Electrochemistry of Glass and Ceramics

Ceramic Transactions
Volume 92

Electrochemistry of Glass and Ceramics

Edited by
S.K. Sundaram
Pacific Northwest National Laboratory

Dennis F. Bickford
Westinghouse Savannah River Company

E.J. Hornyak Jr.
Libbey, Inc.

Published by
The American Ceramic Society
735 Ceramic Place
Westerville, Ohio 43081

Proceedings of the Electrochemistry of Glass and Ceramics Symposium, held at the 100th Annual Meeting of The American Ceramic Society in Cincinnati, Ohio, May 3–6, 1998.

COVER PHOTO: "SEM images of the surface of composite Ni/SiC films annealed at 600°C in air atmosphere for six hours," is courtesy of M. Hepel and C. Mahoney, and appears as figure 5b in their paper, "Electrodeposition and Characterization of Nickel/Silicon Carbide Films," which begins on page 303.

Library of Congress Cataloging-in-Publication Data
A CIP record for this book is available from the Library of Congress.

For information on ordering titles published by The American Ceramic Society, or to request a publications catalog, please call 614-794-5890.

Printed in the United States of America.

1 2 3 4–02 01 00 99

ISSN 1042-1122
ISBN 1-57498-062-9

Contents

Electrochemical Devices and Sensors

Processing

Preface

Electrochemistry may be considered to have originated from chemistry, but it involves far more than knowledge of the chemistry of substances and the physics of electric fields. It is an interdisciplinary area. A symposium on electrochemistry of glass and ceramics took place in Cincinnati, OH, May 3–6, 1998. The symposium was held in conjunction with the 100th Annual Meeting of The American Ceramic Society and was co-sponsored by the Nuclear and Environmental Technology, Glass and Optical Materials, and Electronics Divisions.

The primary objectives of the symposium are to 1) present the major advances in the electrochemistry (both fundamental as well as applied) research on glass and ceramic systems, 2) revive the multidisciplinary aspects of electrochemistry (applied to glass and ceramics), and 3) bring the researchers and industrial leaders in the field together to recognize and appreciate the progress made in this field. We hope that the symposium, first in this series, rekindles a sustained interest in this field. This volume contains manuscripts of a number of papers that were presented at this symposium.

We are grateful to all who contributed to the success of this symposium. We would like to thank the authors and reviewers for their time, expertise, and dedication. These contributions greatly enhanced the quality of this volume. We would also like to acknowledge the contributions of the session chairs and invited speakers: Glenn W. Hollenberg, Henry D. Schreiber, Susan L. Schiefelbein, Thomas O. Mason, Srikanth Gopalan, and John P. Cronin. We appreciate the travel support and participation of Solartron.

S. K. Sundaram
Dennis F. Bickford
E. J. Hornyak Jr.

Thermodynamics and Redox

PHASE EQUILIBRIA AND THERMOCHEMISTRY OF COMPLEX OXIDES BY ELECTROCHEMICAL TITRATION

Turgut M. Gür
Center for Materials Research
Stanford University
Stanford, CA 94305-4045

ABSTRACT

Due to their complex structures and composition, determination of phase relations and fundamental thermodynamic properties of multicomponent oxides using conventional analytical techniques present major experimental challenges and demand laborous procedures. However, solid state electrochemical tools may mitigate many of these difficulties and simplify procedures. One powerful and versatile method that will be described here is the solid state oxygen coulometric titration technique. It not only reduces substantially the number of samples that are needed for mapping out phase diagrams and establishing stability limits of multicomponent oxides but also provides precise and accurate determination of thermochemical parameters of individual phases.

INTRODUCTION

Although oxides are among the oldest materials known to man, the recent decades have seen an increased degree of structural and compositional complexity in these materials. Inspired partly by the discovery of high Tc superconductivity in multicomponent oxides and driven by an intense search for novel and interesting scientific phenomena and useful properties, many new oxide materials with multiple cations are being synthesized and developed with increasingly complicated structures. Multicomponent oxides containing up to six cations in their crystal structures are not uncommon. Many such oxides are interesting for their superconducting, magnetic, ferroelectric, magneto-optic, piezoelectric, nonlinear optical, ionic and mixed ionic-electronic conducting properties. Naturally, these multicomponent oxides display complex phase relations and equilibrium diagrams over a wide range of temperatures and oxygen pressures. The situation is complicated further by the fact that many of these complex oxides display wide ranges of oxygen nonstoichiometry. Moreover, the desired phases of these oxides are usually synthesized or made at elevated temperatures. Unfortunately, detailed information about their phase relations, stability regimes and fundamental thermodynamic properties is scarce at best, and in many cases it is simply nonexistant. This poses serious difficulties in choosing the proper conditions for the selective synthesis of the desired phase and the potential utilization of these oxides for technological applications. Part of the problem is the multitude of experimental difficulties one often encounters during comprehensive determination of phase equilibria and relations in such multicomponent complex oxide systems and accurate measurement of their thermochemical properties. This is not an easy task.

There are naturally a variety of analytical techniques such as x-ray diffraction (XRD), neutron diffraction, electron microprobe analysis (EMPA), electron diffraction, electron microscopy, thermal gravimetric analysis (TGA) and differential scanning calorimetry (DSC) available to study phase equilibria and relations in solids. Unfortunately, for multicomponent oxides any one of these conventional analytical procedures would be quite cumbersome, time consuming and tedious since it generally requires a very large number of samples with incrementally varying atomic ratios in the composition field to be prepared under carefully controlled conditions and then structurally examined and analyzed in order to construct the phase diagram and obtain thermochemical information. This is primarily a batch process that generally involves proper annealing and quenching operations since samples are usually synthesized or made at elevated temperatures but analyzed often at room temperatures. Even at high temperatures, one must be mindful that not all solid state reactions may yield equilibrium phases but may instead form metastable phases. Equally important is the fact that one may not be able to preserve the high temperature phase during quenching to room temperature. Furthermore, there may be issues related to the appropriateness of sensitivity and resolution limits of the conventional analytical technique used. For example, it is difficult to detect by x-ray diffraction a minority phase which is less than a few per cent. Similarly, one may be limited by the sensitivity of the microbalance in TGA or the atomic number distribution in the sample in EMPA. Above all, in an N-component oxide system these traditional techniques allow only 1,...., (N-1) phase regions to be accessible for analysis. It is usually very difficult to obtain the invariant point where N-phases coexist in equilibrium.

In contrast, a solid state electrochemical technique described below simplifies many of the laborious experimental procedures and difficulties posed by conventional analytical tools. In combination with analytical tools such as x-ray diffraction analysis, this solid state electrochemical tool provides an opportunity to achieve relatively easier, expedient, and more comprehensive determination of high temperature phase relations in these complex oxide systems using a significantly smaller number of samples.

This paper will present the salient features of the solid state oxygen coulombic titration technique that uses an oxide ion conducting solid electrolyte. It will demonstrate the efficacy of this technique by examples of phase studies in ternary as well as quaternary oxide systems and describe the process for the determination of phase stability limits and extraction of thermochemical information. It will also discuss how this useful tool can be employed to either fix or measure the oxygen nonstoichiometry in multicomponent oxides.

OXIDE-ION CONDUCTING SOLID ELECTROLYTES

Solid oxide electrolytes that selectively conduct oxide ions have direct relevance and importance in oxygen coulometric titration, so will be briefly reviewed below. They usually require elevated operating temperatures due to the relatively large activation energies (about 1 eV) they exhibit for ionic transport. On the other hand, many remain electronically insulating over wide ranges of temperature and oxygen pressure. Hence, these oxides serve as excellent solid electrolytes for selective transport of oxide ions.

In this work, the solid oxide electrolyte of choice for oxygen coulometric titration experiments was yttria stabilized zirconia (YSZ). Predominant oxide ion conduction in this material was first reported by Nernst at the turn of the century [1]. It is one of many in the rich family of stabilized zirconia solid-solution compounds. Pure ZrO_2 itself is an insulator and exists in three polymorphs, namely, monoclinic, tetragonal and cubic where the cubic phase, which has the fluorite structure, is stable only above 2200 °C. These phase transformations are accompanied by large volume changes, which are naturally not desirable for practical use, such as thermal insulation for furnace refractory applications. Coincidentally, zirconia has a large solid solubility for many alkali and rear earth oxides. Heavy doping, up to 15 mole % of MgO, CaO, Y_2O_3 or Sc_2O_3 stabilizes the cubic fluorite structure to room temperature and eliminates volume and phase changes. Interestingly, this results in the di- or trivalent cations substituting for the tetravalent zirconium ions in the lattice and generating a very large concentration (up to 10^{22} cm^{-3}) of oxygen vacancies in order to maintain charge neutrality of the solid. This large vacancy concentration is primarily responsible for the observed high oxygen conductivity in these materials [2,3]. The electrical conduction is entirely due to the migration of oxygen vacancies and, hence, predominantly ionic in nature. In fact, the ionic transference number for oxygen ions in this material is near unity (i.e., $t_i > 0.99$) [4]. In other words, more than 99 % of the total current through this solid electrolyte is carried by oxygen ions.

Stabilized zirconia electrolytes have very wide range of oxygen activities in which they maintain predominantly ionic conduction. Thus, they can be used safely in a large variety of electrochemical systems under both oxidizing and reducing conditions. This range of activities (i.e., oxygen pressures) extends from about 10^5 atm down to unusually low values of 10^{-28} atm at elevated temperatures [5]. Hence, it is suitable to employ as the solid electrolyte for oxygen coulometric titration.

OXYGEN COULOMETRIC TITRATION TECHNIQUE

Solid state electrochemical measurements were successfully used in the past to determine thermodynamic quantities such as Gibbs free energy of formation of solids [6-8] and solubility of oxygen in molten metals [9,10]. However, it was the pioneering paper by Wagner [11] that first successfully demonstrated the electrochemical coulometric titration technique for thermodynamic studies in Ag_2S. This is followed by several studies reported for binary and ternary oxide systems [12-14]. The first successful application of solid state oxygen coulometric titration technique to quaternary systems involved a comprehensive study of both the phase relations and thermochemistry of the high Tc superconducting oxide $YBa_2Cu_3O_{7-x}$ [15,16].

Solid state oxygen coulometric titration technique involves changing the composition of the sample by adding or deleting a small but precisely controlled amount of oxygen. Figure 1 illustrates this process for a hypothetical A-B-C-O quaternary oxide system schematically represented by a tetrahedron with the cations

Fig. 1. Tetrahedral phase diagram of an A-B-C-O quaternary system. The arrow indicates the direction of oxygen titration from a sample with a starting composition "x" on the pseudoternary phase plane (shaded). The arrow also traces the nominal composition of the sample as oxygen is titrated.

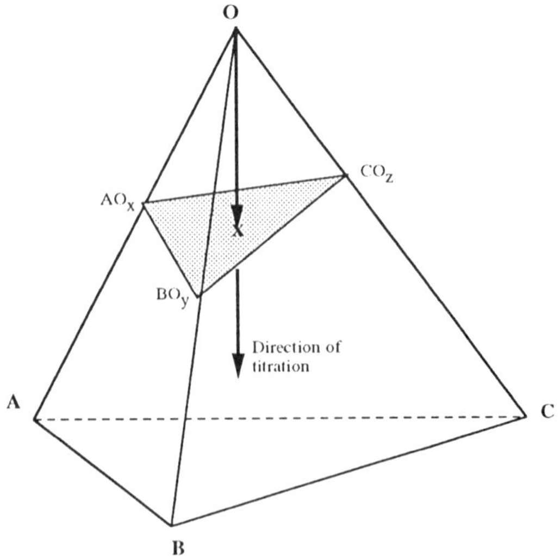

at the corners of the base, and oxygen appropriately at the apex. The titration line indicates the loci of all compositions accessible during coulometric titration steps as oxygen is added to or deleted from this quaternary oxide of initial nominal composition X. The cation ratios in individual samples remain unchanged during oxygen titration. It is obvious that irrespective of the initial nominal composition of the sample, all titration lines necessarily need to pass through the oxygen apex since the samples only either gain or lose oxygen during titration. Consequently, one can in principle traverse the entire range of phase fields and access all compositions that lie along the titration line with just one starting sample. This is of major practical advantage over conventional analytical techniques which require vast numbers of samples carefully and separately prepared for each and every incremental composition.

Generally, a solid state electrochemical cell employing a YSZ electrolyte is used in oxygen coulometric titration experiments for both transporting (i.e., titrating) oxygen to actively alter the oxygen ambient of the sample and for monitoring the ambient oxygen activity after the sample is allowed to relax to its new equilibrium state with the new oxygen pressure ambient. The compositional variation of thermodynamic properties is achieved by a stepwise titration process with equilibration allowed between each individual titration step.

The titration process involves application of a small DC bias, either galvanostatically or potentiostatically, imposed across the YSZ solid electrolyte which serves as an oxygen "pump" that selectively transports a small but precise amount of oxygen in or out of the sample. The amount of oxygen transported in this manner, Δm, is given by Faraday's law and is directly related to the amount of electrical charge, Q, flowing in the external circuit. This is shown by

$$\Delta m = MQ/zF \qquad (1)$$

where M is the atomic weight of oxygen, z is the charge number of the oxide ion, and F denotes Faraday's constant. Since, even extremely small amounts of charge can precisely be measured, the oxygen coulometric titration method offers superior precision compared to all other conventional techniques. For example, transport of about $8x10^{-11}$ g of oxygen corresponds to 10^{-6} C of charge that can readily be measured with great precision.

After the titration step is completed and the DC bias is removed, the equilibration of the sample with the new oxygen activity is then monitored by measuring the open circuit potential across the YSZ electrolyte, which now acts as an oxygen "sensor" where the oxygen activity, a_{O2}, is related to the open circuit potential, E, by the Nernst equation;

$$E = - RT/4F \ln (a'_{O_2}/a''_{O_2}) \qquad (2)$$

where R is the universal gas constant, T denotes the temperature, and the superscripts indicate oxygen activities on either side of the electrolyte. Assuming that ideal gas behavior is obeyed at the elevated operating temperatures and the oxygen activity on one side of the electrolyte can be fixed by using a suitable reference, Eqn (2) can be expressed as

$$E = - RT/4F \ln (P_{O_2}/P_{ref O_2}) \qquad (3)$$

where $P_{ref O_2}$ is the reference oxygen pressure on one side of the electrolyte (e.g., for air, $P_{ref O_2} = 0.21$ atm) and P_{O_2} denotes the unknown oxygen pressure on the other side.

The open circuit potential eventually decays to a constant value when the sample reaches equilibrium with the new oxygen activity in the ambient. This equilibrium oxygen activity value is plotted against the amount of oxygen transported. The complete procedure is repeated for a new titration step until a sufficiently wide compositional range is traversed. The resulting coulometric titration plots, either as E vs Q or P_{O_2} vs Δm, are equivalent to activity-composition diagrams where the flat plateaus of the plots indicate the invariant points in the phase diagram where all N phases coexist and all intensive variables are fixed at constant temperature and oxygen pressure. Invoking Gibbs' phase rule, say for a ternary oxide system,

$$p + f = c + 2 \qquad (4)$$

where p is the number of phases, f the degrees of freedom, and c the number of independent components, a plateau in the oxygen coulometric titration curve corresponds to an invariant reaction at that temperature where three solid phases coexist in equilibrium with the oxygen activity in the gas phase.

The open circuit potential of the electrochemical cell at this plateau is related to the equilibrium constant, K, and hence, the standard Gibbs free energy change, ΔG^o, that accompany this invariant reaction by the expression

$$\Delta G^o = -4FE = -RT \ln K \qquad (5)$$

As long as the three solid phases coexist on this plateau at constant temperature and pressure, the activities of all components are fixed and independent of the compositional change due to oxygen titration. Thus, the cell voltage remains constant. Studying the temperature dependence of the cell voltage at the plateau provides valuable thermochemical information. Standard entropy values can be obtained either from the temperature dependence of the cell voltage or the change in the free energy given by

$$\partial(\Delta G^o)/\partial T = -4F\, \partial E/\partial T = -\Delta S^o \qquad (6)$$

Similarly, the standard enthalpy of the invariant change can be calculated either from Eqn. (7)

$$\Delta G^o = \Delta H^o - T\Delta S^o \qquad (7)$$

or from van't Hoff's equation

$$\partial \ln K/\partial(1/T) = \partial \ln P_{O_2}/\partial(1/T) = -\Delta H^o/R \qquad (8)$$

Thus, the slope of the semilog plot of oxygen pressure versus reciprocal temperature yields the standard enthalpy value. In this manner, all fundamental thermodynamic properties can readily be obtained.

In summary, oxygen coulometric titration technique is a powerful tool to study phase relations and thermodynamic properties of multicomponent oxides. Its advantages weigh heavily over conventional analytical procedures. It is an in situ, dynamic method that requires a relatively small number of samples to map out the complete phase diagram of multicomponent oxides; it is a very sensitive ($< 10^{-10}$ moles of oxygen) and high resolution technique since it involves precise measurement of electrical quantities only; it renders all phase regions 1,...., N accessible for study; it has the ability to even resolve "line" phases with very narrow compositional ranges; and it allows simultaneous determination of stability regimes and fundamental thermochemical quantities of individual phases.

EXPERIMENTAL ASPECTS

The electrochemical cell arrangement generally used for oxygen coulometric titration experiments can be represented as:

$$P_{O_2}\text{, sample / Pt / YSZ / Pt, Air} \qquad (9)$$

Typically a potentiostat is used for applying a small DC bias across the YSZ electrolyte during oxygen titration. A digital coulometer usually monitors the precise amount of the charge passed during the titration step which may be done under either galvanostatic (i.e., constant current) or potentiostatic (i.e., constant voltage) mode. After a small amount of oxygen is titrated, the DC bias is removed. The decay in the open circuit potential of the cell is monitored to observe equilibration of the sample with the new oxygen activity. Depending on the polarity of the applied bias, the equilibration process involves either loss or gain of oxygen by the sample until it establishes a new equilibrium state with the corresponding oxygen activity in the gas phase.

The individual phases participating in the invariant reactions are determined by x-ray diffraction analysis of samples quenched from the titration temperature to room temperature. Prior to quenching, these samples are allowed to equilibrate under electrochemically controlled precise oxygen pressures that are purposefully chosen slightly above and slightly below the titration plateaus. Subsequent x-ray diffraction analysis of the quenched samples reveal the chemical and structural identity of the reactants (i.e., from above the plateau) and products (i.e., from below the plateau) that participate in the invariant solid state reaction at the particular plateau.

Typical schematic design for a solid state electrochemical cell usually consists of an enclosed leak-tight chamber that holds both the sample and the YSZ solid oxide electrolyte. The chamber material may be quartz or ceramic, such as alumina. Porous Pt layers deposited on YSZ surfaces from conductive inks commonly serve as electrodes. One electrode is usually exposed to air to serve as the reference oxygen electrode while the other electrode shares the common ambient with the sample.

EXAMPLES

The versatility and utility of oxygen coulometric titration is illustrated by the following two examples. The first one is the Ba-Fe-O ternary system studied around the $BaFe_{12}O_{19}$ composition [17,18] which potentially is an important material for magnetic recording and microwave applications. The second example involves the Y-Ba-Cu-O quaternary system around the high Tc superconducting $YBa_2Cu_3O_{7-x}$ composition [15, 16]. Details of these studies can be found in the referenced articles.

Ba-Fe-O Ternary System

$BaFe_{12}O_{19}$ powder samples are used as the starting material for the oxygen titration experiments at 700, 800, 900 and 1000 °C in a manner described above. The results of the titration experiments are collectively shown in Figure 2. The titration curves show similar behavior within this temperature regime and feature three well-defined plateaus which correspond to three invariant reactions over a wide range of oxygen pressures. As oxygen is intermittently titrated out of the starting composition $BaFe_{12}O_{19}$, Figure 2 indicates that the oxygen activity monotonically decreases until the oxygen nonstoichiometry in the $BaFe_{12}O_{19}$ phase reaches its limiting value at the foot of plateau I. From the amount of charge passed, the oxygen nonstoichiometry is estimated to be small, about 10^{-5} at 1000 °C which is typical for these materials. At this limiting value, the $BaFe_{12}O_{19}$ phase becomes unstable and starts to decompose along plateau I. The decomposition reaction proceeds until all $BaFe_{12}O_{19}$ is consumed. Although the mere existence of the plateau indicates an invariant reaction where $BaFe_{12}O_{19}$ is decomposed into two

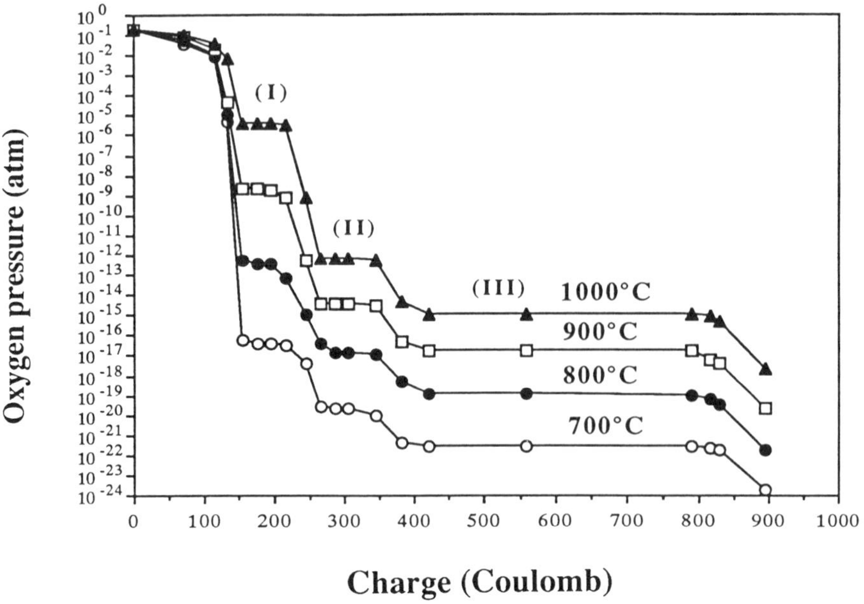

Fig. 2. Oxygen coulometric titration curves for $BaFe_{12}O_{19}$ indicating three well defined plateaus corresponding to three invariant reactions.
"Reprinted from reference [18] J. Mater. Res. 9 (6), 1499 (1994) with kind permission of the Materials Research Society".

other solid phases, it provides no information about their identity. For phase identification by x-ray diffraction analysis, samples are prepared by complete equilibration just slightly above and below the plateau oxygen pressure and subsequently quenched to room temperature. X-ray diffraction results are illustrated in Figure 3 which indicates that the $BaFe_{12}O_{19}$ phase is indeed stable down to plateau I, and magnetite (Fe_3O_4) and orthorhombic barium iron oxide ($BaFe_2O_4$) are the products of the invariant reaction on this plateau I given by

$$BaFe_{12}O_{19} = BaFe_2O_4 + 10/3\ Fe_3O_4 + 5/6\ O_2 \qquad (10)$$

In accordance with Eqn (5), the standard Gibbs free energy for reaction (10) is

$$\Delta G_{10}^{o}\ (J/mol) = 7.23x10^5 - 480\ T \qquad (973K \le T \le 1273K) \qquad (11)$$

while the lower stability limit of $BaFe_{12}O_{19}$ can be obtained from the temperature dependence of the oxygen pressure of plateau I for this decomposition reaction

$$\ln\ (Po_2\ (atm)) = 69.37 - 1.044x10^5/T \qquad (973K \le T \le 1273K) \qquad (12)$$

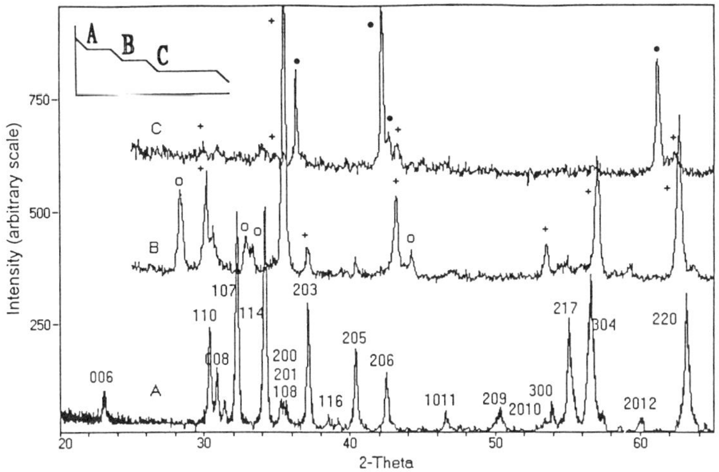

Fig. 3. The x-ray diffraction patterns of quenched samples prepared under electrochemically controlled conditions. (A) Sample is equilibrated at oxygen pressure slightly above plateau I, and diffraction all peaks are due to $BaFe_{12}O_{19}$. (B) Sample is equilibrated between plateaus I and II and peaks indicate the presence of $BaFe_2O_4$ (o) and Fe_3O_4(+). (C) Sample is equilibrated between plateaus II and III. (•) denotes unknown phase.
"Reprinted from reference [18] J. Mater. Res. 9 (6), 1499 (1994) with kind permission of the Materials Research Society".

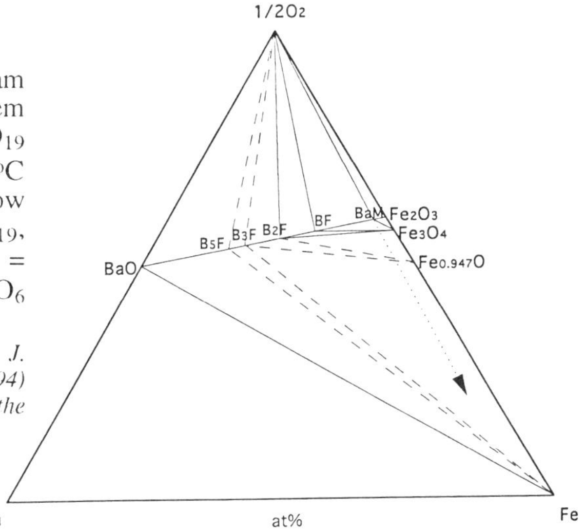

Fig. 4. Partial phase diagram of the Ba-Fe-O ternary system around the BaM = $BaFe_{12}O_{19}$ composition for 700-1000 °C and oxygen pressures below ambient. (BaM = $BaFe_{12}O_{19}$, BF = $BaFe_2O_4$, B_2F = $Ba_2Fe_2O_5$, B_3F = $Ba_3Fe_2O_6$ and B_5F = $Ba_5Fe_2O_8$.
"Reprinted from reference [18] J. Mater. Res. 9 (6), 1499 (1994) with kind permission of the Materials Research Society".

In a similar fashion, thermodynamic information obtained for plateaus II and III can be correlated with structural analysis. Then it becomes possible to construct the Ba-Fe-O ternary phase diagram as shown in Figure 4. The dotted line indicates the direction where the $BaFe_{12}O_{19}$ composition changes as oxygen is titrated out of the sample. The solid lines, experimentally established in this study indicate the tie lines that connect phases in equilibrium with each other.

Y-Ba-Cu-O Quaternary System

Phase relations in quaternary systems are significantly more complex than in ternary systems, primarily because of the added third dimension as depicted in Figure 1. In the early days of the high Tc superconductor $YBa_2Cu_3O_{7-x}$, information about the thermodynamic stability regime of this new quaternary oxide was important and oxygen coulometric titration technique was successfully used for this purpose to obtain rapid results in a relatively short time [15,16].

Using the experimental procedures described above, oxygen coulometric titration curves were obtained for several Y-Ba-Cu-O quaternary samples with different cation ratios around the $YBa_2Cu_3O_{7-x}$ composition. The results shown in Figure 5 clearly indicate the multitudes of distinct plateaus, each associated with an

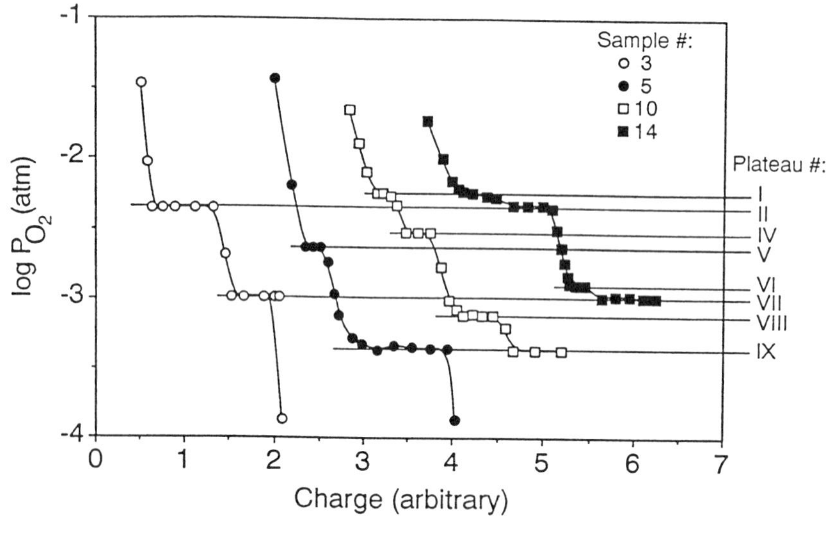

Fig. 5. Oxygen coulometric titration curves for samples No. 3, 5, 10, 14 prepared for various compositions around $YBa_2Cu_3O_{7-x}$ in the Y-Ba-Cu-O quaternary oxide system at 850 °C. Note the multitude of plateaus corresponding to closely spaced invariant reactions.

"Reprinted from reference [19] Physica C 167 529 (1990) with kind permission of Elsevier Science - NL".

invariant point in the phase equilibria. It also indicates that the phase relations around $YBa_2Cu_3O_{7-x}$ depend strongly on oxygen pressure. Following the procedure prescribed above for phase identification, it was possible to determine all phases that participated in the invariant reactions along the plateaus [18]. The results indicated that at 850 °C, the lower stability limit for the $YBa_2Cu_3O_{7-x}$ phase is 4.0×10^{-4} atm of oxygen pressure. Below this oxygen pressure, $YBa_2Cu_3O_{7-x}$ decomposes into Y_2BaCuO_5, $BaCu_2O_2$, and $YBa_3Cu_2O_{6+y}$.

Using coulometric titration, it is also possible to control and change oxygen stoichiometry in the $YBa_2Cu_3O_{7-x}$ phase where the superconducting properties are strongly dependent on oxygen doping. In fact, $YBa_2Cu_3O_{7-x}$ samples prepared by oxygen coulometric titration with precise oxygen nonstoichiometries indicate the presence of a double-plateau behavior with respect to the superconducting transition temperature [19].

CONCLUDING REMARKS

Solid state oxygen coulometric titration technique is a powerful and versatile electrochemical tool that simultaneously provides both thermodynamic and phase equilibria information as well as allowing precise control of oxygen stoichiometry. The latter is particularly important for many nonstoichiometric multicomponent oxides that have interesting electronic, ferroelectric, superconducting, optical, magnetic, piezoelectric, sensing and transport properties.

REFERENCES

[1] W. Nernst, Z. Elektrochem. 6, 41 (1900)

[2] T. M. Gür , I. D. Raistrick and R. A. Huggins, Mat. Sci. Engr. 46, 53 (1980)

[3] T. M. Gür , I. D. Raistrick and R. A. Huggins, Solid State Ionics 1, 251 (1980)

[4] T. M. Gür, I. D. Raistrick, and R. A. Huggins, J. Electrochem. Soc. 127, 2620 (1980)

[5] J. W. Patterson, E. C. Bogren, and R. A. Rapp, J. Electrochem. Soc. 114, 752 (1967)

[6] K. Kiukkola and C. Wagner, J. Electrochem. Soc. 104, 308 (1957)

[7] K. Kiukkola and C. Wagner, J. Electrochem. Soc. 104, 379 (1957)

[8] K. A. Klinedinst and D. A. Stevenson, J. Chem. Thermodynamics 4, 565 (1972)

[9] B. Heshmatpour and D. A. Stevenson, J. Electrochem. Soc. 129 [2], 430 (1982)

[10] B. Heshmatpour and D. A. Stevenson, Metall. Trans. B 13, 53 (1982)

[11] C. Wagner, J. Chem. Phys. 21, 1819 (1953)

[12] R. A. Giddings and R. S. Gordon, J. Electrochem. Soc. 121, 793 (1974)

[13] W. Piekarczyk, W. Weppner, and A. Rabenau, Z. Naturforsh. 34a, 430 (1979).

[14] W. Weppner, C. Li-chuan, and W. Piekarczyk, Z. Naturforsh. 35a, 381 (1980).

[15] B. T. Ahn, T. M. Gür, R. A. Huggins, R. Beyers, E. M. Engler, P. M. Grant, S. S. P. Parkin, G. Lim, M. L. Ramirez, K. P. Roche, J. E. Vazquez, V. Y. Lee, and R. D. Jacowitz, Physica C 153-155, 590 (1988).
[16] B. T. Ahn, V. Y. Lee, R. Beyers, T. M. Gür, and R. A. Huggins, Physica C 167, 529 (1990).
[17] J. Li, T. M. Gür, R. Sinclair, S. S. Rosenblum, and H. Hayashi, Solid State Ionics 73, 1185 (1994)
[18] J. Li, T. M. Gür, R. Sinclair, S. S. Rosenblum, and H. Hayashi, J. Mater. Res. 9, 1499 (1994)
[19] R. Beyers, B. T. Ahn, G. Gorman, V. Y. Lee, S. S. P. Parkin, M. L. Ramirez, K. P. Roche, J. E. Vazquez, T. M. Gür and R. A. Huggins, Nature 340, 619 (1989)

THERMODYNAMIC PROPERTIES OF NANOCRYSTALLINE PRASEODYMIUM-CERIUM OXIDE $Pr_{0.7}Ce_{0.3}O_{2-x}$

P. Knauth * and H. L. Tuller
Massachusetts Institute of Technology, Department of Materials Science and Engineering, Crystal Physics and Electroceramics Laboratory, Cambridge, MA 02139, USA

ABSTRACT

The oxygen deficiency x of a nanocrystalline mixed oxide of composition $Pr_{0.7}Ce_{0.3}O_{2-x}$ was studied by coulometric titration coupled with potentiometric measurements between 739 and 915 K. A single-phase Pr_6O_{11} structure was found by X-ray diffraction. The oxygen pressure-composition isotherms indicate a two-phase region at small relative oxygen deficiency and a single-phase region at larger non-stoichiometry. The oxygen pressure dependence in the homogeneous region can be described by a power law with an exponent -1/6, in accordance with doubly charged oxygen vacancies as majority defects. The enthalpy of reduction amounts to (2.9 ± 0.3) eV (280 kJ/mol).

INTRODUCTION

Mixed ionic-electronic conducting oxide ceramics are finding industrial application in automotive catalysts, membranes for oxygen separation and fuel cells. The function of oxide ceramics in three-way automotive catalysts is the storage and release of oxygen during transients from reducing (fuel-rich) to oxidizing (fuel-lean) environments. A high oxygen-storage capacity is therefore of interest and is related to the ability of the cations to change their valence state. The kinetics of the redox process are obviously important for this kind of application.

Mixed praseodymium-cerium oxides [1-6] present a particularly high oxygen storage capacity, due to the easy valence change of the praseodymium

* On leave from: Laboratoire EDIFIS, CNRS-Université d'Aix-Marseille, F-13397 Marseille Cedex 20, France.

cation (Pr^{4+}/Pr^{3+}). In a previous paper from this laboratory, the oxygen-deficiency of a mixed praseodymium-cerium oxide of composition $Pr_{0.545}Ce_{0.455}O_{2-x}$ was investigated by the coulometric titration technique coupled with potentiometry in a stabilized zirconia solid electrolyte cell [6]. The composition of the mixed oxide is changed by applying a constant current through the solid electrolyte for a short time thereby pumping oxygen into or out of the system. The corresponding equilibrium oxygen partial pressure is calculated using Nernst's equation from the steady-state voltage of the cell following the pumping sequence. A high relative [†] oxygen deficiency ($x - x_0$) was measured for $Pr_{0.545}Ce_{0.455}O_{2-x}$ between 673 and 873 K.

In principle, the preparation of mixed oxides in nanocrystalline form is a means for further improving their properties. First, it has been shown that the oxygen deficiency of nanocrystalline oxides may be enhanced, because of the existence of interface sites with low defect formation energies [7]. This should further increase the already important oxygen storage capacity. Second, the effective oxygen diffusion coefficients may also be enhanced, because of the high density of fast-diffusion paths along grain boundaries in nanocrystalline oxides [8]. This should increase the kinetics of oxygen uptake and release.

In this work, we study nanocrystalline mixed praseodymium-cerium oxide with composition $Pr_{0.7}Ce_{0.3}O_{2-x}$. The isotherms of composition versus oxygen chemical potential are determined by coupling coulometry and potentiometry in a double solid electrolyte cell. The use of a double cell also enables the investigation of the relaxation kinetics from the voltage transient after each titration step [9].

EXPERIMENTAL

The samples were prepared by coprecipitation of cerium nitrate and praseodymium nitrate using oxalic acid solutions [4-6]. Precipitates were calcined at 923 K for 6 h and 1023 K for 2 h. The composition of the oxide was investigated by iodometric titration after dissolution of the samples in hot sulfuric acid. We used a standardized thiosulfate solution and starch solution as indicator.

A double stabilized zirconia cell with platinum paste electrodes was used for the coulometric and potentiometric experiments (galvanostat EG&G 362, multimeter Keithley 197). The amount of electrolyzed oxygen $\Delta n(O_2)$ was calculated from the current I and the time t (F : Faraday's constant) according to:

$$\Delta n(O_2) = It / 4F + (P_f - P_i) V/RT \qquad (1)$$

[†] The term "relative" means that the oxygen deficiency x is indicated as ($x - x_0$) with respect to the value x_0 , which is not necessarily zero, corresponding to the initial equilibrium in air.

The right-hand side of this equation takes into account the so-called dead-volume correction : P_f and P_i are the final and the initial oxygen pressures, respectively, and V/RT is the dead-volume coefficient. Tests of reversibility were performed by applying small currents to the cell (< 1 μA, 10 s) and recording the transient of the open-circuit potential after the current flow. The absence of gas leakage was checked by long-time measurements of the cell voltage (Figure 1).

The mixed oxides were investigated before and after the electrochemical measurements by X-ray powder diffraction (Cu-Kα radiation).

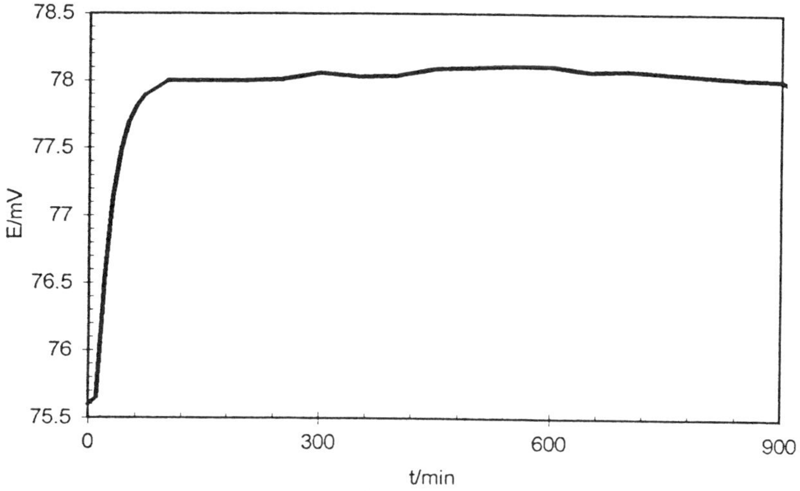

Fig. 1. Typical time dependence of the cell voltage after a coulometric titration step.

RESULTS

Figure 2 shows a typical X-ray diffractogram of the mixed oxide. All diffraction lines correspond to the fluorite structure of β-praseodymia Pr_6O_{11}. An analysis of the mean grain size d using Scherrer's equation gave d = (35 ± 5) nm.

A cerium concentration of (31 ± 1) mol%, identical to the total cerium concentration, was found by iodometric titration. This confirms that cerium is nearly valence-invariant, remaining largely in the Ce^{4+} state. The oxidation state of praseodymium could however not be determined by this technique, because the redox potential of Pr^{4+} ions in aqueous solution is so high (E° ≈ (3.2 ± 0.2) V [10]), that the four-valent state is probably instantaneously reduced by water.

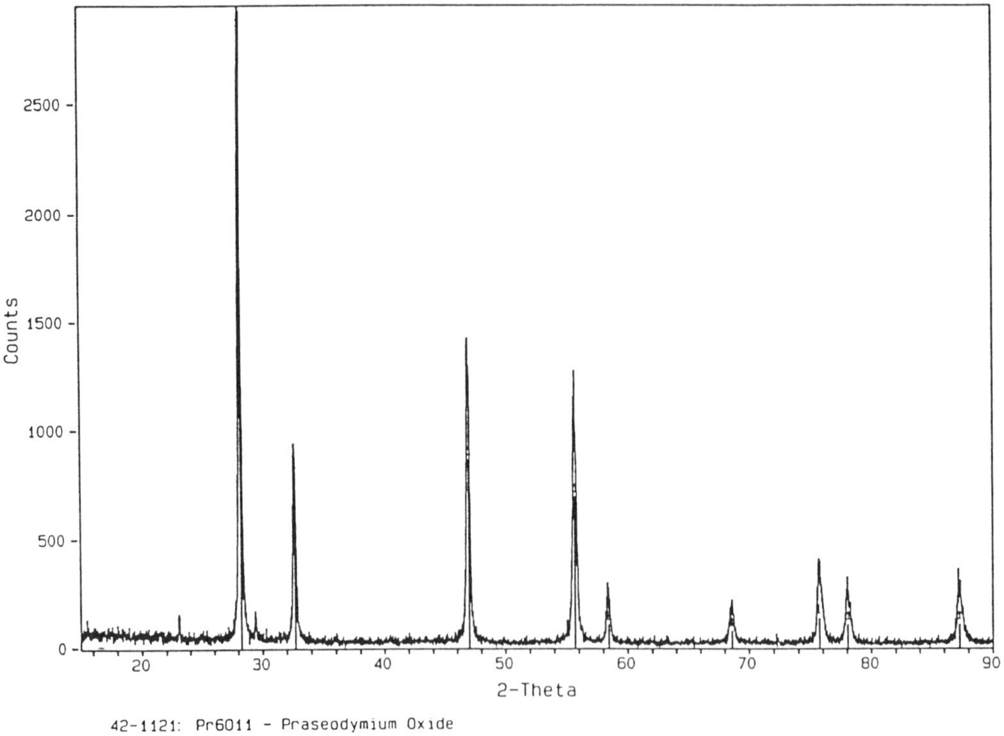

Fig. 2. X-ray diffractogram of nanocrystalline $Pr_{0.7}Ce_{0.3}O_{2-x}$.

Figure 3 is a plot of the oxygen partial pressure versus the relative oxygen deficiency ($x - x_0$) at 739, 830 and 915 K. Large oxygen deficiencies, $x > 0.1$, are indeed observed in the nanocrystalline mixed oxide. In this plot, one recognizes a plateau region at high oxygen pressures, followed by a region with steep change of oxygen pressure for a small variation of composition, corresponding to a single-phase region. The plateau indicates a two-phase domain, because at the temperatures of measurement, the mixed oxide can be considered as a quasi-binary system (the cation ratio is fixed).

Figure 4 shows the corresponding double logarithmic plot. The oxygen deficiency data exhibit a linear oxygen pressure dependence with a slope of -1/n:

$$\log x = A - \log P(O_2) / n \qquad (2)$$

The value of n is approximately 6 at elevated temperatures and about 3 at 739 K.

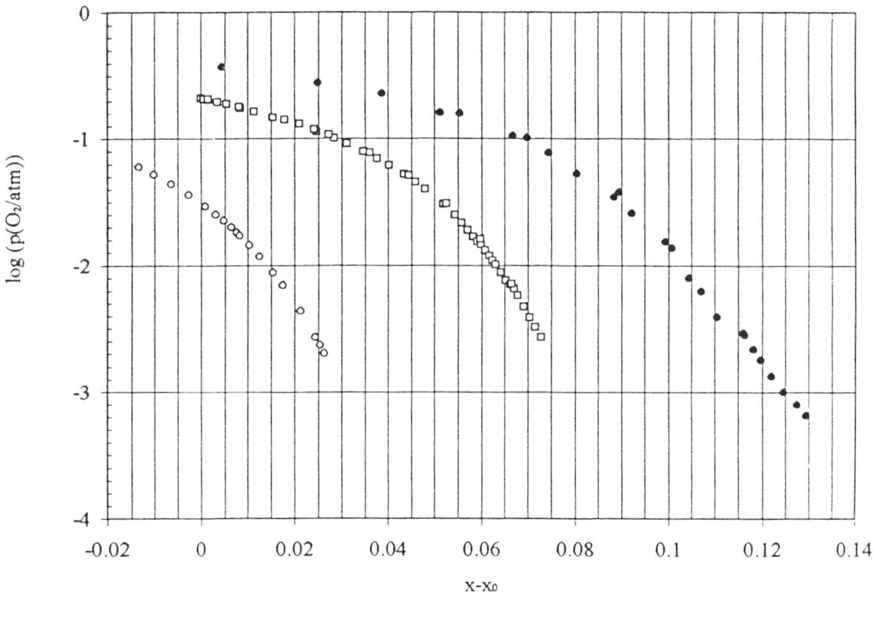

Fig. 3. Dependence of the oxygen partial pressure on the oxygen deficiency x in $Pr_{0.7}Ce_{0.3}O_{2-x}$: ● 915 K, ❑ 830 K, ❍ 739 K.

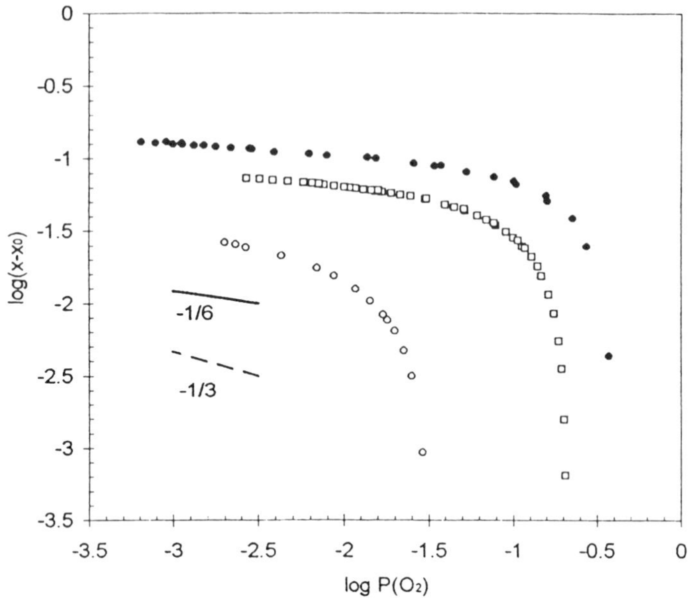

Fig. 4. Oxygen pressure dependence of the oxygen deficiency of nanocrystalline $Pr_{0.7}$ $Ce_{0.3}O_{2-x}$: ● 915 K, ❑ 830 K, ❍ 739 K.

The partial Gibbs free energy variation $\Delta G(O_2)$ can be calculated from the oxygen pressure $P(O_2)$ according to:

$$\Delta G(O_2) = RT \ln P(O_2)/P° \qquad (3)$$

The data are plotted against the relative oxygen deficiency in figure 5. Here, a plateau is clearly visible for $(x-x_0) < 10^{-2}$.

DISCUSSION

Concerning the X-ray diffraction patterns, it is interesting to note that the solid solution appears single-phase and retains the crystal structure of the praseodymium oxide host lattice even with 30 mol% cerium dioxide. This is certainly related to the nanocrystalline microstructure. An enhanced apparent solubility due to grain boundary segregation of the solute is theoretically expected and was previously observed in copper-doped cerium oxide [11]. In conventional materials, Takasu et al. [1] found a single phase CeO_2 fluorite type structure for 70 mol% praseodymium and at higher praseodymium concentrations a two-phase region was observed. On the contrary, Steele and coworkers [3] found a two-phase region between 40 and 60 mol% praseodymium. Logan and Shelef [4] found single-phase ceria for 55 mol% praseodymium and the two phases CeO_2 and Pr_6O_{11} for 83 mol% praseodymia. Clearly, our result differs notably from those of these previous investigations on polycrystalline samples after high temperature annealing treatments.

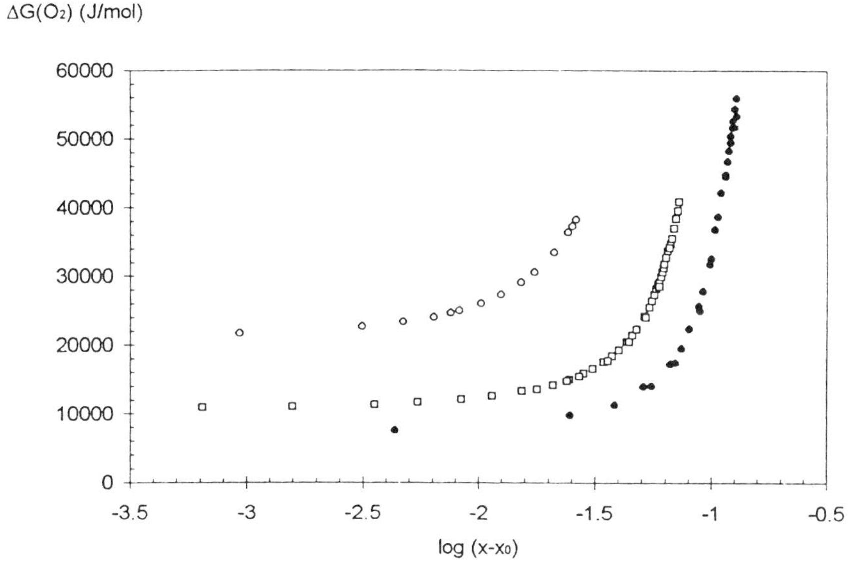

Fig. 5. Partial Gibbs free energy of mixing of oxygen in nanocrystalline $Pr_{0.7}Ce_{0.3}O_{2-x}$: ● 915 K, ❑ 830 K, ❍ 739 K.

The oxygen pressure dependence, with an exponent of -1/6, is in agreement with doubly charged oxygen vacancies as majority point defects. The corresponding reduction reaction can be written in Kroger-Vink nomenclature:

$$O_O \Longleftrightarrow V_O^{\bullet\bullet} + 2\,e' + 1/2\,O_2 \qquad (4)$$

The equilibrium constant K is:

$$K = [V_O^{\bullet\bullet}]\,[e']^2\,P(O_2)^{1/2} = K_0 \exp(-\Delta H/RT) \qquad (5)$$

where ΔH is the enthalpy of reduction. For reduction-controlled behaviour, the electroneutrality condition is:

$$2\,[V_O^{\cdot\cdot}] = [e'] \qquad (6)$$

By definition, the relative oxygen deficiency $(x - x_0)$ is proportional to the mole fraction of oxygen vacancies $[V_O^{\cdot\cdot}]$. We obtain therefore:

$$(x - x_0) \propto [V_O^{\cdot\cdot}] = (K/4)^{1/3}\, P(O_2)^{-1/6} \qquad (7)$$

There is a surprisingly good agreement with the experimental oxygen pressure dependence at 830 and 915 K in spite of the apparently high defect concentrations (Figure 4). One should, however, keep in mind that we are operating in the vicinity of an intermediate ordered compound, in which the predominant part of the oxygen vacancies are ordered and thus should not be considered as defects. Furthermore, these systems exhibit rather high dielectric constants, which thereby significantly reduce Debye-Hückel type defect interactions.

The formation of singly charged and uncharged oxygen vacancies can be considered as an association between doubly charged vacancies and electrons. The corresponding reduction reactions can be written:

$$O_O \Longleftrightarrow V_O^{\cdot} + e' + 1/2\, O_2 \qquad (8)$$

and

$$O_O \Longleftrightarrow V_O + 1/2\, O_2 \qquad (9)$$

A similar derivation shows that the exponent for the oxygen presssure dependence should be -1/4 or -1/2 respectively for these two cases. Experimentally, an exponent of -1/3 was found at 739 K. This may reflect the tendency for defect association between oxygen vacancies and electrons at low temperature, but it can not be excluded that equilibrium is not established at 739 K, the lowest temperature of measurement.

The temperature dependence of K gives an enthalpy of reduction ΔH = (2.9 ± 0.3) eV (280 kJ/mol). This value is lower than the corresponding value for conventional polycrystalline cerium dioxide (4.7 eV [12]), but higher than that for nanocrystalline cerium dioxide (2.3 eV [13]).

CONCLUSIONS

The thermodynamic properties of nanocrystalline mixed praseodymium-cerium oxide were measured by coulometry and potentiometry using a stabilized zirconia solid electrolyte cell. The high relative oxygen deficiencies observed confirm that nanocrystalline oxides are potentially of interest as oxygen storage materials. The oxygen pressure dependence is in agreement with doubly charged oxygen vacancies as majority defects.

REFERENCES

1. Y. Takasu, T. Sugino, Y. Matsuda, "Electrical Conductivity of Praseodymia Doped Ceria", *J. Appl. Electrochem.*, 14, 79 (1984).
2. Ch. Ftikos, M. Nauer, B. C. H. Steele, "Electrical Conductivity and Thermal Expansion of Ceria Doped with Pr, Nb and Sn", *J. Eur. Ceram. Soc.*, 12, 267 (1993).
3. M. Nauer, Ch. Ftikos, B. C. H. Steele, "An Evaluation of Ce-Pr Oxides and Ce-Pr-Nb Oxides Mixed Conductors for Cathodes of Solid Oxide Fuel Cells: Structure, Thermal Expansion and Electrical Conductivity", *J. Eur. Ceram. Soc.*, 13, 493 (1994).
4. D. Logan. M. Shelef, "Oxygen Availability in Mixed Cerium / Praseodymium Oxides and the Effect of Noble Metals", *J. Mater. Res.*, 9, 468 (1994).
5. M. Yu. Sinev, G. W. Graham, L. P. Haack, M. Shelef, "Kinetic and Structural Studies of Oxygen Availability of the Mixed Oxides $Pr_{1-x}M_xO_y$ (M = Ce, Zr)", *J. Mater. Res.,* (1996).
6. O. Porat, H. L. Tuller, M. Shelef, E. M. Logothetis, "Electrical Conductivity and Nonstoichiometry in $Pr_{0.545}Ce_{0.455}O_{2-x}$", in Solid State Chemistry of Inorganic Materials, P. K. Davies, A. J. Jacobson, C. C. Torardi, T. A. Vanderah, Materials Research Soc., Pittsburgh, 1997, p. 531.
7. O. Porat, H. L. Tuller, E. B. Lavik, Y.-M. Chiang, "Coulometric Titration Studies of Nonstoichiometric Nanocrystalline Ceria", in Nanophase and Nanocomposite Materials II, S. Komarneni, J. Parker, H. Wollenberger, Eds., Materials Research Society, Pittsburgh, 1997, p. 99.
8. H. L. Tuller, "Solid State Electrochemical Systems: Opportunities for Nanofabricated or Nanostructured Materials", *J. Electroceramics*, 1, 211 (1997).
9. P. Knauth, H. L. Tuller, in preparation.
10. A. J. Bard, R. Parsons, J. Jordan, *Standard Potentials in Aqueous Solutions*, Marcel Dekker, New York (1985).
11. P. Knauth, G. Schwitzgebel, A. Tschöpe, S. Villain, "E.m.f. Measurements on Nanocrystalline Copper-Doped Cerium Dioxide", *J. Solid State Chem.*, in press.
12. H. L. Tuller, A. S. Nowick, "Defect Structure and Electrical Properties of Nonstoichiometric CeO_2 Single Crystals", *J. Electrochem. Soc.*, 126, 209 (1979).
13. Y. M. Chiang, E. B. Lavik, I. Kosacki, H. L. Tuller, J. Y. Ying, "Nonstoichiometry and Electrical Conductivity of Nanocrystalline CeO_{2-x}", *J. Electroceramics*, 1, 7 (1997).

QUANTITATIVE OXIDATION STATE DETERMINATION OF NICKEL OXYHYDROXIDES BY X-RAY ABSORPTION SPECTROSCOPY

Bahne Cornilsen and Zhiyong Xu
Michigan Technological University
Department of Chemistry
1400 Townsend Dr.
Houghton, MI 49931
and
George Meitzner
Edge Analytical, Inc.
2126 Allen Blvd. #3
Middleton, WI 53562

ABSTRACT

This research demonstrates the power of x-ray absorption spectroscopy to observe, and quantify, the Ni(II) and Ni(IV) contents and to determine the average oxidation state of a mixed-valent, nonstoichiometric material. Nickel oxyhydroxides are the electrochemically "active" material in the positive electrode of Ni-Cd, Ni-H_2, and Ni-MH (MH = metal hydride) batteries. The materials studied in this work are model compounds for this electrode. During cycling the maximum and minimum oxidation states control the electrode capacity; for the various cycles these exhibit non-integer values between *ca.* 2.2 and 3.67. Therefore, accurate measurement of these oxidation states is critical to the understanding and improvement of these electrodes. The position and shape of the nickel K-edge in x-ray absorption spectra provide information about the cation oxidation state. First derivative spectral analyses show unique positions and maxima for Ni(II) and Ni(IV) cations. The presence of the Ni(IV) species is uniquely demonstrated. Oxidation states are calculated from Ni(II) and Ni(IV) standard spectra as a multivariable minimization problem using a constrained optimization function, the non-negative least-squares. The 100% Ni(IV) standard is measured *in situ* in the presence of excess oxidizing agent. Calculated oxidation states are in good agreement with the results of chemical titration for stable materials. The instability of the higher oxidation states and rapid reduction in aqueous media is demonstrated. The presence of both Ni(II) and Ni(IV) is demonstrated, and explains the non-integer average oxidation states.

INTRODUCTION

The nickel K-edge position is dependent upon the oxidation state of the cation. The K-edge for nickel varies between 8338 eV and 8342 eV, for Ni(II) and Ni(IV), respectively (1-3). The principle underlying this analysis is that the K-edge for an intermediate oxidation state, *e.g.* Ni(III), is not composed of a single component at 8340 eV; rather, it is made up of two components, one K-edge from each of the two extreme oxidation states. For a sample of intermediate oxidation state, the two inflection points are clearly seen as maxima in the first derivative spectra (1). X-ray absorption spectroscopy is ideal for the *in situ* study of cation oxidation states because the oxidant and electrolyte are transparent and do not interfere with the measurement.

BACKGROUND

Two electrochemical cycles have been reported in the literature for the nickel electrode (eqns. 1 and 2) (4,5). Interconversion between these cycles occurs in two ways. Overcharge of a β-NiOOH produces a γ-phase material, taking the system into the α/γ cycle. Ageing (*i.e.* standing in KOH electrolyte) transforms α material into β material, returning the system to the β/β cycle (6,7).

$$\alpha\text{-Ni(OH)}_2 \underset{\text{discharge}}{\overset{\text{charge}}{\rightleftharpoons}} \gamma\text{-NiOOH} \qquad [1]$$

α-Ni(OH)₂ ⇓ (?) β-Ni(OH)₂; β-NiOOH ⇑ (overcharge) γ-NiOOH

$$\beta\text{-Ni(OH)}_2 \underset{\text{discharge}}{\overset{\text{charge}}{\rightleftharpoons}} \beta\text{-NiOOH} \qquad [2]$$

The four materials, *i.e.* end-members, in these two cycles were initially defined using powder x-ray diffraction (XRD) by analogy with chemically precipitated materials that were previously named α, β, and γ (5,8). For both of these electrochemical cycles, the discharged materials were initially identified as close-packed, with ...ABAB... (oxygen layer) stacking, and related to β-Ni(OH)$_2$ which exhibits the Brucite structure. The discharged β-material was thought to be the same as ordered β-Ni(OH)$_2$. The α-phase was defined with β-type NiO$_2$ layers separated by water molecules (5,9). The γ-NiOOH structure has had two different structures proposed for it, a simple R$\bar{3}$m structure or a more complex Cm cell (5,10,11). The β-NiOOH structure has not been defined, but it is thought to be similar to β-Ni(OH)$_2$. However, none of the electrochemical or chemical materials (except for ideal, ordered β-Ni(OH)$_2$) exhibit empirical formulae which are consistent with these ideal formulae. Barnard *et al.* reported precise empirical formulae for these four materials, none of which have Ni/O/H ratios of 1/2/1 or 1/2/2 (4). Since the discharged materials were thought to be close packed, and the charged γ-phase is non-close packed, it has generally been accepted that there is a phase change between the charged-γ and discharged-α materials and between the β/β and α/γ cycles. Barnard *et al.* also defined a thermodynamic model to relate these four materials; in so doing they predicted that these four materials are unique phases (4).

In contrast with this *multi-phase* model, a *nonstoichiometric, solid solution* (NSS) structural model was proposed in 1986 (12). Raman spectra of the active mass materials showed that each had the same selection rules and, therefore, the same structure. Powder XRD, IR, and EXAFS analyses of these materials confirmed this result; the discharged materials are not close packed β-$Ni(OH)_2$ phases and the four Barnard materials have a common, non-close packed structure with the $R\bar{3}m$ space group (6,7,11). Very recently Kostecki and McLarnon came to a similar conclusion (13).

Several electrochemical phenomena are unique to the nickel electrode. These include: a) the formation process (in which electrochemical active mass is changed from a green material, lacking the ability to cycle, to a black material which can be cycled 30,000 cycles), b) a maximum oxidation state of 3.67 for γ-NiOOH, c) a maximum oxidation state of 3.25 for charged β-NiOOH, and d) minimum oxidation states reported between 2.2 and 2.7. The multi-phase structural model does not explain these phenomena, either qualitatively or quantitatively.

In contrast, the nonstoichiometric structural model does explain each of the above phenomena (a-d), in particular how the non-integer oxidation states are accommodated in the crystal, and changes in capacity (1,6,7,12). This model explains in simple fashion, the variations in Ni/O/H ratios, oxidation states, and electrochemical properties. The nonstoichiometry (x of up to 25%) is accommodated by nickel cation vacancy point defects. The unique structural difference between the Barnard-type, 2β/3β and 2α/3γ cycles is the magnitude of the nickel vacancy content, *not* different phases. The discharged materials are not β-like; these differ from the chemically prepared materials. All four materials exhibit the same non-close packed structure with ...ABBCCA... layer stacking (11,12).

The Bode diagram (eqns. 1-2) has been modified and expanded to include changes in nonstoichiometry which occur during the ageing process (eqns. 5), the formation processes (eqns. 3), cycling (eqns. 1-2), and overcharging (eqn. 4) (1,7).

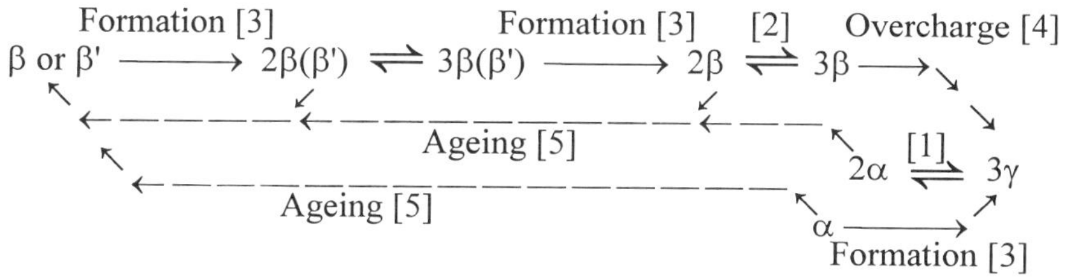

[3-5]

This modified diagram elucidates the structural differences between chemical and electrochemical (active mass) α and β materials. These equations define the relationship of the Barnard *active mass* materials, 2β/3β/2α/3γ, to each other and to precursor materials, α, β', and β, where β' = the nonstoichiometric, disordered β-nickel hydroxide (6,11). The coefficients ("2" and "3") are added to indicate that these materials, especially the two discharged materials, differ in structure from precursor α and β phases.

$$[\overset{4+}{Ni}_{1-(n+1)a-2b-c-d}\ \overset{3+}{Ni}_{(na+b)}\ (nH)_a\ (K)_b\ (Co)_c\ (V_{Ni})_d]\ OO\ H_1 \qquad [6]$$

Nonstoichiometric formulae (equation 6), containing point defects, explain how these variable oxidation states are accommodated (1,6,12). The cations within the left-hand brackets of equation 6 are taken to be on octahedral cation sites within the NiO_2 layers; and the cations to the right of the two oxygen atoms are taken to be on interlamellar sites, between two different NiO_2 layers. Foreign cations, such as potassium or hydrogen, on vacant nickel sites (V_{Ni}) and Ni^{4+} on Ni^{2+} sites are examples of point defects. The presence of higher valent nickel atoms (Ni^{3+} on Ni^{2+} sites of $NiOOH_2$ and Ni^{4+} on Ni^{3+} sites of NiOOH) is required to maintain charge balance for the higher average oxidation states observed. The nonstoichiometry ($x = a+b+c+d$), lower valent cations (K^+ and H^+), and number of protons/Ni site (n) influence the oxidation state of nickel in equation 6. When cobalt is present, it is possible to define the nonstoichiometric formulae with cobalt atoms on nickel sites (Co_{Ni}) (6,7,12). To define a point defect containing, nonstoichiometric structural model, one must refer to the ideal stoichiometry defined by a particular lattice or crystal structure. For these four, active mass materials, this is the non-close packed structure described above. The average nickel oxidation state of a material is, therefore, controlled by the crystal structure, the dopant and additive cation oxidation states, point defects, and the crystalline nonstoichiometry. The oxidation state of each cation on a nickel or hydrogen site and the number of each explains the observed oxidation states.

EXPERIMENTAL

γ-nickel oxyhydroxide was oxidized as it precipitated in a concentrated, aqueous KOH/oxidant (NaOCl) solution, using literature methods (1). The x-ray absorption spectrum of this sample was run with the oxidant present to assure a maximum oxidation state. This is denoted as the "oxidized" sample. Some of this precipitate was washed in water to remove excess oxidant and KOH and then scanned (denoted as the "washed" sample). This washed sample was then reduced in ethanol (denoted as the "reduced" sample). A fourth sample from an earlier preparation was scanned and is denoted as the "dried" sample. Raman spectra of these materials indicated that these are non-close packed γ- and α-type materials that are isostructural with battery electrode active mass (1). A recrystallized β-$Ni(OH)_2$ was scanned and is used as the Ni(II) standard (1).

There are several advantages to using chemical redox rather than electrochemical redox. First, incomplete oxidation due to imperfect electrical contact with the current collector is avoided. Secondly, there is no nickel plaque to interfere with quantitative nickel determination. Finally, incomplete reduction, due to poor electrical contacts, is avoided.

The Ni(II) and Ni(IV) spectra have been used as standards, and the spectra for samples with intermediate oxidation states have been fit to them using the

NNLS (Non-Negative Least-Squares) procedure. The proportion of 2+ and 4+ spectra composing each sample spectrum is then used to calculate the oxidation state of each. This has been implemented as a multivariable minimization problem using the NNLS constrained optimization function in MATLAB™.

Nickel K-edge x-ray fluorescence spectra were scanned at the National Synchrotron Light Source at Brookhaven National Laboratories, Upton, NY, beamline X10C. The I_f/I_o ratios are proportional to concentration (I_f = fluorescence intensity, I_o = incident intensity).

RESULTS AND DISCUSSION

The nickel K-edge spectra for 5 compounds of varying oxidation state are presented in Figure 1. The "half-heights at the edge-jump" (HHEJ) positions, E(HHEJ), are given in Table I and indicate the general trend of increasing E with higher oxidation state. The sample oxidized *in situ* has the highest oxidation state. Comparison with literature edge positions for two other Ni(IV) containing compounds, $BaNiO_3$ and $KNiIO_6$, shows that the oxidation state of this nickel oxyhydroxide is 4+ (1-3). The edge position for recrystallized β-$Ni(OH)_2$ is the lowest; it contains only Ni(II). These two samples shall be used as 2+ and 4+ standards to calculate the oxidation states of the other three compounds.

The other materials, including the washed sample, the reduced sample and a dried sample, exhibit intermediate edge positions as measured by the E(HHEJ) positions (Table I). It is interesting to note that the washed sample underwent immediate reduction upon removal of excess oxidant. The 4+ oxidation state is not stable in aqueous media. The dry sample is even further reduced; presumably by greater exposure to water and air during drying. The reduced material has the lowest edge position of these three. This material is expected to be at the minimum oxidation state allowed by the defect structure (x), and therefore, to be stable. The defect structure prevents reduction to the 2.00 oxidation state.

The derivative spectra for these compounds (Figure 2) show two maxima which represent inflection points in the K-edge. (The third, weaker maximum observed previously has been found to be spurious (1).) The 2+ maximum (E2) is at 8339.3 eV. The 4+ maximum (E4) is at 8343.8 eV. There is a contribution from the 2+ sample at 8343.8 eV which underlies the 4+ maximum. As each edge is mixed with the other, the maxima shift toward the center slightly (Table I), as expected for a weaker band on the side of a stronger band. The presence of the two maxima are direct proof that these compounds contain Ni(II) and Ni(IV). A single phase compound containing cations in two oxidation states is completely consistent with the nonstoichiometric structural model proposed previously (1,11), explaining the non-integer empirical oxidation states.

The washed, dried, and reduced sample K-edges have been fit using the non-negative least squares fitting procedure (see Figure 3). The oxidation states have been calculated from the ratios of the 2+ and 4+ contributions These are given in Table I. Rapid, time dependent changes in oxidation state are documented for the washed and dried samples. The value calculated for the washed sample from the K-edge fitting is 3.26. Reduction from the titrated value of 3.62 (1) occurred after oxidant removal. The dried sample was further reduced to 3.12 (Table I) with drying and time. The oxidation state measured for the reduced compound is 2.66, not 2.00. This material retains some Ni(IV).

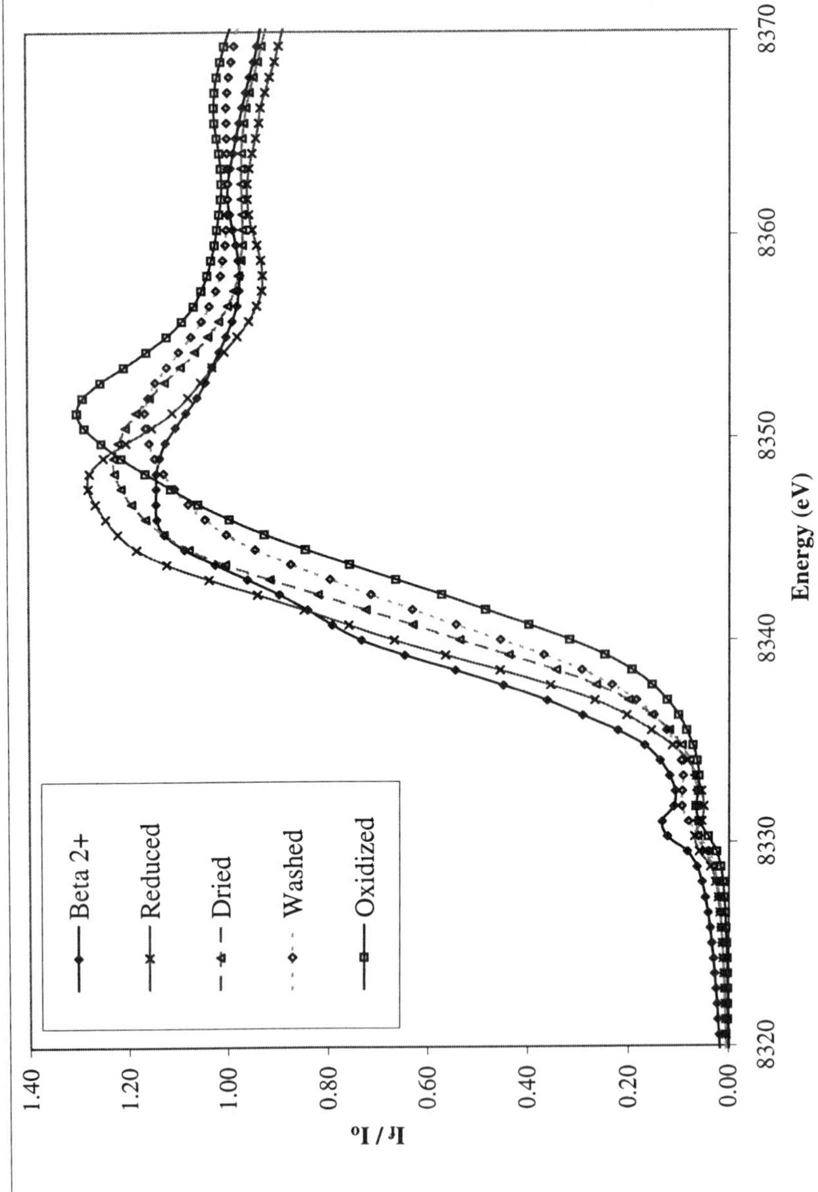

Figure 1. Normalized Nickel K-Edge X-ray Fluorescence Spectra

Table I. K-edge Positions and Calculated Oxidation States, Comparing Derivative Maxima at E2 and E4 with the Half-Height at the Edge Jump (HHEJ).

Sample	E(HHEJ)	E2	E4	Calculated Oxidation State
Oxidized	8341.72	8340.80	8343.80	(4.00)*
Washed	8340.42	8340.80	8343.00	3.26
Dried	8339.79	8340.00	8343.00	3.12
Reduced	8338.82	8339.30	8343.00	2.66
Beta 2+	8338.17	8339.30	8343.80	(2.00)*
Ni Foil	8330.40	———	———	———

* standards

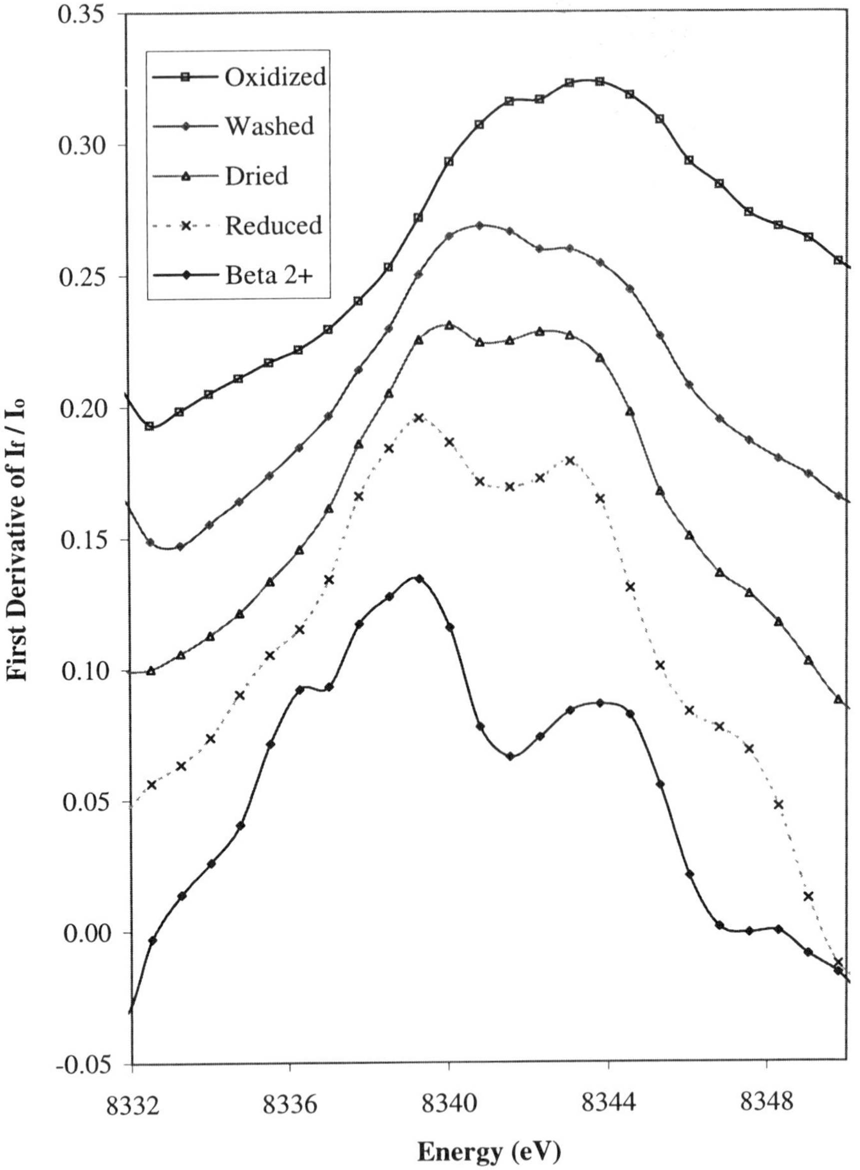

Figure 2. Derivatives of K-Edges

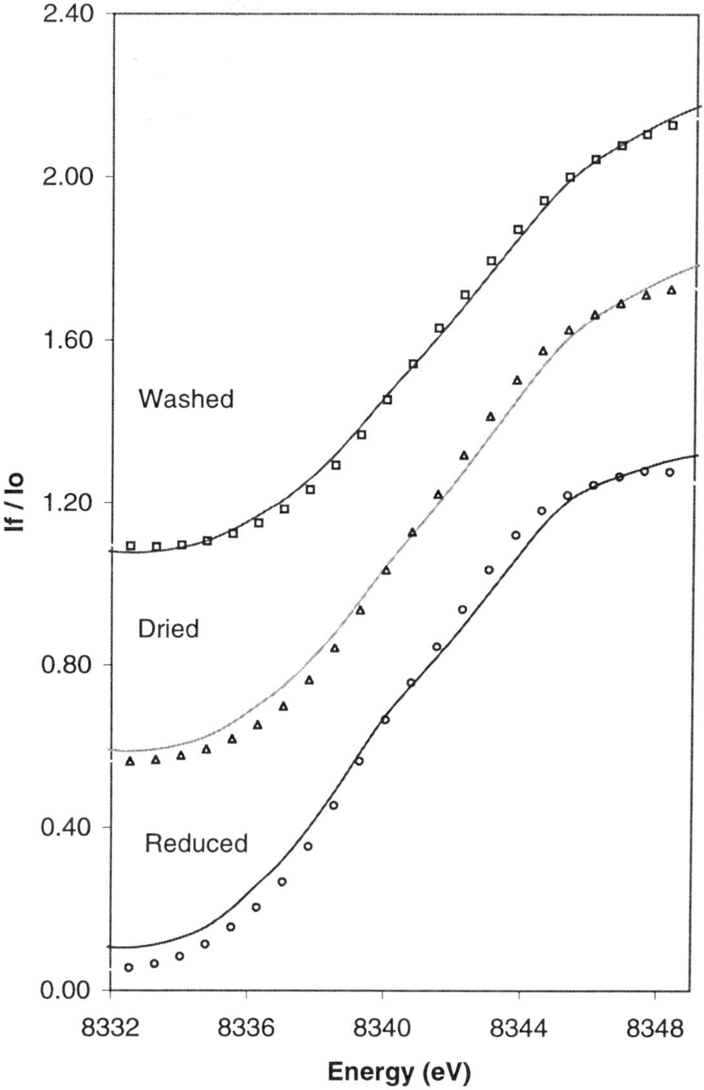

Figure 3. Comparison of Experimental K-Edges with the Non-negative Least Squares (—) Fits Using the 2+ and 4+ Standards.

The oxidation state of the reduced compound can be predicted on the basis of the oxidation state and nickel nonstoichiometry of the high oxidation state material. Immediate titrations of a washed material give an average oxidation state of 3.62 ($\pm 0.03 = 1\sigma$) (1). The x value corresponding to this oxidation state is 0.237 (see equation 7, with n = 1). The oxidation state of the reduced material is then predicted assuming x does not change appreciably during reduction. This assumption has been verified by empirical analyses (6). Using equation 7 (with n'=0) and the washed material oxidation state (3.62), the minimum oxidation state for this reduction is predicted to be 2.62 (see equation 8). This value is in excellent agreement with a titrated value of 2.59 (± 0.19) (1) and a value of 2.66 calculated from the edge fitting (Table I). The reduction does not return the system to a Ni(II) state; it does not reduce to $[Ni_{(0.763)}(2H_{Ni})_{(0.237)}]_1OOH_2$.

$$[Ni_{(1-x)}(n'H)_x]_1OOH_2 + [1+x(n'-n)]\,OH^- \underset{\text{discharge}}{\overset{\text{charge}}{\rightleftharpoons}} [Ni_{(1-x)}(nH)_x]_1OOH_1 + [1+x(n'-n)]\,H_2O + [1+x(n'-n)]\,e^- \qquad [7]$$

$$\overset{2.62+}{[Ni_{(0.763)}}(V_{Ni})_{(0.237)}]_1OOH_2 + 0.763\,OH^- \xleftarrow{\text{reduction}} \overset{3.62+}{[Ni_{(0.763)}}(1H_{Ni})_{(0.237)}]_1OOH_1 + 0.763\,H_2O + 0.763\,e^- \qquad [8]$$

CONCLUSIONS

The results presented exemplify the import of an *in situ* measurement technique in that these allow the observation of the maximum oxidation state in the presence of oxidant. The immediate changes in oxidation state when the oxidant is removed are documented, *i.e.* reduction (to 3.67 or 3.26), as is the further reduction that occurs as the material is dried (to 3.12). These observations explain why the "maximum" empirical average oxidation states measured in the literature vary over such a wide range. The *in situ* measurement of the oxidation state while oxidant is present demonstrates that an oxidation state substantially higher than 3.67 is attainable (4.0). This observation of essentially 100% Ni(IV), followed by reduction to 3.67, allowed one to understand, for the first time, why the maximum nonstoichiometry (x) is 0.25; the x value produced during Ni(IV) production controls the maximum observed oxidation state when oxidant is removed (1). Now it has been demonstrated that this x value also controls the minimum oxidation state upon reduction. A nonstoichiometric reduction mechanism to a stable product predicts a minimum oxidation state of 2.62, which is consistent with the two experimental measurements of the oxidation state. This success adds further support for the nonstoichiometric structural model.

ACKNOWLEDGMENTS

We are grateful to the National Synchrotron Light Source at Brookhaven National Laboratories, Upton, NY, beamline X10C, for x-ray beam time.

REFERENCES

1. B.C. Cornilsen, X. Cai, R. Tanbug, and G. Meitzner, "Nickel Electrode Oxidation States, the Presence of Ni(IV), and One-electron Transfer in Oxidized and Reduced Materials;" pp. 88-98, in *Aqueous Batteries*, PV 96-16, Edited by P.D. Bennett and S. Gross. The Electrochemical Society, Pennington, NJ, 1997.

2. A.N. Mansour, C.A. Melendres, M. Pankuch, and R.A. Brizzolara, *J. Electrochem. Soc.*, **141**, L69 (1994).

3. W.E. O'Grady, K.I. Pandya, K.E. Swider, and D.A. Corrigan, *J. Electrochem. Soc.*, **143**, 1613 (1996).

4. R. Barnard, C.F. Randell, and F.L. Tye, *J. Appl. Electrochem.*, **10**, 109 (1980).

5. H. Bode, K. Dehmelt, and J. Witte, *Electrochim. Acta*, **11**, 1079 (1966) and, *Z. anorg. allg. Chemie*, **366**, 1 (1969).

6. B.C. Cornilsen, X. Shan, and P.L. Loyselle, *J. Power Sources*, **29**, 453 (1990).

7. B.C. Cornilsen, X. Shan, and P.L. Loyselle, "The Structure and Reactions of the Nickel Electrode," p. 82 in *Nickel Hydroxide Electrodes*, PV 90-4, Edited by A. Zimmerman and D. Corrigan. The Electrochemical Society, Pennington, NJ, 1990.

8. P. Oliva, J. Leonardi, J.F. Laurent, C. Delmas, J.J. Bracconier, M. Figlarz, F. Fievet, and A. de Guibert, *J. Power Sources*, **8**, 229 (1982).

9. R.S. McEwen, *J. Phys. Chem.*, **75**, 1782 (1971).

10. O. Glemser and J. Einerhand, *Z. anorg. allg. Chemie*, **261**, 43 (1950).

11. B. C. Cornilsen, P. J. Karjala, and P. L. Loyselle, *J. Power Sources*, **22**, 351 (1988).

12. P. L. Loyselle, P. J. Karjala, and B. C. Cornilsen, "A Point Defect Model for Nickel Electrode Structure," p. 114 in *Electrochemical and Thermal Modeling of Battery, Fuel Cell, and Photoenergy Conversion Systems*, PV 86-12, Edited by J.R. Selman and H.C. Maru. The Electrochemical Society, Pennington, NJ, 1986.

13. R. Kostecki and F. McLarnon, *J. Electrochem. Soc.*, **144**, 485 (1997).

APPLICATIONS OF REDOX CHEMISTRY IN GLASS MANUFACTURING

Emery J. Hornyak, Jr., PhD
Libbey Inc.
940 Ash Street
Toledo, OH 43611

INTRODUCTION

There are a number of the processes involved in glass manufacturing in which redox chemistry plays a significant role. Among these are fining, i.e. the elimination of the seeds formed during melting, decolorizing of glass, production of certain types of colored glass, and, at times, the formation of undesirable defects, particularly in some specialty glasses. Also redox impacts the ferrous iron concentration which affects high temperature conductivity and, consequently, furnace operations

The glass manufacturing process is a continuous one, so seldom is equilibrium attained. Thus, although equilibria data are useful in indicating what the final state of the system would be, the kinetics of the reactions are most important for attaining the final quality of the glass.

In working with the manufacturing process, it is important to understand both the thermodynamics and kinetics of the reactions involved. In this report will be presented examples of data from studies which were used to solve problems and/or produce products of acceptable quality.

GENERAL BACKGROUND

In the production of commercial glasses, both the thermodynamic and kinetic properties of redox couples may be important. For the large scale melting of glasses, e.g. container or TV compositions, kinetics of the reactions may be predominant. However, on a smaller scale, e.g. specialty glasses, both the thermodynamics and kinetic behavior of the system may be important.

The redox characteristics of a particular element will be a function of the glass composition, temperature, oxygen fugacity, and redox state of the glass as influenced by the addition of oxidizing or reducing agents to the batch.

Based on the data obtained from equilibrium studies of several redox couples, Johnston (1) proposed a model which describes the change in concentration of the ratio of two valence states as a function of oxygen partial pressure at constant temperature. This model has been successful in correlating the changes in the ratio with oxygen pressure in terms of the number of electrons transferred between ions. For the generalized ionic reaction in glass

$$\frac{4}{n} M^{(n+x)+}_{(glass)} + 2O^{2-}_{(glass)} \rightleftarrows \frac{4}{n} M^{x+}_{(glass)} + O_{2_{(g)}} \qquad [1]$$

where n is the number of electrons transferred, $M^{(n+x)}$ and M^{x+} are the oxidized and reduced forms of the element, respectively, and assuming essentially constant oxygen ion activity (for a fixed glass composition) and little variation in the activity coefficients of the ions as a function of concentration, the equilibrium constant may be written as:

$$k = \left[\frac{[M^{x+}]}{[M^{(n+x)+}]} \right]^{4/n} \bullet p_{o_2} \qquad [2]$$

However, the oxygen activity is a function of the glass composition and, as the latter is modified, the equilibrium ratio of the redox couple will change. A number of studies have indicated that the ratio of the reduced to the oxidized species decreases with increasing basicity of the solvent glass. Therefore, these shifts in equilibria may make it necessary to modify process conditions, particularly for "smaller" scale melts, as the composition is changed. An example of this will be discussed.

Also of significant importance is the temperature dependence of the redox equilibria. This, together with kinetic considerations, may dictate what process conditions must be used to successfully obtain a product meeting the quality criteria.

FINING

In glass manufacturing, fining is the process by which bubbles formed by the gases released from the chemical reactions of the batch material (typically CO_2) and air entrapped during the melting process (N_2) are removed. The acceptable level of bubbles - also called "seeds" - is dependent upon the quality requirements of the product.

For most glass systems, fining is the most time consuming part of the glass preparation process. In addition, the compounds normally used to accelerate fining are either hazardous materials to handle and/or contribute to environmental concerns.

Therefore, industry has continually focused on attempting to improve the fining process through a combination of operational changes and optimization of fining agent concentrations.

To assist in the fining process, chemical compounds having multivalent ions are added to the batch. The most used compounds today are antimony oxide, arsenic oxide, and sulfates (generally with carbon).

Antimony and Arsenic Oxides

In order to understand the mechanism of fining by antimony and arsenic oxide, it is necessary to know the thermodynamic behavior of the systems. The relevant reactions for these systems are:

$$Sb^{5+}_{(glass)} + O^{2-}_{(glass)} \rightleftharpoons Sb^{3+}_{(glass)} + \tfrac{1}{2}O_{2(g)} \qquad \boxed{3}$$

and

$$As^{5+}_{(glass)} + O^{2-}_{(glass)} \rightleftharpoons As^{3+}_{(glass)} + \tfrac{1}{2}O_{2(g)} \qquad \boxed{4}$$

The process by which the two compounds influence the fining process can be described as follows.

As the temperature increases during the melting phase, reactions (3) and (4) will proceed to the right, i.e. to the reduced state, resulting in the formation of O_2. The physical solubility of O_2 is low and, therefore, the melts become supersaturated with respect to it. Because of this, the O_2 will either form bubbles which will coalesce with other seeds formed during melting or the O_2 will diffuse into seeds formed during the melting phase. In both cases, the seed size increases and the rate of rise of the bubbles, or the fining rate, is accelerated because of this. This is called the primary phase of fining.

Subsequently, after the batch reactions and primary fining have been completed, the temperature of the melt is decreased to condition the glass for forming. As this occurs, reactions (3) and (4) will want to proceed to the left, i.e. to the oxidized state. Thus, the remaining seeds, which should be principally O_2, will react with the Sb^{3+} or As^{3+} and be absorbed. The latter is referred to as the secondary phase of fining, or refining.

The effectiveness of these fining agents is clearly dependent upon the thermodynamics of the system. This is illustrated in Figure (1) which presents thermodynamic data for the arsenic equilibrium in a soda-lime-silica glass, as determined by Baak (2) and the antimony equilibrium in a sodium disilicate glass, reported by Baak and Hornyak (3). As can be seen, pentavalent antimony (Sb^{5+}) is easily reduced at lower temperatures while pentavalent arsenic is considerably more stable to higher temperatures. This data appears to verify statements made over the years that antimony is an excellent fining agent for "lower" melting temperature glasses while arsenic is preferred as the fining agent for "higher" melting temperature glasses.

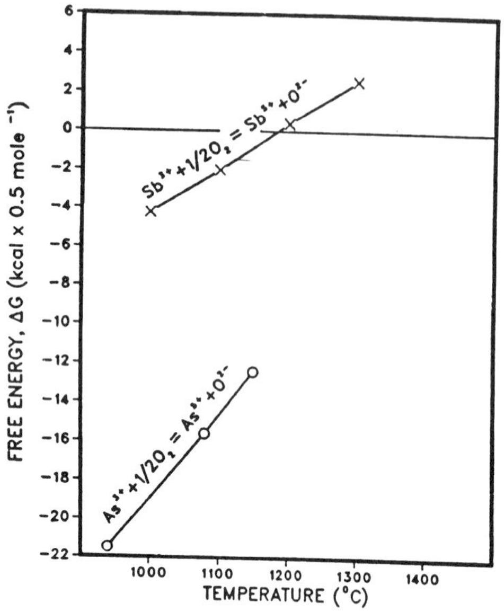

Figure 1: Free Energy vs. Temperature for Arsenic and Antimony in Glass

Following are data obtained in laboratory studies on a commercial glass composition in which both antimony and arsenic oxides were used as fining agents. The antimony was added as sodium antimonate (Sb^{5+}), the arsenic as the trioxide (As^{3+}), and nitrates were used in the batch. The temperature of these studies was ≈1370°C which mathematical modeling studies indicated was the maximum temperature attained by the majority of the glass in the melting and fining phase.

In Figure (2) is depicted the effect on fining of variations in the Sb^{5+} concentration in the batch. The data are compared to a batch having no antimony or arsenic oxides. In addition, the indicated borderline refining batch was a production one having what was determined to be the minimum concentrations of antimony and arsenic oxides to produce an acceptable seed level in the product.

As can be seen, the degree of refining, as expected, improves as the concentration of Sb^{5+} is increased. However, there is an indication that, as the Sb^{5+} concentration increases, the incremental rate of increase in the refining rate decreases. This is consistent with published results of other investigators. For example, Beerkens, et. al. (4) concluded from their mathematical model that up to 0.6 wt.% antimony oxide, the fining process was accelerated by an increase in the fining agent concentration. Above that concentration, there was minimal improvement in the fining.

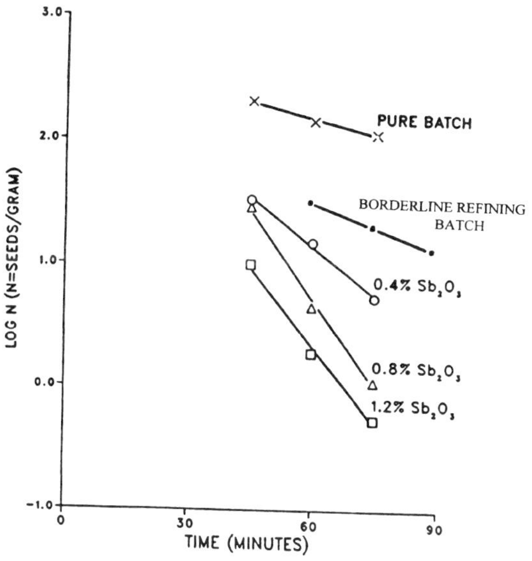

Figure 2: Effect of Antimony Level on Degree of Refining

The results obtained from an investigation of the effect of the arsenic trioxide concentration on the fining of this composition is illustrated in Figure (3). The data are the seed counts from the longest time samples (75 minutes) of the tests. As can be seen, significant increases in the concentration of the % As_2O_3 in the batch had little effect on the final seed count. That this would be the case is evident from the data in Figure (3) from which it can be concluded that the reaction

$$Sb^{5+}_{(glass)} + As^{3+}_{(glass)} \rightarrow Sb^{3+}_{(glass)} + As^{5+}_{(glass)} \quad \boxed{5}$$

is favored. Therefore, the As^{5+} is stabilized and would have little effect on the fining.

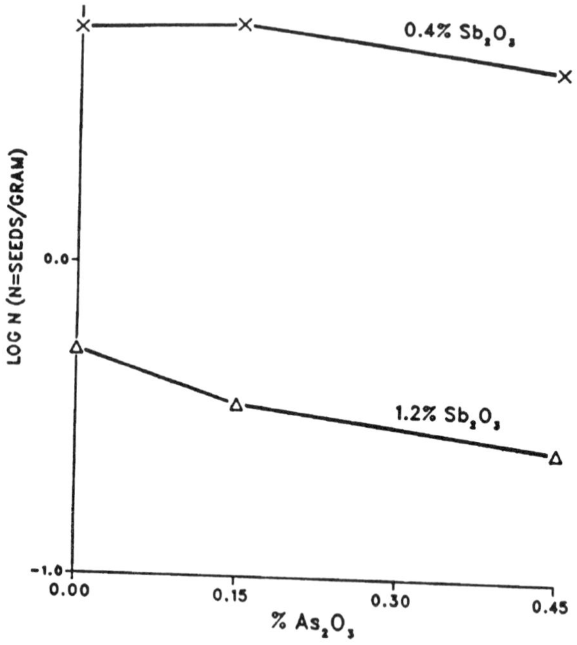

Figure 3: Effect of As_2O_3 Content on Refining

In Figure (4) is depicted the necessity of using nitrates when antimony is added to the batch in the trivalent state as Sb_2O_3. As can be seen, when no nitrates are added, no refining takes place. The seed counts of the batch are essentially the same as those of the pure batch with no fining agents (see Figure 2). However, when nitrates are added, the Sb^{3+} is oxidized and, therefore, the Sb_2O_3 containing batch refines just as well as the one in which the antimony is added as Sb^{5+}.

As a result of these studies, several modifications were made to the fining system of the production operation. These included the following:

1. A significant reduction of the amount of antimony oxide used in the batch.
2. The elimination of arsenic trioxide.

3. Some modification of the level of nitrate.

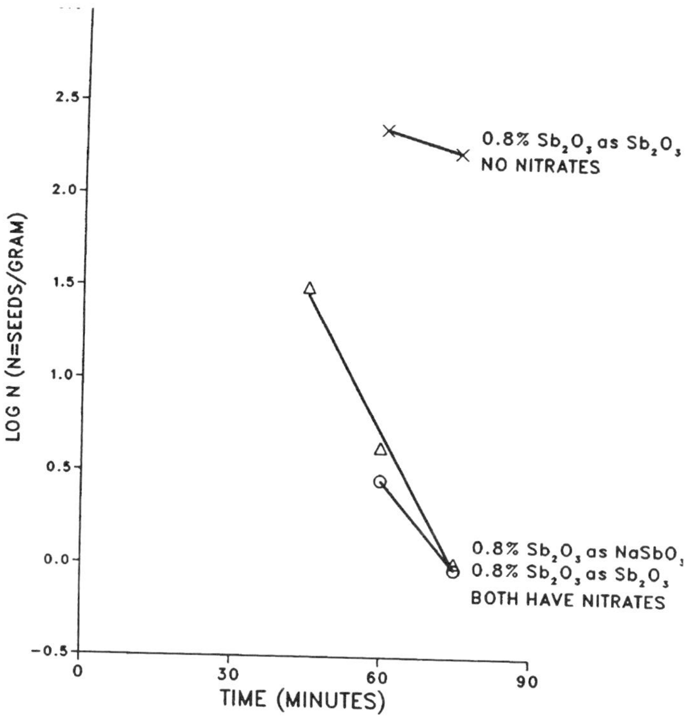

Figure 4: Effect of Antimony Valence State on Refining

Sulfates as Fining Agents

Sulfates, added principally as Na_2SO_4 and sometimes $CaSO_4$, are used as the fining agents for glass container, flat glass, and tableware compositions. Thus, on a tonnage basis, sulfates are the most widely used fining agents in the industry.

The chemical reaction defining the action of sulfates is as follows:

$$SO^{2-}_{4_{(glass)}} \rightleftarrows SO_{2_{(g)}} + \tfrac{1}{2}\, O_{2_{(g)}} + O^{2-}_{(glass)} \qquad \boxed{6}$$

Both the SO_2 and, as stated previously O_2, have limited solubilities in the molten glass. The oxygen may react with reducing agents, thus leaving SO_2 as the gas available for fining.

Equilibrium solubility data for reaction (6) have been determined by Holmquist (5), in sodium-disilicate glass, and Papadopoulos (6), in soda-lime-silica glasses. In these studies, samples of the glasses were equilibrated in atmospheres having different $p_{SO_2} \bullet p_{O_2}^{1/2}$ ratios.

The results of these studies indicated that the solubility of sulfate increased with a higher $p_{SO_2} \bullet p_{O_2}^{1/2}$, increasing Na_2O concentration in the glass, and decreasing temperature. The data imply that both temperature and the redox state are important parameters related to the efficiency of fining with sulfates.

One process which may be used to refine with sulfates is to heat the melt to sufficiently high temperatures that (6) shifts significantly to the right. When the internal pressure of the gases generated (SO_2 + O_2) exceeds ~1.2 bar, they will diffuse rapidly into the existing seeds, thus enlarging them which increases the rate of bubble rise by buoyancy effects, and, subsequently, leading to more rapid removal of the bubbles from the melt.

A second method is to modify the oxidation state of the batch by adding a reducing agent such as carbon to it. This initiates the decomposition reaction at a lower temperature which starts the fining earlier in the melting process and also accelerates the dissolution rate of silica. Thus, for a given production rate, this generally allows for running at lower temperature, or conversely, to increase the production rate at a given operating temperature.

The concept of the redox number of batches, originally devised by Manring and Hopkins (7), and modified by others over the years, has proven to be quite valuable in the sulfate fining of glasses. This is an empirical approach to determining the effectiveness of oxidizing and reducing agents in the glass melting process, principally relative to fining and control of glass color. Relative redox values are assigned to different batch materials which have a role in fining or color development. The concentrations of the different materials can be varied to give a particular redox number to the batch which then results in a glass having the desired characteristics e.g. low seed levels and desired flint (crystal) or amber glass color.

In most instances, the carbon-sulfate ratio, or redox number, must be optimized to obtain both the desired quality and color and evermore stringent pollution requirements. In addition the concentrations of sulfate in the batch, and the resulting concentration in the glass, must be consistent with both the sulfate solubility limits and the operating parameters of the commercial operation to avoid both foaming and blister problems.

PROBLEMS OF MELTING LASER GLASS IN PLATINUM (8)

The glass lasers produced in this program were used in the initial development of systems for nuclear fusion research. Because of the high power applications, the glass had to meet very stringent requirements of high purity, homogeneity, and have very low seed and inclusion levels.

Because of the quality required, platinum melters were used. However, other problems surfaced which were the result of using platinum, and thus, considerable process development was required in order to meet all quality criteria.

The first generation of glasses used for this program was silicate based. In this system, at times it was found that micron and sub-micron size platinum inclusions were formed during the melting process. These defects would absorb energy from the laser pulse, either expand or vaporize, and cause permanent damage in the form of small cracks in the laser component. The results of research aimed at understanding and solving this problem will be discussed.

Preliminary Experiments

From a review of the literature, it was suggested that there were several probable sources of platinum contamination. Therefore, attention was devoted initially to gaining a more detailed understanding of the mechanisms of platinum particle formation.

An experimental approach was devised in which the laser glass was exposed to volatile oxides of platinum formed by exposing the latter to either O_2 or air atmospheres. No platinum particles were found in either case. However, upon switching from air or O_2 to a neutral atmosphere (N_2), very rapid growth of particles occurred. This work indicated that solution and precipitation may be the predominant mechanism. It further substantiated the fact that atmosphere control was a necessary requirement of the process.

This mechanism might occur via the following reaction sequence:

$$Pt_{(s)} + \frac{n}{4}O_2 \rightleftarrows \mathrm{Pt}^{n+}O_{n/2_{(g)}} \qquad \boxed{7}$$

$$\mathrm{Pt}^{n+}O_{n/2}\ (\mathrm{g}) \leftrightarrows Pt^{n+}_{(glass)} + \frac{n}{2}O^{2-}_{(glass)} \qquad \boxed{8}$$

$$Pt^{n+}_{(glass)} + \frac{n}{2}O^{2-}_{(glass)} \rightleftarrows \mathrm{Pt}^{0}_{(s)} + \frac{n}{4}O_{2_{(g)}} \qquad \boxed{9}$$

This indicates that the dissolved platinum content is a function of the O_2 partial pressure of the system and that ionic Pt will be retained in solution if p_{O_2} and the oxidation state of the glass are maintained at sufficiently high levels.

Thus, one necessary phase of this investigation was a thermodynamic description of the equilibrium state of the system, specifically,

1. the equilibrium solubility of Pt as a function of O_2 partial pressure;
2. the oxidation state of the glass as reflected in the equilibrium Ce^{3+}-Ce^{4+} ratio as a function of p_{O_2}

Equilibrium Studies

The isothermal equilibria to be established were:

1. Pt, O_2 (p_{O_2} = const.), glass (T = 1450°C)
2. $2Ce^{4+} + O^{2-} \rightarrow 2Ce^{3+} + \frac{1}{2}O_2$ (T = 1450°C, 1232°C)

The equilibrium Pt solubilities were determined at 1450°C using: (a) a closed system for $p_{O_2} \geq 0.21$ atm. and (b) an open system for $p_{O_2} < 0.21$ atm. In the latter case, periodic mass spectrometer analyses of the atmosphere were conducted. Long time equilibration periods were required to achieve equilibrium. Commercial ED-2 glass was used for the higher O_2 pressures, whereas for the lower O_2 pressures, special high purity glasses were required since the Pt content of the commercial glasses was higher than the equilibrium solubility.

After the desired time at temperature, the glasses were removed, cooled, and carefully inspected microscopically for evidence of metallic particles in the bulk and on the surface. Since the ionic Pt content was wanted, only samples free of metallic particles were analyzed.

The Ce^{3+} - Ce^{4+} equilibrium was studied at 1450°C and 1232°C in the open system. The samples, after equilibration, were crushed (with special care taken to avoid iron contamination) and the Ce^{4+} was analyzed by potentiometric titration with Fe^{2+}. The total Ce content was determined, after complete oxidation to the tetravalent state, by Fe^{2+} titration. Trivalent cerium was determined by difference.

The equilibrium Pt solubility in ED-2 as a function of O_2 pressure at 1450°C is shown in Figure (5). It is seen to vary between $\approx$3-4 ppm at a p_{O_2} of $\approx 10^{-3}$ atm. to $\approx$38 ppm at 1 atm. The isothermal plot of log $[Pt^{n+}]$ vs. log p_{O_2} should have a slope characteristic of the Pt valence. Evaluation of the slope gives a value of n=1.6 which would indicate that platinum exists in the divalent state in ED-2 at this temperature. (Using only the data points for p_{O_2} = 1 atm. and 0.21 atm. gives n=1.9 which substantiates this.) This conclusion is in agreement with that from an extensive study of the solubility of Pt in sodium-silicate glasses which also was conducted in our laboratories.

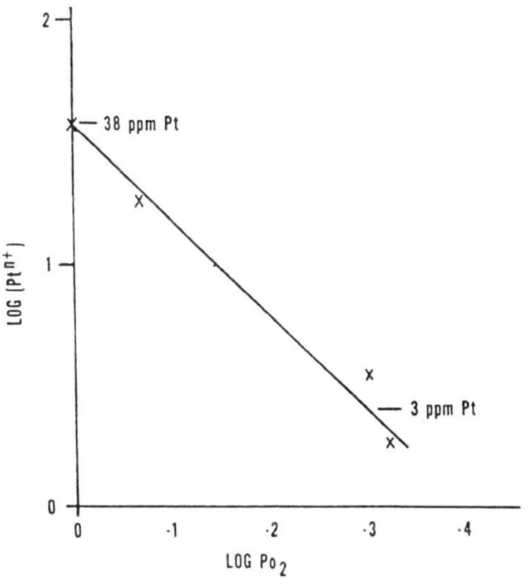

Figure 5: Equilibrium Pt Solubility as a Function of Oxygen Partial Pressure in Ed-2 at 1450°C

The equilibrium Ce^{3+}/Ce^{4+} ratios at 1450°C as a function of O_2 partial pressure are presented in Figure 6. Once again, the log-log plot is linear and, using the previously described ionic model, a value of n=0.96 is obtained which is that expected for the oxidation-reduction reaction which involves a one-electron change.

Since the log of the equilibrium relationships of both systems are linear functions of log p_{O_2}, they can be combined into a plot of the type depicted in Figure 7 in which the equilibrium Pt solubility can be expressed in terms of the Ce^{3+}/Ce^{4+} ratio. This was a useful working plot which can be applied directly to interpreting the results of measurements made on an operating system.

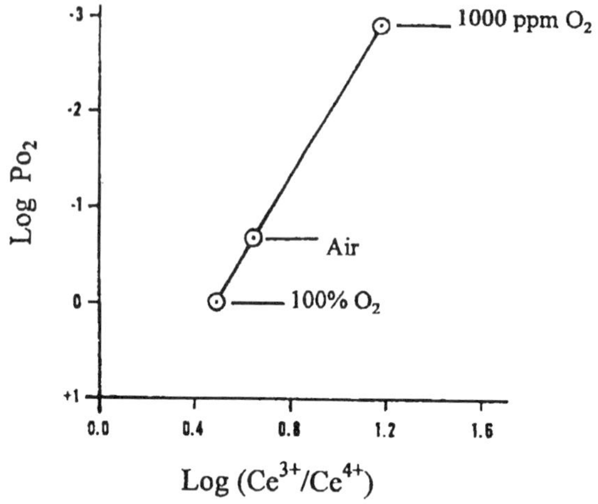

Figure 6: Equilibrium Ce^{3+}/Ce^{4+} Ratio as a Function of Oxygen Partial Pressure in ED-2 at 1450°C

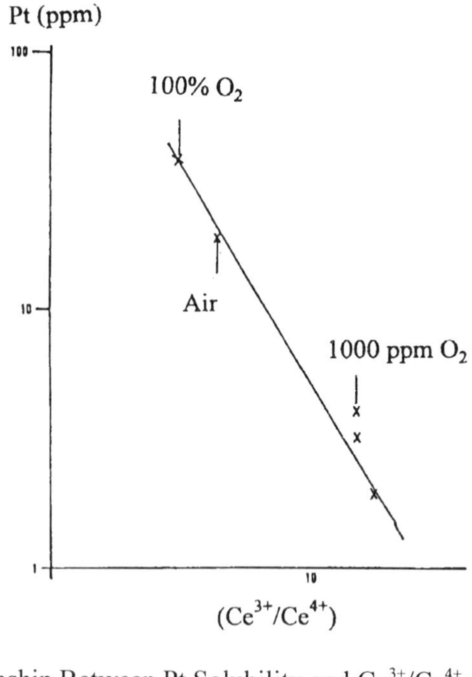

Figure 7: Relationship Between Pt Solubility and Ce^{3+}/Ce^{4+} Ratio for ED-2 at 1450°C

Application to ED-2 Production

The commercial production of ED-2 was done in platinum melters having provisions for maintaining a controlled atmosphere. Two melters having the same surface area but with the volume of one being double that of the other were used.

The second phase of this study, which was done concurrently with the equilibrium studies, involved the determination of the kinetics of the process, i.e., the rate of platinum pick-up and the change in oxidation state of the melt during the melting, refining, and homogenization cycles. This was accomplished by sampling the system at certain times during the melt cycle and analyzing the samples for Pt content and the Ce^{3+}/Ce^{4+} ratio.

The typical change in oxidation state of the melt for the two different units, as reflected by changes in the Ce^{3+}/Ce^{4+} ratio, is presented in Figure 8. This data, coupled with the working plot relating equilibrium Pt solubility to the Ce^{3+}/Ce^{4+} ratio, could be used to determine the approach to equilibrium for the systems, as is depicted in Figures 9 and 10.

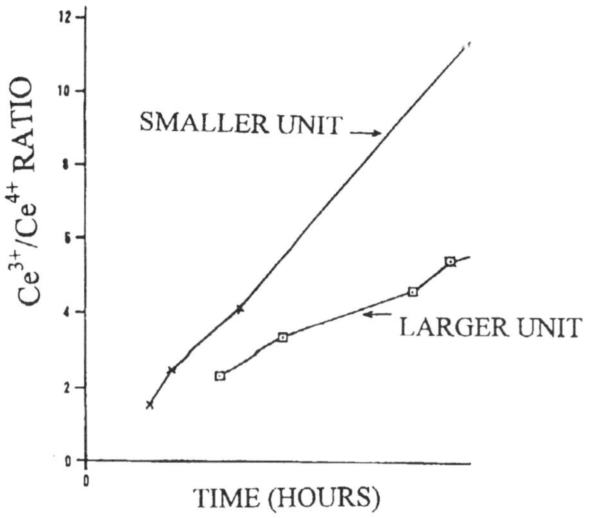

Figure 8: Change of Oxidation State as a Function of Time for the Two Different Size Melters

In the case of the smaller melter, it is seen that the majority of the platinum pick-up (≈70%) occurs during the charging and melting cycle with only a small

increase in concentration during the remaining refining and homogenization part of the cycle. The equilibrium Pt solubility of the melt at different times during the cycle, as inferred from the Ce^{3+}/Ce^{4+} ratio, is also depicted and is seen to have a marked decrease with time. In fact, for this particular cycle, there was a crossover near the end of the cycle and this melt did have Pt particles in it.
The same trend was found for the larger melter, as shown in Figure 10. In this case, the Pt content is approximately twice that of the smaller unit because of a doubling of the charging time. Again, the majority of the Pt solution occurs during this phase of cycle. However, because of the higher oxidation state of

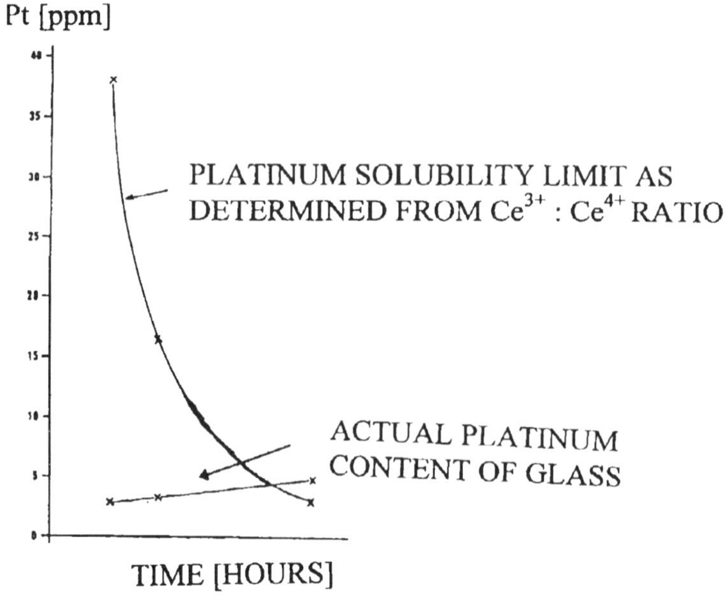

Figure 9: Actual Platinum Content and Solubility Limit for ED-2 Produced in the Smaller Unit

this system, the equilibrium platinum solubility does not decrease at as rapid a rate, and therefore, the system is not as critical with respect to potential for platinum precipitation. The fact that the larger melter has twice the volume but the same surface area as the smaller unit results in a significant modification of the kinetics of reduction of the melt. In practice, it was noted that if half the melt were cast at the end of the normal cycle and the other half a few hours later, the first cast was free of Pt particles while the second had several thousand Pt inclusions. The additional time and smaller volume of glass was sufficient to bring the melt to a critical state relative to precipitation of Pt.

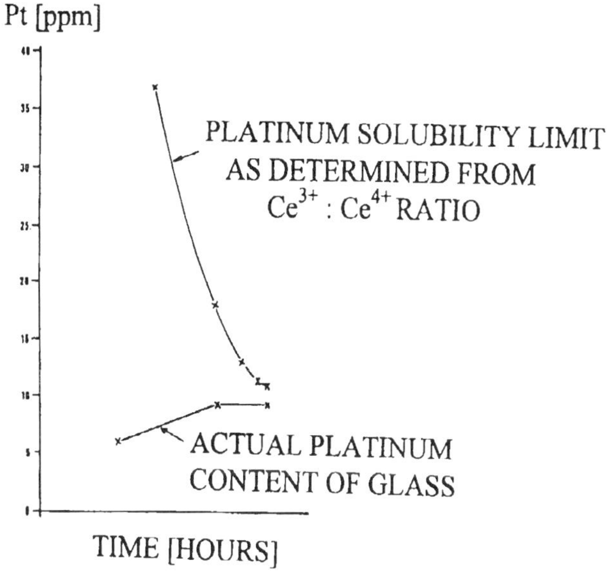

Figure 10: Actual Platinum Content and Solubility Limit for ED-2 Produced in the Larger Unit

Additional monitoring of the process included following the change in the O_2 content of the atmosphere above the melt during the cycle by use of an oxygen analyzer. Typical behavior of the two melters is depicted in Figure 11. The higher oxygen levels during the charging cycle are the result of the rapid rate of decomposition of $CeO_2(2CeO_2 \rightarrow Ce_2O_3$ (glass) + ½ O_2) and the introduction of air during charging. It will be recalled that it was during this period that the maximum platinum pick-up also occurred. This substantiates the conclusions of the laboratory studies relative to the relationship between Pt content and O_2 pressure in the system.

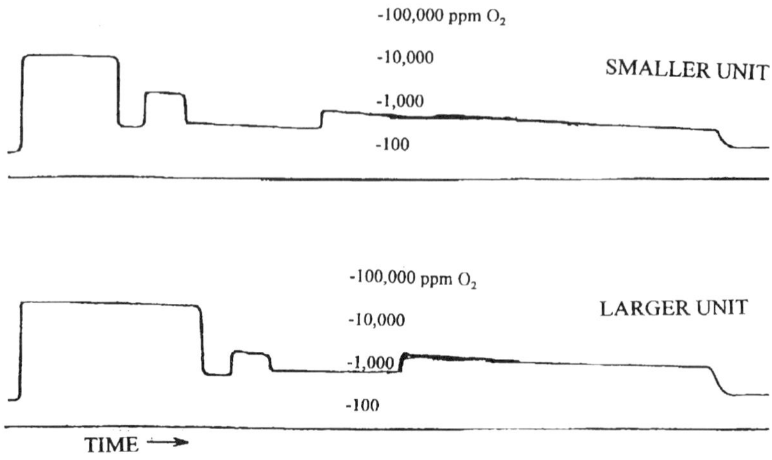

Figure 11: Melter Atmosphere Oxygen Content

Throughout the rest of the cycle a continuous O_2 evolution from the melt is noted and the O_2 pressure of the system decreases with time. The O_2 evolved is the result of the reaction

$$2Ce^{4+}_{(glass)} + O^{2-}_{(glass)} \rightarrow 2Ce^{3+}_{(glass)} + \tfrac{1}{2}O_{2_{(g)}} \qquad \boxed{10}$$

which is occurring in the glass. This is substantiated by the fact that the total O_2 evolved during a part of the cycle, determined by integrating the O_2-time curve, is almost exactly that predicted from the change in Ce^{4+} content as determined by analysis of samples taken at the start and end of that part of the melt cycle.

Thus, on the basis of the data obtained in this study, it was clear that, for a given process, one method of avoiding Pt precipitation in glasses was to manipulate the kinetics of the process such that the equilibrium state of the system is never attained in the time required to melt, refine, and homogenize the glass.

Applications to Other Silicate Glasses

Similar but less detailed investigations were done to either understand or modify process cycles required for silicate laser glasses having compositions different from that of ED-2. Such composition changes were instituted to improve the lasing properties of the system

One such composition modification involved the substitution of 10 mole % CaO for SiO_2 (ED-8). This was found to significantly affect the oxidation-reduction characteristics of the glass as is depicted in Figure 12 where the equilibrium Ce^{3+}/Ce^{4+} ratio for the modified composition is compared to that for ED-2. The data show that, for a given p_{O_2}, the modified composition has a significantly higher Ce^{4+} content at equilibrium (much lower Ce^{3+}/Ce^{4+} ratio) than ED-2. This implied that, if the kinetics of the melting process were the same, the modified system would attain equilibrium sooner and Pt precipitation would occur. Indeed, it was found that significant modifications to the melt cycle were required in order to produce glass free of Pt inclusions.

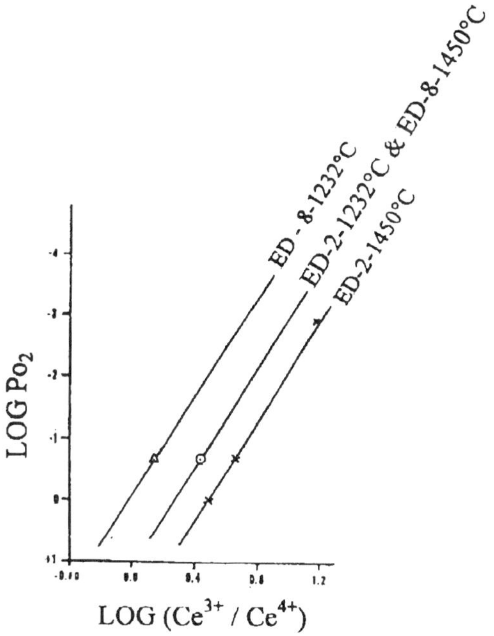

Figure 12: Comparison of Equilibrium Ce^{3+}/Ce^{4+} Ratios as a Function of Oxygen Partial Pressure for the ED-2 and ED-8 Compositions

A second composition modification involved the removal of CeO_2 from ED-8 and the substitution of Sb_2O_3 as the solarization inhibitor. In contrast to the previous efforts, for this modification we were asked to predict what process cycle would be required to produce a glass free of Pt inclusions. Therefore, equilibrium Sb^{3+}/Sb^{5+} ratios were determined and compared to the Ce^{3+}/Ce^{4+} ratios for the same composition. The results are presented in Figure 13. One striking feature of this plot is the observation that the Sb^{3+}/Sb^{5+} ratio is more sensitive to changes in both p_{O_2} and temperature. Thus, if the Sb^{3+}/Sb^{5+} ratio of the glass at melt temperatures was greater than that at a lower temperature, lowering the temperature might result in a strong tendency for oxidation of Sb^{3+} to occur. One way in which this could happen is via the reaction

$$Sb^{3+}_{(glass)} + Pt^{2+}_{(glass)} \rightarrow Sb^{5+}_{(glass)} + Pt^{0}_{(s)} \qquad \boxed{11}$$

Based on the equilibrium data, it was predicted that, in order to avoid Pt particle formation, the process would have to be isothermal, i.e. melting and casting would have to be done at the same temperature. This recommendation was not followed and the first melt used the conventional cycle, i.e. melt at 1450°C and cast at 1232° C.

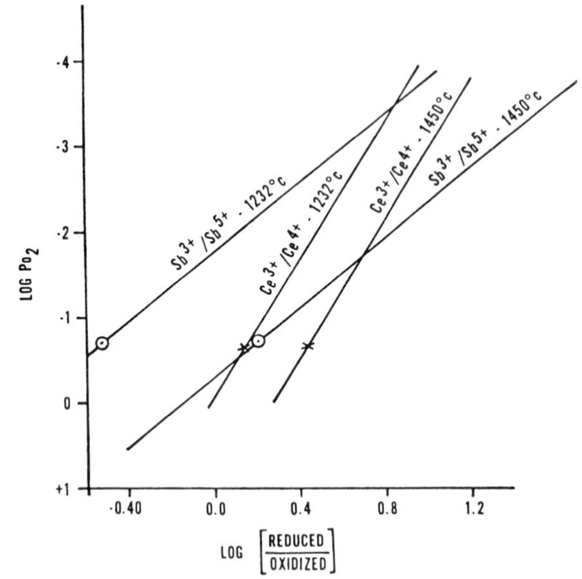

Figure 13: Cerium and Antimony Equilibrium Data for ED-8

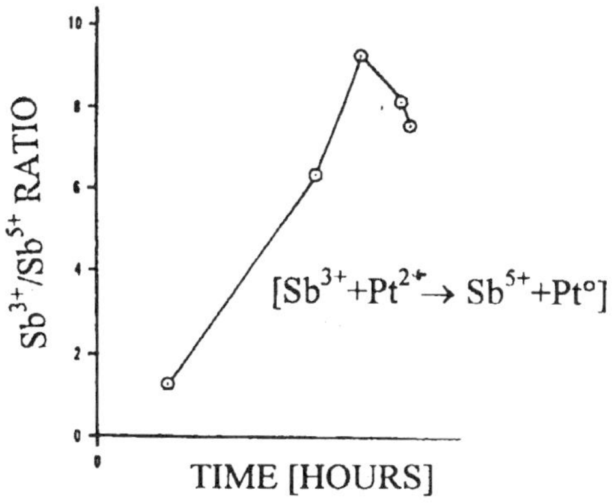

Figure 14: Sb^{3+}/Sb^{5+} as a Function of Time for an ED-8 Melt

In Figure 14 is depicted the Sb^{3+}/Sb^{5+} ratio as a function of the melt cycle. As can be seen, a maximum in the Sb^{3+}/Sb^{5+} ratio was obtained. A component cast at the maximum was inclusion-free, whereas one cast later at the lower temperature was found to contain thousands of Pt particles. This verified the prediction, based on the equilibrium relationships, that reaction (11) would take place.

Once again, significant changes in the melt cycle were required to produce glass of this composition which did not contain Pt inclusions. Specifically, it was necessary to melt and cast the glass at the same temperature.

COLORED GLASSES

Transition metal and rare-earth elements are important in the manufacturing of colored glasses. Many of these can exist in more than one oxidation state.

There are several of these elements in which one oxidation state is preferred in glass. For example, cobalt exists primarily in the +2 state in glass and imparts a blue color. The intensity of the color is easily varied just by changing the concentration of the element.

However, there are times when the multivalent character of the element is important and the redox chemistry and kinetics of the reaction must be manipulated to obtain the desired color.

The latter will be illustrated by a discussion of the steps which were taken in the development of selenium and manganese based colors.

It should be mentioned that these colors were produced by the addition of color concentrates, manufactured by Ferro Corp., to crystal glass in a forehearth. Thus, all melting, fining, and mixing of the color concentrate had to take place within the colorant section of the forehearth.

Selenium Pink

A request was received to produce a "peach" color for Marketing. After consultation with Ferro Corp. and some preliminary melting studies at their facility, the decision was made to use a color concentrate, containing selenium as the principal colorant, for the production of this color.

It should be noted that it is elemental selenium, Se^o, which gives the desired color. Elemental selenium is quite volatile and, therefore, in order to improve its retention in the glass, the selenium is added as the selenite $(SeO_3)^{2-}$ in the color concentrate. Thus, in order to obtain the color, the Se^{4+} must be reduced, i.e.,

$$Se^{4+}_{(glass)} \rightarrow Se^{0}_{(glass)} + 4e^{-} \qquad \boxed{12}$$

When put into production, some difficulty was encountered in maintaining the desired color, particularly at different facilities. After eliminating production variables as factors, it was concluded that perhaps the glass itself was a key element.

Several samples of peach were analyzed for Se^o. Indeed it was found that the Se^o varied from 3.6 ppm for a pink glass to 9.5 ppm for a dark orange sample. In addition, certain "peach" samples, as well as the base crystal glass, were analyzed for the %FeO both spectroscopically and by conventional techniques.

The results of these analyses are presented in Table I.

Table I. Selenium and FeO Contents of Selected Peach Glasses

% FeO		Elemental Selenium	
Base Glass	Peach	(ppm)	Color
0.0090	ND	9.5	Dark Orange
0.0068	0.0038	4.4	Orange
0.0060	0.0046	3.6	Pink

ND = Not determined

From the data in Table I, it can be seen that the %FeO is lower in the "peach" than in the original crystal glass. This implies that one of the reactions taking place leading to the formation of Se^{o} is the following:

$$Se^{4+}_{(glass)} + 4Fe^{2+}_{(glass)} \rightarrow Se^{0}_{(glass)} + 4Fe^{3+}_{(glass)} \qquad \boxed{13}$$

The data indicated that approximately 60% of the elemental selenium is the result of the above reaction. Thus, it was concluded that the %FeO and therefore, the redox state of the base glass was a significant factor for control of this color.

The understanding obtained here was also important in attempting to maintain consistency with another color in which selenium was a significant component. In this case, while producing this color, programs were also initiated to improve the "crystal" quality of the base glass by reducing the total $\%Fe_2O_3$ and, consequently, the % FeO.

As the latter program progressed, it became very difficult to maintain the desired color because of an inability to obtain the required concentration of Se^{o}. Thus, modifications were made to the color concentrate formulation to alleviate this problem.

Manganese "Purple"

A request was received to manufacture a "dark purple" product. Again, with the assistance of Ferro Corp. personnel, it was determined that a manganese based color concentrate should be used. It should be noted that Mn^{3+} is the oxidation state which gives the purple color. Laboratory melts and their experience dictated that a relatively high feed rate was required.

The product of the proper color intensity obtained from the first trial contained many blisters which subsequently were found to contain only oxygen, O_2. In addition, an analysis of the glass indicated that only ~10% of the manganese added was present as Mn^{3+} in the glass.

The data indicated that the following reactions occurred:

$$2MnO_2 \rightarrow Mn_2O_3 + \tfrac{1}{2}O_2(g) - very\ rapid \qquad \boxed{14}$$

$$Mn_2O_3 \rightarrow 2MnO + \tfrac{1}{2}O_2(g) - rapid \qquad \boxed{15}$$

Thus, it was obvious that a high concentration of MnO_2 was required to obtain a small amount of Mn_2O_3.

At this point, attention was focused on attempting to determine if it would be possible to stabilize Mn^{3+} in the glass. In Figure 15, are plotted the equilibrium curves for (Mn^{2+}, Mn^{3+}) and (Cr^{3+}, Cr^{6+}) as determined by Schreiber (9). The data imply that, if the higher oxidation state was present, it would oxidize Mn^{2+} that was formed, thus acting as a buffer, and a higher concentration of Mn^{3+} would be maintained in the glass.

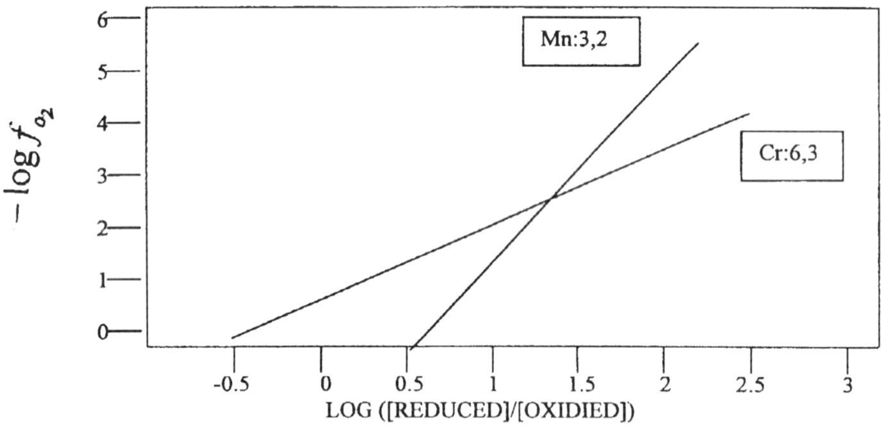

Figure 15: Percentage of Redox Couple in the Reduced State [after Schreiber (9)]

This was posed to Ferro Corp. personnel who indicated that low concentrations of Cr_2O_3 had in fact been added to manganese containing color concentrates to obtain that effect. Presumably, the reactions which occur are as follows:

$$2MnO_2 \rightarrow Mn_2O_3 + \tfrac{1}{2}O_2 \qquad \boxed{16}$$

$$Cr_2O_3 + \tfrac{3}{2}O_2 \rightarrow 2CrO_3 \qquad \boxed{17}$$

$$2CrO_3 + 6MnO \rightarrow 3Mn_2O_3 + Cr_2O_3 \qquad \boxed{18}$$

The addition of low concentrations of Cr_2O_3 to the initial manganese containing color concentrate did in fact allow for a significant reduction in the feed rate and the obtaining of commercially acceptable product.

CONCLUSIONS

It has been shown that there are a number of processes in which redox chemistry plays a significant role in glass manufacturing. Increasingly tighter specifications with respect to product quality requires that a better understanding of the fundamentals of these processes is required.

It has been shown that data obtained in the laboratory can be usefully applied to large scale production. For example, in the fining area, the data were used to modify the fining agent concentrations without affecting glass quality. In applying laboratory data to a production process, it must be recognized that it is the trends which are important, and not necessarily the absolute numbers.

It is encouraging to see that, in the fining area, some agreement has been noted between laboratory data and a mathematical model of fining.

It was also pointed out that laboratory equilibrium data, together with that obtained on samples from the manufacturing process, can be used to both solve problems as well as predict process parameters or composition modifications which can be implemented to obtain the desired quality or color.

It should also be noted that the redox relationships obtained for one glass system appear to be general enough that they can be applied to other compositions. Thus, in the absence of data on a given system, one can use with a certain degree of confidence redox data which has been obtained on another glass composition.

ACKNOWLEDGMENTS

The author is grateful to Libbey Inc., Techneglas Inc., Owens-Illinois, Inc., and Ferro Corp. for giving permission to publish information presented in this report. I also am deeply appreciative to Ms. Mary Clare Powers and Ms. Marguerite Erpelding of Libbey Inc. for their effort in preparing this manuscript.

REFERENCES

[1]W. D. Johnston, "Oxidation-Reduction Equilibria in Molten $Na_2O{\bullet}2SiO_2$ Glass," *Journal of the American Ceramic Society*, **48** [4], 184-90 (1965).

[2]T. Baak, "Thermodynamics of Arsenic in Glass," *Journal of the American Ceramic Society*, **42** [1], 27-29 (1959).

[3]T. Baak and E. J. Hornyak, "Antimony in Glass." *The Glass Industry*, 374-78 (1966), 418-20, 455-57 (1966).

[4]R. Beerkins, E. Muijsenberg, G. Peters and O. Verheyen, "Modeling Explores Bubble Behavior in Molten Glass," *The Glass Researcher*, **6** [2], 19-23, 25 (1997).

[5]S. Holmquist, "Oxygen Ion Activity and the Solubility of Sulfur Trioxide in Sodium Silicate Melts," *Journal of the American Ceramic Society*, **49** [9], 467-73 (1966).

[6]K. Papadopoulos, "The Solubility of SO_3 in Soda-Lime-Silica Melts, *Physics and Chemistry of Glasses*, **14** [3], 60-65 (1973).

[7]W. H. Manring and R. W. Hopkins, "Use of Sulfate in Glass," *Glass Industry*, **39** [3], 139-142, 170 (1958).

[8]E. J. Hornyak and R. P. Abendroth, "Problems of Melting Laser Glass in Platinum," Presented at the 80th Annual Meeting of the American Ceramic Society (1978).

[9]H. D. Schreiber, G. B. Balazs, B. E. Carpenter, et. al., "An Electromotive Force Series in a Borosilicate Glass-Forming Melt," *Journal of the American Ceramic Society*, **67** [6], C106-C108 (1984).

DEPENDENCE OF THE REDOX STATE OF A POLYVALENT ELEMENT ON ITS CONCENTRATION IN GLASS

Henry D. Schreiber, Nicholas R. Wilk Jr., and Charlotte W. Schreiber
Department of Chemistry, Virginia Military Institute, Lexington, VA 24450-0304

ABSTRACT

Ever since the 1930s, it has been generally accepted that the iron redox ratio, Fe^{2+}/Fe^{3+}, increases as the total iron concentration, [ΣFe], decreases in soda-lime-silicate glasses. This trend has been reported as especially pronounced when [ΣFe] is less than 1 wt%. Similar trends for redox ratios as a function of total polyvalent element concentration have been reported for Cr^{3+}-Cr^{6+} and As^{3+}-As^{5+} redox couples in glass. More recently, this concentration effect has been questioned by some, but confirmed by others. This study investigated Mn^{2+}-Mn^{3+}, Ce^{3+}-Ce^{4+}, and Ti^{3+}-Ti^{4+} redox equilibria in a soda-lime-silicate glass. Separate series of glasses were prepared containing a single polyvalent element over a concentration range of 0.05 to 5 wt%. The apparent stabilization of Fe^{2+}, Mn^{2+}, and Ce^{4+} at low [ΣFe], [ΣMn], and [ΣCe] respectively is shown to be an artifact of the experimental method used to analyze the redox ion concentrations. The redox ratio of a polyvalent element is generally constant in a soda-lime-silicate glass, as long as the polyvalent element concentration is less than about 5 wt%.

INTRODUCTION

When a polyvalent element is introduced into a glass-forming system, redox equilibrium is established between that element's oxidized and reduced states. For example, equation 1 represents the redox equilibrium for iron in glass.

$$4Fe^{3+} + 2O^{2-} \leftrightarrow 4Fe^{2+} + O_2 \qquad (1)$$

In order to adjust this redox equilibria to produce more Fe^{2+} at the expense of Fe^{3+} in a glass-forming system, one could make the base composition more acidic, synthesize the glass at a lower oxygen fugacity or a higher temperature, add less total iron content, or include reducing agents during processing [1]. However, to explain more clearly the compositional dependence of the redox equilibrium, equation 1 is often written as,

$$4FeO_2^- \leftrightarrow 4Fe^{2+} + 6O^{2-} + O_2 \qquad (2)$$

which indicates that the redox ions are not present as isolated ions in the silicate solvent but as oxide or coordination complexes within the silicate network [2].

The concentration of the polyvalent element establishing its redox equilibrium in an ideal solvent-solute system should not affect the value of the equilibrium constant or, in other words, the relative concentrations of the oxidized and reduced states of the polyvalent element. Such an ideal situation usually exists when the concentration of the solutes (redox ions of the polyvalent element) are sufficiently dilute for ionic activities to be approximated by their respective concentrations. However, experimental studies have shown a significant concentration effect with respect to the redox ratio of a polyvalent element [3], especially at the low concentration limit. Evidently, redox ions have dramatically changing activity coefficients, even when the ions are present at low concentrations. At the other extreme, it is not too surprising that the redox ratio of a polyvalent element varies as a function of polyvalent element concentration at high concentrations; under such conditions, the polyvalent element can contribute to a changing oxide ion activity of the base composition [4] or the redox ions can begin to cluster [5].

Historically, Densem and Turner [3] as well as Fuwa [6] in the 1930s showed that the iron redox ratio (Fe^{2+}/Fe^{3+}) increased as the iron concentration decreased in soda-lime-silicate glasses. This increase in redox ratio was especially dramatic at low iron concentrations. In fact, there was more of an increase below 0.35 wt% total iron than from 10 to 0.5 wt% total iron, as shown in figure 1. These results remain ingrained within the overall understanding of redox chemistry in silicate glasses, even though their analytical methods were considered suspect and equilibration of their samples questioned [7]. However, these early results compared favorably with the later reports of Paul [4] who studied the redox chemistry of iron in simple alkali silicates as a function of the total iron content in the glass. Once again, rather dramatic increases in the iron redox ratio were observed as the total iron concentration decreased in a concentration range below 0.5 wt%. Paul [4] also reports similar behavior for chromium and cerium redox equilibria when glass composition, melt temperature, and furnace atmosphere were all kept constant. The increasing redox ratio, or selective destabilization of the oxidized redox ion in the glass, suggested that Fe^{3+} and Ce^{4+} exist in different coordination complexes within the glass and these coordination equilibria are dependent on the concentration of the redox ion [4]. Because a coordination change is linked to a major change in the nature of bonding between the central redox ion and the surrounding ligands, the activity coefficients of the redox ions change correspondingly. Specifically, the origin of the effect of total iron concentration on the Fe^{2+}/Fe^{3+} ratio may be attributed to Fe^{3+} preferring octahedral coordination more and more as its concentration becomes less and less (even at extremely low concentrations). However, these observations of Paul [4] were disputed by Goldman [8], who questioned whether the systems were really at equilibrium and who did not find any variation in the iron redox ratio when the iron content was varied from 0.09 wt% to 0.5 wt%.

Figure 2 shows the results of a recent study of the dependence of iron redox ratio on total iron content in a soda-lime-silicate glass [9]. Although the general trend of increasing redox ratios with decreasing iron concentrations, when iron is included at relatively low iron concentrations, was observed, the rapid rise was now shifted to below about 0.1 wt% total iron. That the concentration range over which the effect is observed gets lower and lower with time may arouse suspicions about correlations to analytical methods for redox measurements getting better and better with time.

The dependence of iron redox ratio on iron's concentration in two glasses is also compared in figure 3 [10,11]. The results in this figure indicate that iron becomes a contributing factor to a glass' overall acidity at relatively high iron contents and that the magnitude of this effect varies with the actual glass composition. Once again, however, whereas this is a relatively small change in redox ratio at iron concentrations greater than 1 wt%, much larger changes in redox ratio occur at lower total concentrations of iron. Figure 4 shows the results of a recent study on the redox chemistry of arsenic and chromium in a soda-lime-silicate glass [10]. Even at relatively low total concentrations of arsenic and chromium, there is a trend to the more reduced state of each polyvalent element as the concentration decreases. This was again attributed to changing coordination or structure of the polyvalent element oxyanions as a function of the concentration, even at these low concentrations.

OBJECTIVE

The purpose of this study was to determine whether the redox equilibria of polyvalent elements depend on the polyvalent element's concentration, especially at relatively low concentration, in a soda-lime-silicate glass. If such an effect was found to be real, a secondary goal was then to explain the underlying cause in the chemistry of the redox ions at low concentrations. Alternatively, the dependence reported in previous studies may be found to be an artifact of the methods used to synthesize or analyze the glass samples --- and merely be a reflection of the limitations of accurately determining small concentrations of a specific redox ion in glass.

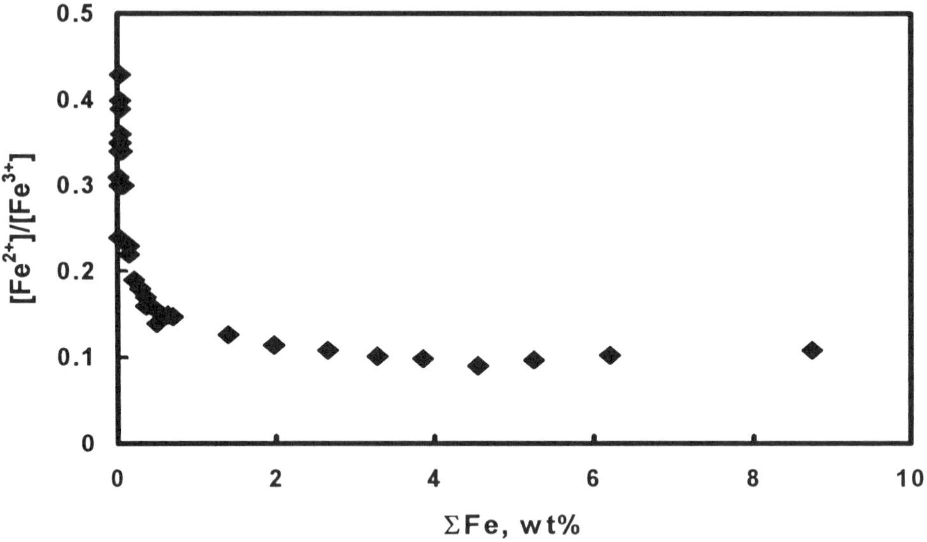

Figure 1. The effect of iron content on iron redox ratio in glass, according to [3].

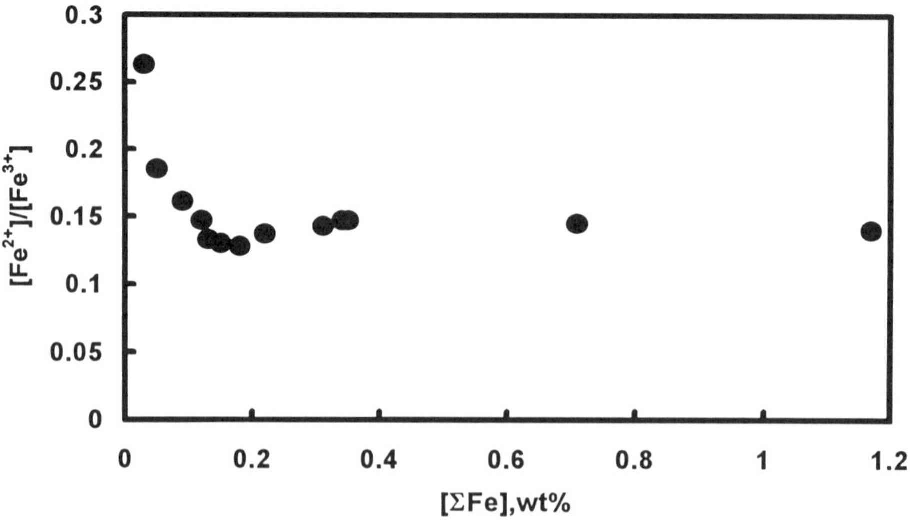

Figure 2. The effect of iron content on iron redox ratio in glass, according to [9].

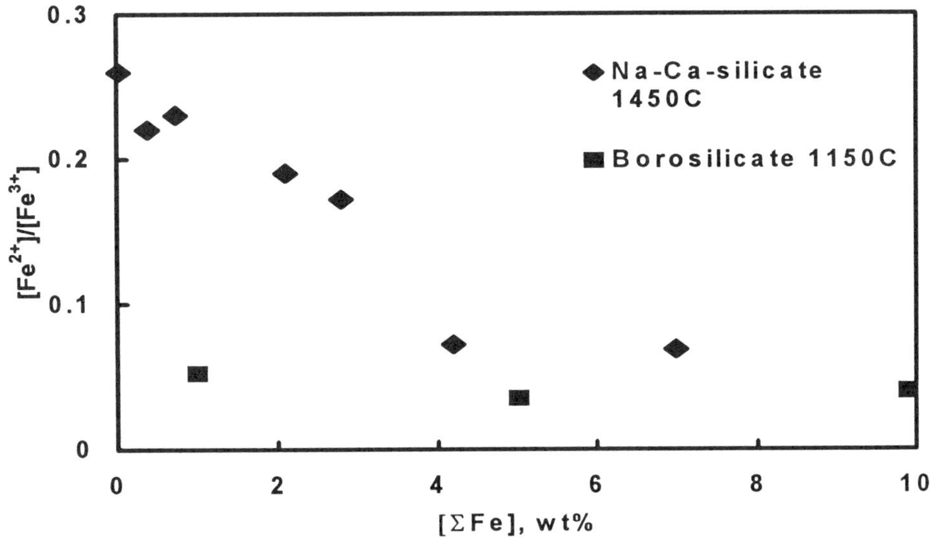

Figure 3. The effect of iron content on iron redox ratio in glass, according to [10,11].

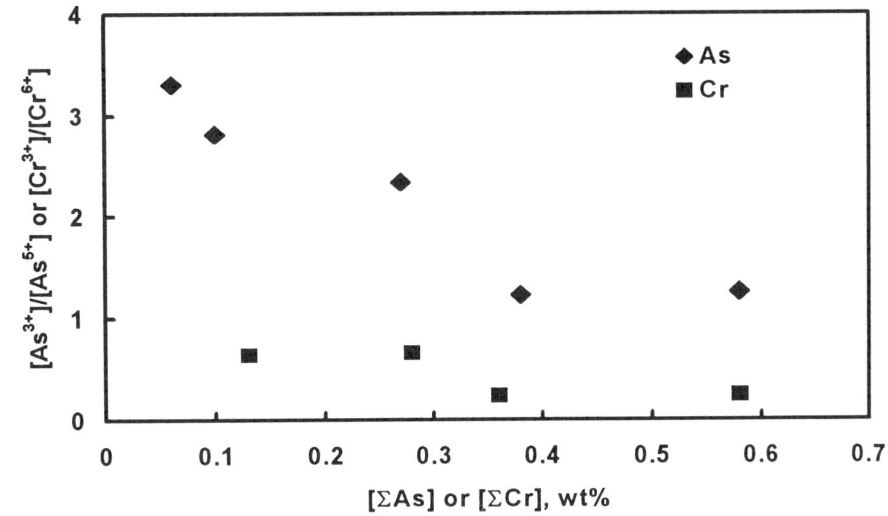

Figure 4. The effect of total concentration on redox ratio in glass, according to [10].

EXPERIMENTAL METHODS

The base composition of the glass is a soda-lime-silicate of 17 mol% Na_2O, 11 mol% CaO, and 72 mol% SiO_2. It was prepared by melting stoichiometric amounts of ultrapure Na_2CO_3, CaO, and SiO_2 at 1400°C, rapidly quenching to a glass, and finely powdering to a frit. A series of several glasses was made by adding MnO so that the concentration of manganese in the glasses varied from 5 wt% to 0.05 wt%. Other series were made with cerium, added as CeO_2, and with titanium, added as TiO_2, over the same concentration ranges.

All glasses within a particular series were prepared of the same base composition and processed at the same melt temperature and sample atmosphere. Syntheses were accomplished in a high-temperature, controlled-atmosphere furnace. Manganese- and cerium-containing glasses were synthesized in Pt capsules of about 1 mL capacity suspended vertically in the furnace held at 1400°C in an air atmosphere. Titanium-containing glasses were synthesized in graphite capsules of the same capacity and at the same temperature, but in a constant atmosphere of pure CO. Samples were equilibrated under processing conditions for about 24 hours, a time previously shown to be sufficient to establish redox equilibrium for this sample configuration. Samples were quenched by removal from the furnace.

Manganese- and titanium-containing glasses were mounted within an acrylic mold and cut into 1 mm thick sections, which were subsequently ground and polished on both sides. Spectra of these samples were obtained on a Shimadzu 3400 UV/vis/NIR spectrophotometer from 800 nm to 300 nm. Absorbances were monitored for the Mn^{3+} peak at 460 nm and for the Ti^{3+} peak at about 500 nm. A glass of the base composition without any additive served as a reference to define the baseline at each of these wavelengths, in order to obtain the net absorbance provided by each redox ion. The Mn^{3+} absorbance had previously been calibrated to Mn^{3+} concentration in this glass [12].

All samples were subsequently powdered for chemical redox titrations with standard ceric sulfate, as adapted from the classic methods developed by Johnston [13] and Close and Tillman [14]. Both H_2SO_4 and HF were diluted 50/50 with water, then treated with $KMnO_4$ and heated to remove any residual oxidants or reductants. To dissolve the manganese- and cerium-containing glasses, a solution of ferrous sulfate in the treated sulfuric acid was used; while for titanium-containing glasses, a solution of ferric sulfate in the treated sulfuric acid was required. The powdered glass samples were digested in the acids [13,14] under an inert argon atmosphere. Excess HF was then complexed with boric acid. The resulting solutions were titrated against standardized 0.00125*M* ceric sulfate, using o-phenanthroline as a visible indicator. These indirect titrations determined the amount of Fe^{2+} remaining in the digestion solution in the case of manganese- and cerium-containing samples, and of Fe^{2+} produced in the case of titanium-containing samples. Such quantities could then be related to Mn^{3+}, Ce^{4+}, and Ti^{3+} contents, respectively, in the samples. The other redox ion, Mn^{2+}, Ce^{3+} or Ti^{4+}, in each redox couple was calculated by difference with the total element concentration.

RESULTS

Figures 5, 6, and 7 provide the results of this study, showing the redox ratios of manganese, cerium, and titanium respectively as a function of their total concentration. The redox ratio of manganese, Mn^{2+}/Mn^{3+}, was found to be independent of the concentration of manganese in the glass, as long as the redox ion concentrations were fixed by the spectral determination of Mn^{3+} in the glass. No change in Mn^{2+}/Mn^{3+} was observed from 0.05 to 5 wt% ΣMn within experimental error. However, a dramatic increase in Mn^{2+}/Mn^{3+} was determined for low ΣMn, similar to previously reported iron redox trends [3,4,9], if analyses of ionic concentrations by redox titrations were used. The results in figure 5 are averages of triplicate analyses, with very good precision for each sample's analysis. The results attributable to titration analyses appear as though a systematic effect is operational on the redox ratio, one that is accentuated at the lower manganese concentrations. In other words, there appears to be a constant small effect that is negligible at 1 wt% ΣMn, but is significant at 0.1 wt% ΣMn. The results for titanium are similar to these for manganese, albeit not as pronounced. Cerium redox ratios were determined only by redox titrations, with no spectral counterparts. Below 1 wt% ΣCe, the poor precision demonstrates the inherent difficulty in analyzing small concentrations of an individual redox ion. Nevertheless, the trend leads only to the likelihood that systematic analytic error leads to the apparent stabilization of Ce^{4+} at low cerium concentrations.

DISCUSSION

That there is no change in redox ratio of a polyvalent element upon changes in the polyvalent element concentrations, when such concentrations are relatively low, in glass is satisfying in that coordination changes of the redox ions do not have to be postulated at such low concentrations. In fact, such coordination changes would be more likely to occur at high, not low, concentrations. Although such a conclusion is proven when Mn^{2+}/Mn^{3+} is determined by direct analysis of the glass by spectral methods, it could also be extrapolated to other polyvalent elements, in particular iron. An unchanging iron redox ratio with total iron concentration is consistent with a report that the spectral absorptivity of Fe^{2+} follows Beer's Law as a function of total iron concentration [15].

Results for polyvalent redox ratio determinations differ between the titration and spectral analyses only when done for the low concentration samples.

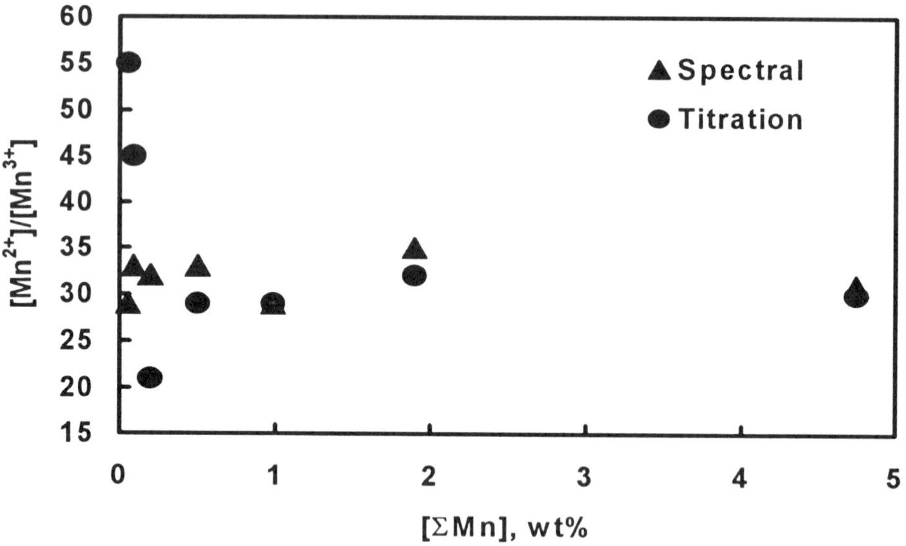

Figure 5. Manganese redox ratio as a function of total manganese concentration in glass.

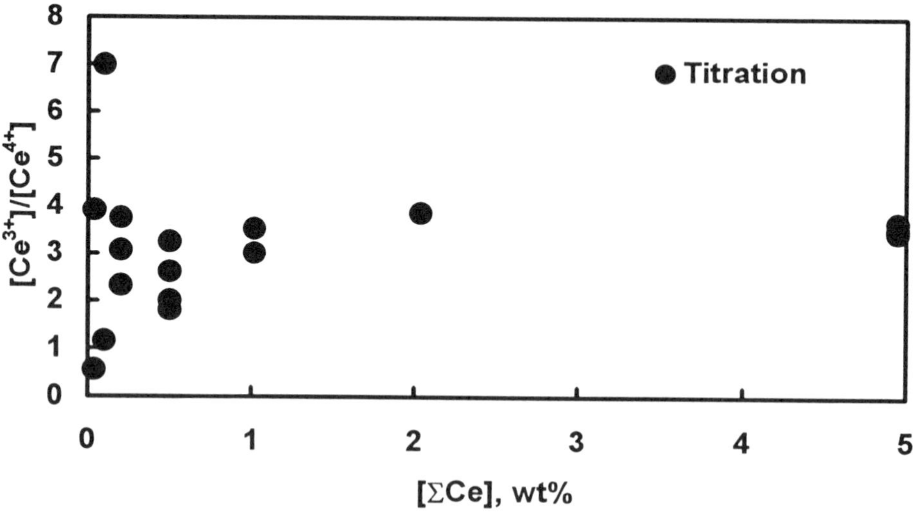

Figure 6. Cerium redox ratio as a function of total cerium concentration in glass.

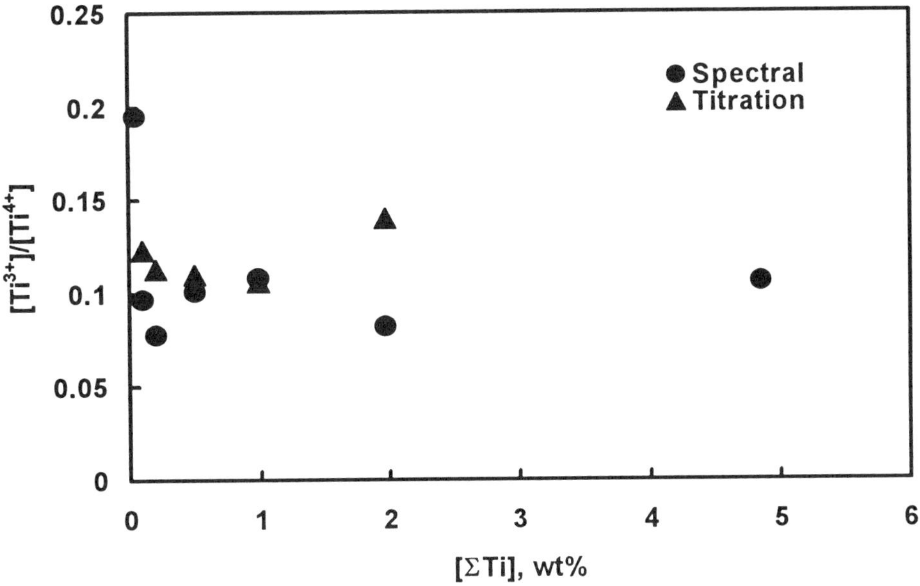

Figure 7. Titanium redox ratio as a function of total titanium concentration in glass.

Close and Tillman [14] stated that whereas the determination of 0.5 wt% Fe^{2+} in glass was not that difficult, measurement of 10 ppm Fe^{2+} required sophisticated techniques. This is illustrated by the lower limit of analyzed Fe^{2+} content, when measured by redox titration, being about 0.004 wt% Fe^{2+} for a 2 gram glass sample [14]. Further, there is inherently a 3% error in Fe^{2+} analyses for both titration [13,14] and colorimetric [16] procedures, but this error may go up to about 10% when $[Fe^{2+}] < 0.1$ wt%. One assumption in such analytical procedures to measure iron (or other polyvalent element) redox ratios is that the redox state concentrations are not disrupted when the glass is digested in acids (or even upon powdering the glass). Precautions such as removing oxidizing and reducing agents from the acids used to dissolve the glass and dissolving the glass under a protective inert atmosphere are part of the analytical procedure. However, if there is some disruption in the redox state upon dissolution of the glass into the acidic medium, and if this disruption is a small constant amount, it would greatly affect the analyses at low concentrations but be negligible at higher concentrations. This is indeed consistent with the results of this study. The previous measurements from the 1930s [3,6] to the 1990s [9,10], shown in figures 1 through 4, may simply be an indication of the quality of redox analyses getting better with time.

CONCLUSIONS

When a polyvalent element is doped into a soda-lime-silicate glass at concentrations less than about 5 wt%, concentration has little effect on the redox equilibrium of that polyvalent element. Previous reports in the literature of a dramatic effect of increasing redox ratio as a function of polyvalent element concentration, especially at very low concentrations, can be attributed to analytic limitations, most likely a small but constant disruption of the redox ratio upon digestion of the glass.

ACKNOWLEDGMENTS

This research was supported by the NSF-industry-university Center for Glass Research, centered at Alfred University.

REFERENCES

[1] H.D. Schreiber and M.T. Coolbaugh, "Solvations of redox ions in glass-forming silicate melts", *J. Non-Cryst. Sol.* **181** (1995) 225-230.

[2] K.H. Karlsson, K. Fröberg, and B. Skrifvars, "Chemical approach to medium range order in glass", *Glastech. Ber.* **63** (1990) 37-47.

[3] N.E. Densem and W.E.S. Turner, "Equilibrium between ferrous and ferric oxides in glass, *J. Soc. Glass Technol.* **22** (1938) 372-389.

[4] A. Paul, *Chemistry of Glasses*, 2nd ed. Chapman & Hall, London, 1990.

[5] C. Rüssel, "Iron oxide-doped alkali-lime-silica glasses", *Glastech. Ber.* **66** (1993) 68-75.

[6] K. Fuwa, "Glasses containing iron oxide (report 1)", *J. Jap. Ceram. Assocn.* **43** (1935) 489-495.

[7] T. Baak and E.J. Hornyak, "The iron-oxygen equilibrium in glass: Effect of platinum on the Fe^{2+}/Fe^{3+} equilibrium", *J. Amer. Ceram. Soc.* **44** (1961) 541-544.

[8] D.S. Goldman, "Oxidation equilibrium of iron in borosilicate glass," *J. Amer. Ceram. Soc.* **66** (1983) 205-209.

[9] B. Hahn, P. Buhler, and R. Weissman, "Influence of the total iron content on the redox ratio", *Rivista d. Staz. Speriment. del Vetro (supplement)* **23** (1993) 205-210.

[10] L. Ortmann, D. Höhne, and G. Nölle, "Equilibrium constant - determination and influence on redox reactions in soda-lime-silica glass melts", *Glastech. Ber. Glass Sci. Technol.* **69** (1996) 235-241.

[11] H.D. Schreiber, S.J. Kozak, R.C. Merkel, G.B. Balazs, and P.W. Jones, "Redox equilibria and kinetics of iron in a borosilicate melt", *J. Non-Cryst. Sol.* **84** (1986) 186-195.

[12] H.D. Schreiber, L.J. Peters, J.W. Beckman, and C.W. Schreiber, "Redox chemistry of iron-manganese and iron-chromium interactions in soda lime silicate glass melts", *Glastech. Ber. Glass Sci. Technol.* **69** (1996) 269-277.

[13] W.D. Johnston, "Oxidation-reduction equilibrium in iron-containing glasses," *J. Amer. Ceram. Soc.* **47** (1964) 198-201.

[14] W.P. Close and J.F. Tillman, "Chemical analysis of some elements in oxidation-reduction equilibria in silicate glasses", *Glass Technol.* **10** (1960) 134-146.

[15] J-P. Traverse, T. Toganidis, and C. Ades, "Spectrophotometric analysis of ferrous, ferric and total iron content in soda-lime-silica glass", *Glastech. Ber.* **65** (1992) 201-206.

[16] D.R. Jones, W.C. Jansheski, and D.S. Goldman, "Spectrophotometric determination of reduced and total iron in glass with 1,10-phenanthroline", *Anal. Chem.* **53** (1981) 923-924.

REDOX REACTIONS AND FOAMING BEHAVIOR IN GLASS MELTS

Haochuan Jiang and William C. LaCourse
New York State College of Ceramics at Alfred University
2 Pine Street
Alfred, NY 14802

ABSTRACT

The foaming behavior of soda-lime-silica glass melt was studied by an electrical probe method. The glasses were melted in an atmosphere-controlled tube furnace and the foam volume was measured. There are two distinct foaming processes for the glasses containing the following fining agents: sodium sulfate, antimony oxide, and arsenic oxide: low temperature foaming which occurs from about 1100°C to 1200°C, and high temperature foaming which occurs at above 1350°C for Na_2SO_4 (reboiling) and from 1200°C to 1300°C for As_2O_3 and Sb_2O_3 + CeO_2 fining agents. The low temperature foaming is caused by the decomposition of carbonates in the batch after the formation of viscous liquidous phase. Redox reactions of the fining agents and the relevant oxygen release are the key factors for the high temperature foaming. The foam volume is closely related to the amount of fining agents used. This also suggests that the quantity of fining agents in glass melting should be carefully controlled for the purpose of foaming reduction.

INTRODUCTION

The foam formed on a glass melt surface can cover about one third of the melting tank. It reduces the heat transfer from flame to glass melt, causing over-heating in the combustion chamber, leading to increased refractory corrosion and energy consumption. Additionally, the lack of heat in the glass melt can give rise to higher defect concentrations such as bubbles and inclusions. As many furnaces have been changed from air-firing to oxy-fuel-firing, the foaming pattern has also been changed. In order to reduce the foam formation, it is critical to understand the fundamental mechanisms of the foaming formation and stability in glass melting.

There are diverse redox reactions in glass melts. The most important ones are those of the fining agents such as sodium sulfate, antimony oxide, cerium oxide, and arsenic oxide. These reactions occur through the whole glass melting process and contribute to the acceleration of gas bubble removal form the melts. This process is called fining and refining. The mechanisms of fining have been extensively studied. At lower temperature, the fining agent ions are oxidized into higher valence state, e.g. Sb^{5+}, As^{5+}, Ce^{4+}, and SO_4^{2-}, but as the temperature increases, the redox equilibriums shift and the above ions or ion groups are reduced into lower valence state, e.g. Sb^{3+}, As^{3+}, Ce^{3+}, and SO_2(gas).

These reactions release oxygen or SO_2 that can help the bubble growth and removal. When the glass melts are cooled, the redox reactions shift back and some of the tiny oxygen and SO_2-rich bubbles will be absorbed.

Jiang and LaCourse[1] studied the foaming behavior of soda-lime-silica glass melts with antimony and cerium oxide fining agents. There were two distinct foaming processes for both fining agents. The low temperature foaming was caused by the decomposition of carbonates. The high temperature foaming was caused by the redox reaction of fining agents and thus the oxygen release. Cable and his coworkers[2-4] studied the reboiling in premelted silicate glasses. They found only the glasses containing sulfate could foam and reboiling occurred only in oxidizing atmospheres. The reboiling was enhanced when water vapor was present. Goldman et al[5-6] studied the foaming of simulated nuclear waste glass. They proposed that foaming was caused by the release of water vapor and oxygen from the melt. The oxygen was supposed to come from the redox reactions of species other than iron. The foam stability was greatly decreased in reducing atmospheres. Kim and Hrma[7,8] studied the foaming of soda-lime-silica glass under both ramp heating and isothermal conditions. The foaming temperature was compared with the sulfate solubility of the glasses. Laimböck and Beerkens[9] showed that at above 1460°C the release of SO_2 and O_2 due to the decomposition of Na_2SO_4 caused a large volume of foam. When carbon was added into the batch, the foaming temperature was decreased considerably because of the reduction of sulfate to sulfide. The addition of carbon could reduce the foam volume at high temperature, but also increased the batch blanket thickness.

Here we report on the foaming processes in soda-lime-silica glasses with sulfate, arsenic oxide, antimony oxide and cerium oxide fining agents under either dry or "wet" CO_2. Effects of atmosphere will be reported elsewhere. The foaming behavior was discussed on the basis of the redox reactions of the relevant fining agents.

EXPERIMENTAL PROCEDURES

The compositions of the base glasses in mol% are $75SiO_2$-$15Na_2O$-$10CaO$ (SLS) and $73SiO_2$-$15Na_2O$-$10CaO$-$2Al_2O_3$ (SLAS). Batches for 40 g glass were prepared from silica sand (40-100 mesh), Na_2CO_3, $CaCO_3$, Al_2O_3, and fining agents Na_2SO_4, As_2O_3, Sb_2O_3 and CeO_2. $NaNO_3$ were used when needed. Batches were reacted in clear cylinder fused silica crucibles (inner diameter of 34 mm) in an atmosphere-controlled tube furnace. The apparatus and the principles of the foam measurements were described elsewhere[1]. The heating schedule for the melting of arsenic oxide, antimony oxide, and cerium oxide fined glasses is shown in Fig. 1.

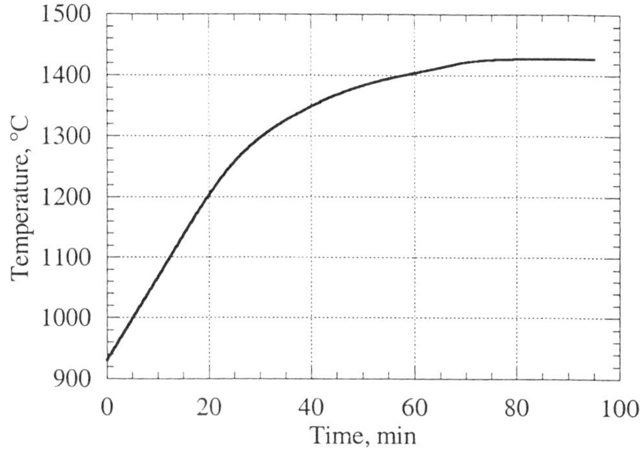

Fig. 1. The heating schedule for the melting of SLAS glasses with As_2O_3, Sb_2O_3, and CeO_2.

RESULTS AND DISCUSSIONS

Sodium Sulfate Fining Agent

Fig. 2 shows the maximum high-temperature foam volume formed during the melting of SLS glasses containing different amount of Na_2SO_4. This foam occurred at temperatures above 1400°C, depending on the heating rate, furnace atmosphere, and glass composition. The furnace atmosphere was CO_2 with 30% moisture for Fig. 2. It clearly indicates that as the content of Na_2SO_4 in the batch increases, the maximum foam volume increases, in agreement with previous studies[8,9].

Fig. 3 shows the foam volume change of SLAS glass containing Na_2SO_4 fining agent. The furnace atmosphere was dry CO_2. It gives the same trend in relationship between the maximum foam volume and the sulfate content. There are two distinct foaming processes in Na_2SO_4 fined glass melts: low temperature foaming and high temperature foaming. The low temperature foam occurring at about 1100°C is just simply the batch expansion caused by the decomposition of carbonates. This foam decays quickly, usually in several minutes, and is therefore not of particular importance for the melting process. Our study shows that the maximum foam volume of the low temperature foam is not obviously related to the original sulfate content in the batch.

The high temperature foam shown in Fig. 2 and Fig. 3 is caused by the decomposition of Na_2SO_4:

SO_4^{2-}(melt) fi O^{2-}(melt) + SO_2(g) + $1/2O_2$(g) (1) This foam can have a lifetime as long as one hour, depending on the temperature, atmosphere, and many other factors. This foaming process is very important because it can considerably influence the melting and fining process of glass and cause a lot of defects such as bubbles and inclusions. The chemistry of the reaction and gas evolution were described in detail by previous investigators[9]. As temperature increases, the solubility of sulfate or SO_3

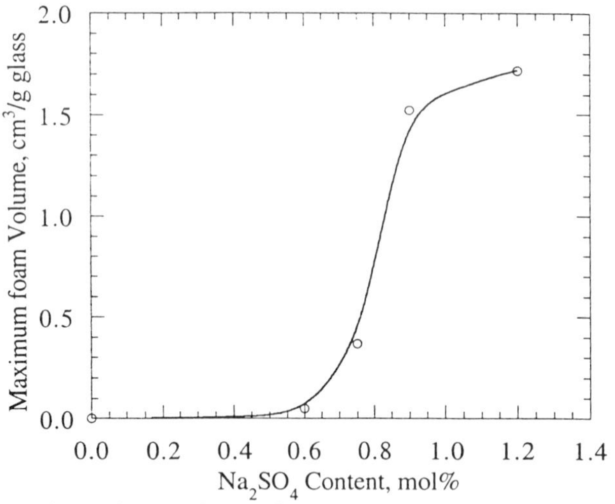

Fig. 2. The maximum foam volume of SLS glass vs. Na_2SO_4 content in the batch.

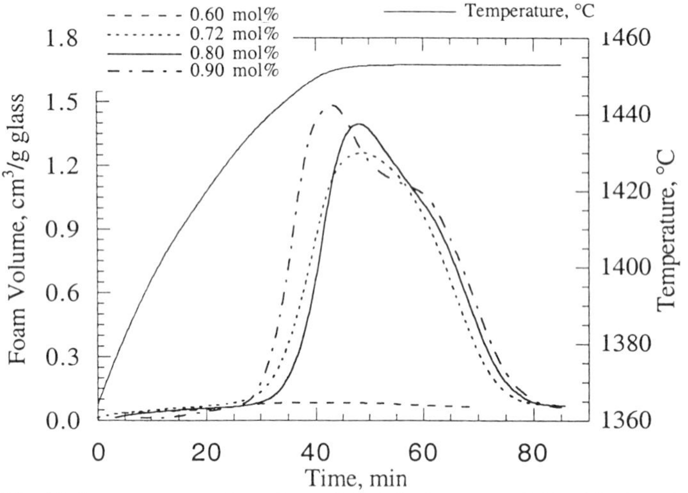

Fig. 3. The high temperature foaming of SLAS glass containing Na_2SO_4 (mol%).

decreases. At certain temperature (foaming temperature), some of the oversaturated sulfate will decompose by reaction (1). The gas ($SO_2 + O_2$) released will be also oversaturated in the glass melt and form bubbles. The first stage of the bubble formation , nucleation, can take place at the many interfaces in the glass melt, such as the surface of undissolved silica sand grains, refractory surface, and the pre-existing bubbles. As more and more bubbles form and grow, a layer of foam will be formed on the surface of the glass melt. According to this mechanism, the solubilities of both sulfate and gases are critical.

Oversaturation of sulfate is a required condition for foam formation. Kim and Hrma[8] compared the foaming temperature with different sulfate content and the previously reported solubility data of sulfate in glasses and found they fit well. Based on the fact of solubility and oversaturation, one can not expect a linear relationship between the maximum foam volume and the original sulfate content. This is confirmed by both Fig. 2 and Fig. 3. When the sulfate content is below 0.6mol % (on the glass base), the foam volume is very small. But at higher concentration, the foam volume rapidly increases, thus showing a threshold. Practically this is very important for determining the fining agent content in glass melting. If the sulfate content is lower than the threshold value, it is possible to considerably reduce the foam volume.

When 2 mol % SiO_2 was substituted by Al_2O_3, the foaming temperature was considerably reduced. For example, the foaming temperature of the SLS glass containing 0.9 mol% Na_2SO_4 is about 1440°C[10], but the foaming temperature of SLAS glass with same amount of Na_2SO_4 is decreased to about 1365°C. This change can be explained by the role of Al_2O_3 in the melt structure.[11] In the case of melt containing sulfate, we assume that sulfur exists in the glass melt as $[SO_4]^{2-}$ or $[SO_4]_n^{2n-}$ groups, weakly bonded with Na^+ or Ca^{2+} in the channels. Due to the relatively large size of the sulfur ion groups, their content is limited by the number of non-bridging oxygens and the channel sizes. As Al_2O_3 is added into the SLS glass melt (when the ratio is below certain value) it eliminates some non-bridging oxygens and forms AlO_4 tetrahedra, thus reducing the total number of non-bridging oxygen and the free volume, therefore reducing the solubility of sulfate (and other gasses). Even with same original sulfate content, the oversaturation of SLAS glass occurs at about 1365°C while the oversaturation of SLS glass occurs at much higher temperature, about 1440°C.

Arsenic Oxide Fining Agent

Fig. 4 shows foaming of SLAS glass with different As_2O_3 content. The furnace atmosphere is CO_2. For all glasses, 1 mol% of Na_2O was from $NaNO_3$. There are two distinct foaming processes for As_2O_3 fined glass melts, similar to that of Sb_2O_3 fined glass[1]. The primary foaming occurred at about 1050 to 1200°C, almost same as that of Sb_2O_3 fining glass[1]. The starting temperature of the primary foaming is lower than that of the SLS glasses with sodium sulfate fining agent which is about 1130°C[11]. The primary foaming is caused by the decomposition of carbonates in the batch. As indicated by Kim and Hrma[12], there are three necessary conditions for foam formation: 1. a large amount of gas is released, 2. viscous melt is formed, 3. the melt is sufficiently interconnected. At above 400°C CO_2 starts to release and reaches its maximum rate at about 900°C. The eutectic liquid phase of the alkali and alkaline earth oxides and carbonates formed is highly basic and rapidly dissolves silica, increasing the viscosity of the liquid phase. When $NaNO_3$ and As_2O_3 was added, the temperature at which liquidus phase forms is reduced, and the dissolution of silica sand is accelerated. The temperature at which the liquid phase is sufficiently viscous for foam formation is closer to that at which the gas release rate is maximum. Therefore the primary foam forms at lower temperature and its maximum increases compared to the glass with sulfate.

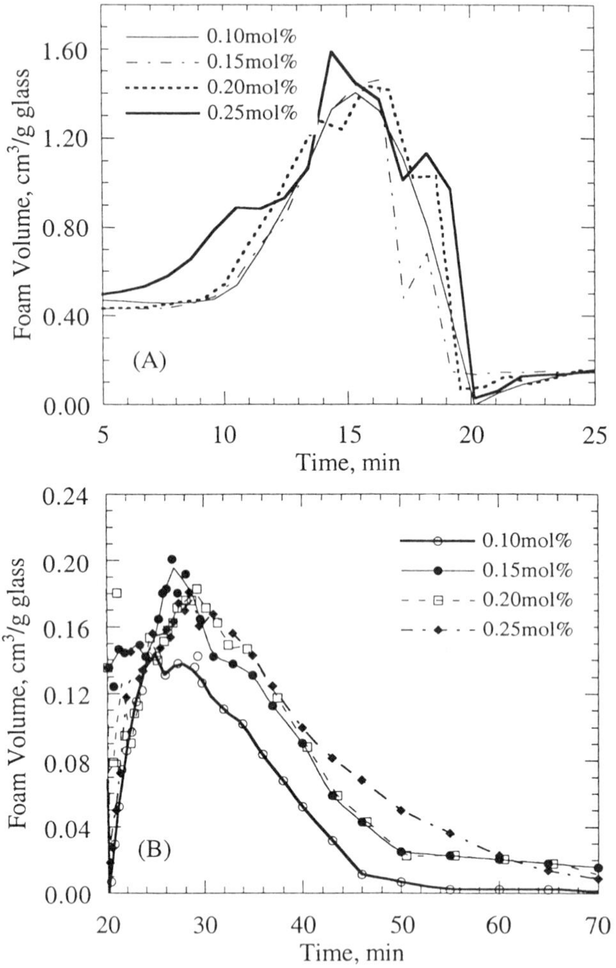

Fig. 4. (A) The low temperature, (B) the high temperature foaming of SLAS glass with different amount of As_2O_3 in the batch. Note scale differences.

The high temperature foaming occurred above 1200°C. It has been shown that there was no high temperature foam above 1200°C with only $NaNO_3$[1]. The foam increased in both volume and decay time with increasing As_2O_3 content. However, the maximum foam volume is much lower compared to the glasses with same amount of Sb_2O_3. As proposed for the Sb_2O_3 fining agent by the authors[1], the high temperature foaming is caused by the oxygen release from the redox reaction of As_2O_3. At lower temperatures, the following reactions occur:

$$2NaNO_3 = 2NaNO_2 + O_2 \quad (1)$$

and

$$4NaNO_2 = 2Na_2O + 2N_2 + O_2 \quad (2)$$

The reactions are strongly oxidizing, and the As_2O_3 will be oxidized as follows:

$$2NaNO_3 + As_2O_3 = As_2O_5 + Na_2O + NO + O_2 \quad (3)$$

The fining behavior of As_2O_3 is from the following redox equilibrium:

$$As^{5+}(melt) + O^{2-}(melt) ¤ As^{3+}(melt) + 1/2O_2 \quad (4)$$

As the temperature increases (fining stage), the reaction shifts to the right side, a lot of O_2 gas evolves and forms bubbles, thus accelerating the bubble removal from the glass melt; As the melt is cooled (refining stage), the reaction shifts back to the left side, and some of the tiny bubbles of O_2 can be reabsorbed. The equilibrium constant of reaction (4) at temperature T is:

$$K(T) = \frac{a_{As^{3+}} \cdot (a_{O_2})^{1/2}}{a_{As^{5+}} \cdot a_{O^{2-}}} \quad (5)$$

Since the content of As is very small, its activity can be replaced by its concentration. Oxygen ion
activity in the glass melt can also be assumed to be a constant, thus the equilibrium constant becomes:

$$K(T) = \frac{[As^{3+}] \cdot a_{O_2}^{1/2}}{[As^{5+}]} \quad (6)$$

This constant depends strongly on temperature,

$$K(T) = \exp(-\frac{\Delta H}{RT} + \frac{\Delta S}{R}) \quad (7)$$

The reported data are[13]: at DH° = 230 kJ mol^{-1}, DS° = 137 J mol^{-1}K^{-1}. At T = 1000°C, K = 0.005; at T = 1300°C, K = 0.33. The formation of 1 mole As^{3+} will result in the evolution of 0.5 mole of O_2, or approximately 11.2 liters of O_2. The volume of the oxygen gas evolved from the above reduction is proportional to the concentration of As^{3+} in the glass melt. The equilibrium constant can be also expressed as:

$$K = \frac{[As^{3+}] \cdot a_{O_2}^{1/2}}{[As_T] - [As^{3+}]} \quad (8)$$

So $[As^{3+}]=[As_T]k/(1+k)$, where $[As_T]$ is the concentration of total antimony, $k = K/a_{O_2}^{1/2}$. Thus the volume of oxygen gas evolved $V = k_1[As_T]$, where k_1 is a constant. Below 800°C, all the arsenic ions are oxidized into As^{5+}. As the temperature increases, more and more arsenic is reduced into trivalent state and oxygen gas forms. The large amount of oxygen nucleates on the surface of the undissolved silica sand grains or diffuses into the pre-existing bubbles and the bubbles grow. The rapid reaction and gas evolution cause foam formation. Since foam volume is related to the volume of oxygen evolved one must strictly control the fining agent concentration.

Antimony and Cerium Oxide Fining Agent

Fig. 5 shows the foaming processes of SLAS glass melts with Sb_2O_3 and CeO_2 fining agents. The furnace atmosphere was dry CO_2. There are also two distinct foaming peaks: low temperature foaming from 1100°C to 1200°C, and high temperature foaming above 1200°C. The redox equilibriums for antimony and cerium ions in the glass melt are:

$$Sb^{5+}(melt) + O^{2-}(melt) = Sb^{3+}(melt) + 1/2O_2 \quad (9)$$

and

$$Ce^{4+}(melt) + 1/2O^{2-}(melt) = Ce^{3+}(melt) + 1/4\,O_2 \quad (10)$$

The relationship between glass foaming and the above redox reactions were analyzed by the authors[1]. Both oxygen releases from reactions 9 and 10 contribute to the high temperature foaming. Fig. 5 also shows that when Sb_2O_3 is substituted by CeO_2 (equal amount based on the theoretical oxygen yield), the high temperature foam volume decreases. The overall foam volume is an accumulated result of the relevant single fining agents according to the previous experimental results[1].

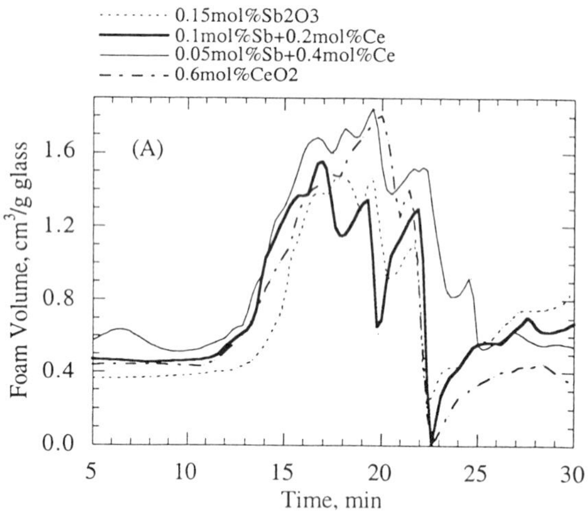

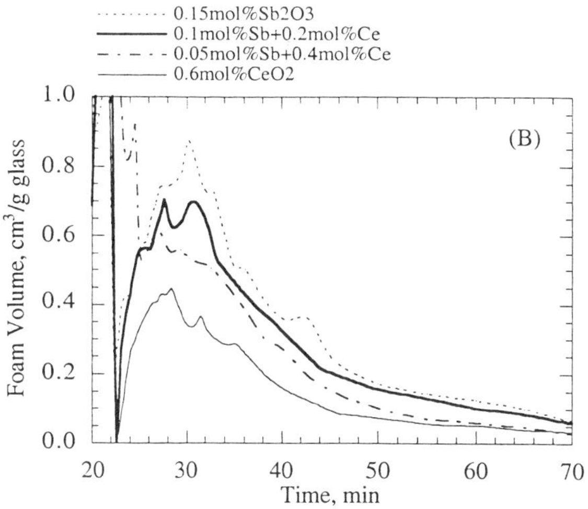

Fig. 5. The foaming of SLAS glass with Sb_2O_3 and CeO_2 fining agents.

SUMMARY

There are two distinct foaming processes for SLAS glasses with sodium sulfate, arsenic oxide, and antimony oxide and cerium oxide fining agents: low temperature foaming and high temperature foaming. Low temperature foaming is caused by decomposition of carbonates in the batch after the formation of viscous liquidous phase. The low temperature foaming of glasses containing arsenic or antimony and cerium occurs at lower temperatures than that of the glasses containing sodium sulfate. The high temperature foaming which is more important in practical glass melting is related to the redox reactions of the fining agents. For glasses with sodium sulfate, the high temperature foaming or reboiling occurs at above 1350°C and is caused by the decomposition of sulfate after oversaturation. For the glasses with other fining agents, the high temperature foaming is caused by the release of oxygen gas due to the reduction of As^{5+}/As^{3+}, Sb^{5+}/Sb^{3+}, and Ce^{4+}/Ce^{3+} ions. The maximum foam volume generated by As_2O_3 is much lower that that generated by equal amounts of Sb_2O_3. When Sb_2O_3 is substituted by an equal amount (based on oxygen yield) of CeO_2, the high temperature foam volume decreases considerably.

ACKNOWLEDGMENTS

This study was funded by the NSF Industry-University Center for Glass Research at Alfred University and the US Department of Energy.

REFERENCES

[1]H. Jiang and W.C. LaCourse, "Foaming in the Soda-Lime-Silica Glass Melts With Antimony and Cerium Oxide Fining Agents," Proceedings of the 18th International Congress on Glass, San Francisco, July 5 - 10, 1998.

[2]C.G. Rasul and M. Cable, "Spontaneous Bubble Formation in Silicate Melts at High Temperatures," Journal of the American Ceramic Society, **49**[10]568-571(1966).

[3]M. Cable and C.G. Rasul, "Spontaneous Bubble Formation in Silicate Melts at High Temperatures: II, Effect of Wet Atmospheres and Behavior of Mixed Alkali Glasses," Journal of the American Ceramic Society, **50**[10]528-531(1967).

[4]M. Cable, C.G. Rasul, and J. Savage, "Laboratory Investigation of Foaming and Reboil in Soda-Lime-Silica Melts," Glass Technology, **9**[2]25-31(1968).

[5]D.S. Goldman, D.W. Brite, and W.C. Richey, "Investigation of Foaming in Liquid-Fed Melting of Simulated Nuclear Waste Glass," Journal of the American Ceramic Society, **69**[5]413-417(1986).

[6]D.S. Goldman, "Melt Foaming, Foam Stability and Redox in Nuclear Waste Vitrification," Journal of Non-Crystalline Solids, **84**[1]292-298(1986).

[7]D.S. Kim and P. Hrma, "Foaming in Glass Melts Produced by Sodium Sulfate Decomposition Under Isothermal Conditions," Journal of the American Ceramic Society, **74**[3]551-555(1991).

[8]D.S. Kim and P. Hrma, "Foaming in Glass Melts Produced by Sodium Sulfate Decomposition Under Ramp Heating Conditions," Journal of the American Ceramic Society, **75**[11]2959-63(1992).

[9]P.R. Laimböck and R.G.C. Beerkens, "Foaming Behavior of Glass Melts," pp. 16-21 in Properties of Molten Glass, Proceedings of the Philips-TNO Seminar, Eindhoven, the Netherlands, April, 1996.

[10]H. Jiang and W.C. LaCourse, unpublished results.

[11]W.C. LaCourse and A.N. Cormack, "Glasses with Transitional Structure,: 5th International Conference on Advances in the Fusion and Processing of Glass, July 27-31, 1997, Toronto, Canada. Ceramic Transactions, Am. Ceram. Soc. (In press)

[12]D.S. Kim and P. Hrma, "Volume Change During Batch to Glass Conversion," American Ceramic Society Bulletin, **69**[6]1039-43(1990).

[13]C. Rüssel and E. Freude, "Voltammetric Studies of the Redox Behavior of Various Multivalent Ions in Soda-Lime -Silica Glass Melts," Physics and Chemistry of Glasses, **30**[2]62-68(1989).

THE EFFECT OF Pt ALLOYING REACTIONS ON ELECTROCHEMICAL MEASUREMENTS IN GLASS MELTS

Susan L. Schiefelbein
Corning, Inc.
HP ME 02 38
Corning, NY 14831

ABSTRACT

Since electrochemical measurements are made at the measuring electrode/melt interface, the desired bulk properties of the melt can be measured only when the measuring electrode is inert. For certain glass melts, no metals qualify as measuring electrodes — even Pt reacts strongly with these melts. The fact that Pt reacts with glass melts is hardly news; the interesting new result is that the effects of these reactions on the redox state of the adjacent glass melt can be very large and long lasting. The results of in-glass P_{O_2} measurements of several commercial glasses are reported. It was found that in-glass P_{O_2} measurements are especially difficult in viscous melts with low oxygen mobility. In these melts, the absolute value of measured P_{O_2} is strongly influenced by factors other than the bulk P_{O_2} of the melt, namely: (i) thickness of the measuring electrode; (ii) concentration of alloyable constituents in the glass melt; and (iii) composition of the measuring electrode. These effects are explained by alloying of the measuring electrode with melt constituents.

INTRODUCTION

The objective of many electrochemical techniques is measurement of a bulk property (*e.g.* redox ratio of a polyvalent species). However, since electrochemical techniques are interfacial by nature, any reaction between the electrode and electrolyte can prevent measurement of the desired bulk property. In the case of Pt measuring electrodes in glass melts, alloying of the Pt electrode with glass melt constituents results in oxidation of the adjacent glass melt and, therefore, loss of the bulk condition at the measuring electrode interface. For example, consider the equilibrium $Sb_2O_3 \leftrightarrow 2Sb[Pt] + \frac{3}{2}O_2$; the initially pure Pt measuring electrode is alloyed with Sb from Sb_2O_3 in the glass melt, and oxygen is generated causing the P_{O_2} of the melt in contact with the electrode to rise. In viscous melts with low oxygen diffusivities, it was found that this shift in the local oxidation state can be very large and long lasting.

In this paper, in-glass P_{O_2} measurements are used to show the effect of several variables on the magnitude and duration of Pt alloying reactions and the consequent change in oxidation state of the adjacent glass melt. Most measurements were made at 1430°C in an alkaline earth aluminoborosilicate melt with <0.01wt% FeO, and 0-2wt% Sb_2O_3.[1] At this temperature, this glass melt is quite viscous, $\eta \approx 2{,}000$ poise, and has a particularly low oxygen diffusivity, $D_O^* \approx 5\text{x}10^{-7}$. It should be stressed that the results reported in this paper are of a general nature, restricted to neither the particular electrochemical technique employed nor the particular glass studied.

Briefly, in-glass P_{O_2} measurements are based on the Nernst equation [1]:

$$E = \frac{RT}{nF} \ln\left(\frac{P_{O_2}^{ref}}{P_{O_2}^{melt}}\right) \qquad \{1\}$$

which relates the P_{O_2} of the melt at the measuring electrode, $P_{O_2}^{melt}$, to the measured EMF between the measuring and reference electrodes, E, the temperature, T, and the P_{O_2} of the reference gas, $P_{O_2}^{ref}$. R is the ideal gas constant, n is the number of electrons transferred, and F is the Faraday constant. Figure 1 is a schematic of the basic setup. In the configuration shown, the reference electrode is a closed-one-end yttria-stabilized zirconia tube with an inner Pt wire and flowing reference gas, and the measuring electrode is a Pt tab. It should be noted that the Nernst equation applies only if the melt is both isothermal and homogeneous [2].

[1] Under test conditions, FeO and Sb_2O_3 were the only reducible constituents in the glass melt.

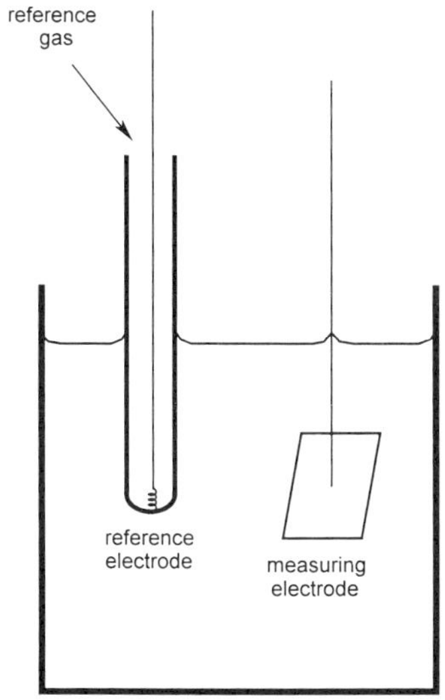

Figure 1: Basic setup for in-glass P_{O_2} measurements.

RESULTS

There were three interesting and important results concerning the validity and interpretation of in-glass P_{O_2} measurements. First, the absolute value of measured P_{O_2}, $P_{O_2}^{meas}$, is strongly influenced by factors other than the bulk P_{O_2} of the glass melt, $P_{O_2}^{melt}$, namely: (i) thickness of the measuring electrode; (ii) concentration of alloyable constituents in the glass melt; and (iii) composition of the measuring electrode. Second, alloying of the measuring electrode with melt constituents, and transport of reaction products away from the measuring electrode, lead to initial transients and drift in $P_{O_2}^{meas}$. Third, despite these problems, there is strong evidence that thermodynamic values derived from $P_{O_2}^{meas}$ *vs.* T data are reliable even for melts with low D_O^*.

The absolute value of $P_{O_2}^{meas}$ is strongly influenced by factors other than $P_{O_2}^{melt}$, namely: (i) thickness of the measuring electrode; (ii) concentration of alloyable constituents in the glass melt; and (iii) composition of the measuring electrode.

Remarkably, it was found that $P_{O_2}^{melt}$, as set by the melting temperature, had *no effect* on $P_{O_2}^{meas}$. There was no measurable difference between identical melts containing 0.5wt% Sb_2O_3 melted at different temperatures. Assuming that each melt equilibrated with 1 atm O_2 at its melting temperature, $P_{O_2}^{melt}$ at 1430°C calculated using $\Delta H(Sb_2O_3) = 317$ kJ·mol^{-1} is 0.163 atm and 0.117 atm for melting temperatures of 1580°C and 1610°C, respectively. Figure 2 clearly shows that $P_{O_2}^{meas}$ was a function of tab thickness only; $P_{O_2}^{melt}$, as set by melting temperature, had no effect.

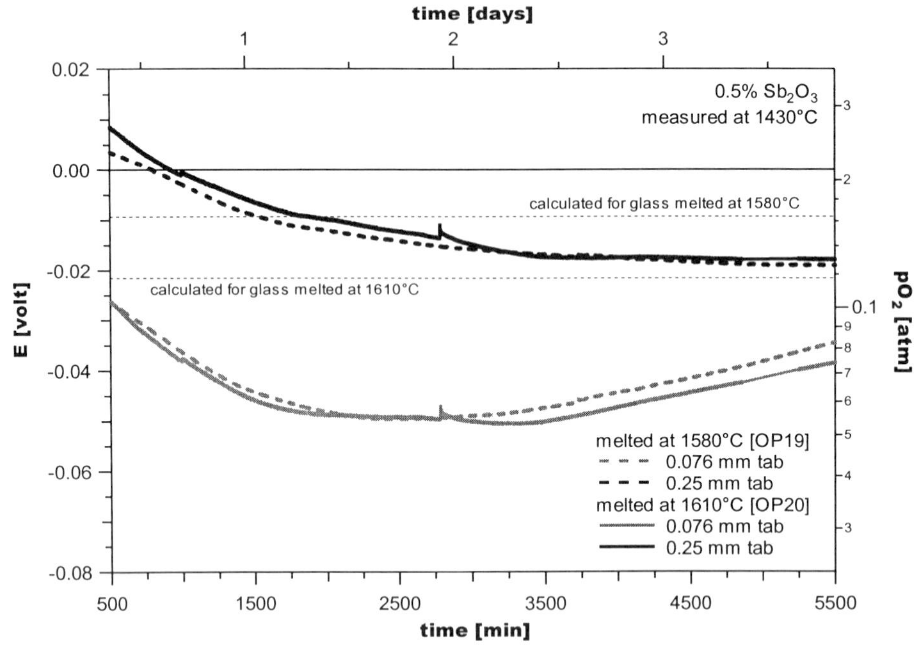

Figure 2: Effect of melting temperature on $P_{O_2}^{meas}$.

Measuring electrode thickness had perhaps the most profound effect on $P_{O_2}^{meas}$. As shown in Figure 3, $P_{O_2}^{meas}$ increased with electrode thickness, with $P_{O_2}^{meas}$ [0.25 mm] $\approx 5\, P_{O_2}^{meas}$ [0.025 mm] (this corresponds to an EMF difference of ~50mV at 1430°C). It's clear that alloying reactions at the Pt measuring electrode control the P_{O_2} of the adjacent melt — thicker tabs require more Sb for equilibration with the glass melt, and result in a more oxidized adjacent glass melt.

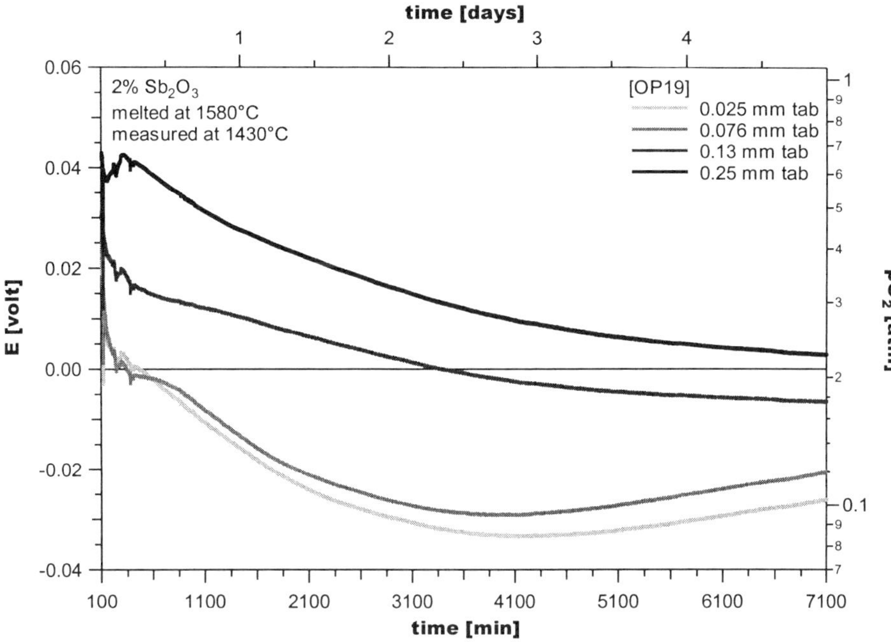

Figure 3: Effect of measuring electrode thickness on $P_{O_2}^{meas}$.

Concentration of alloyable constituents in the glass melt also had a significant effect on $P_{O_2}^{meas}$. It is known from experience and thermodynamics [3] that above a certain minimum concentration, $P_{O_2}^{melt}$ is not a function of oxide fining agent concentration; $P_{O_2}^{melt}$ at any temperature is entirely determined by the type of fining agent

($e.g.$ Sb_2O_3 or As_2O_3) and the melting temperature. The first sets the slope, $\frac{d\left(\ln\left(P_{O_2}^{melt}\right)\right)}{d(1/T)}$,

and the second fixes a single point (T^{melt}, $P_{O_2}^{melt} = 1$ atm). An experiment was conducted on two similar melts melted at 1580°C, one had 0.5wt%, the other had 2wt% Sb_2O_3. As shown in Figure 4, $P_{O_2}^{meas}$ of the 2wt% Sb_2O_3 melt was significantly higher than that of the 0.5wt% Sb_2O_3 melt (reported by tabs of the same thickness). Chemical analysis revealed that the 0.13 mm Pt tab immersed in the 2wt% Sb_2O_3 melt contained twice as much Sb as the 0.13 mm Pt tab immersed in the 0.5wt% Sb_2O_3 melt. Once again, it's clear that alloying reactions at the Pt measuring electrode control the P_{O_2} of the adjacent melt — higher concentrations of Sb_2O_3 shift the equilibrium ($Sb_2O_3 \leftrightarrow 2Sb[Pt] + \frac{3}{2}O_2$) to the right, and lead to increased oxidation of the adjacent melt.

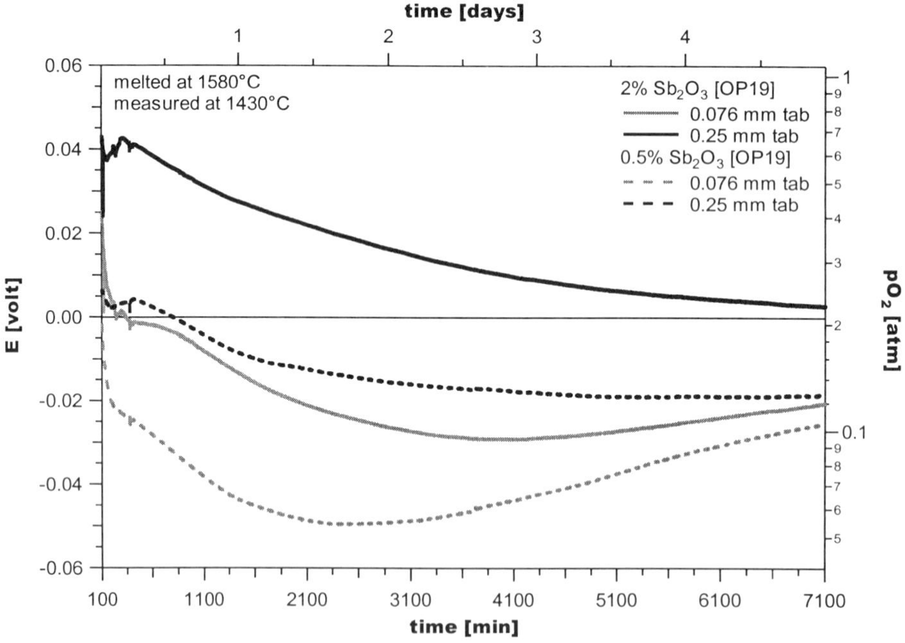

Figure 4: Effect of Sb_2O_3 concentration on $P_{O_2}^{meas}$.

Measuring electrode composition, expressed as either tramp element content or type of metal (Pt *vs.* Ir), also had a notable effect on $P_{O_2}^{meas}$. As shown in Figure 5, even though all four Pt tabs reported the same $P_{O_2}^{meas}$ at long times in a melt containing 0wt% Sb_2O_3, there was an initial wide variation in $P_{O_2}^{meas}$. This initial variation appears to be the result of different tramp element contents (especially Fe, As, and Sb) in the Pt foils. The sums, in ppm, of Fe+As+Sb are given in Figure 5 for three of the four foils. The 0.13 mm tab had the highest tramp element content and initially reported the lowest $P_{O_2}^{meas}$, while the 0.025 mm tab had the lowest tramp element content and initially reported the highest $P_{O_2}^{meas}$. Presumably, the 0.025 mm tab contained too low, and the 0.13 mm tab too high, a tramp element concentration for equilibration with the glass melt at this temperature. The tramp element excess in the 0.13 mm tab lead to reduction, whereas the tramp element deficiency in the 0.025 mm tab lead to oxidation, of the adjacent glass melt.

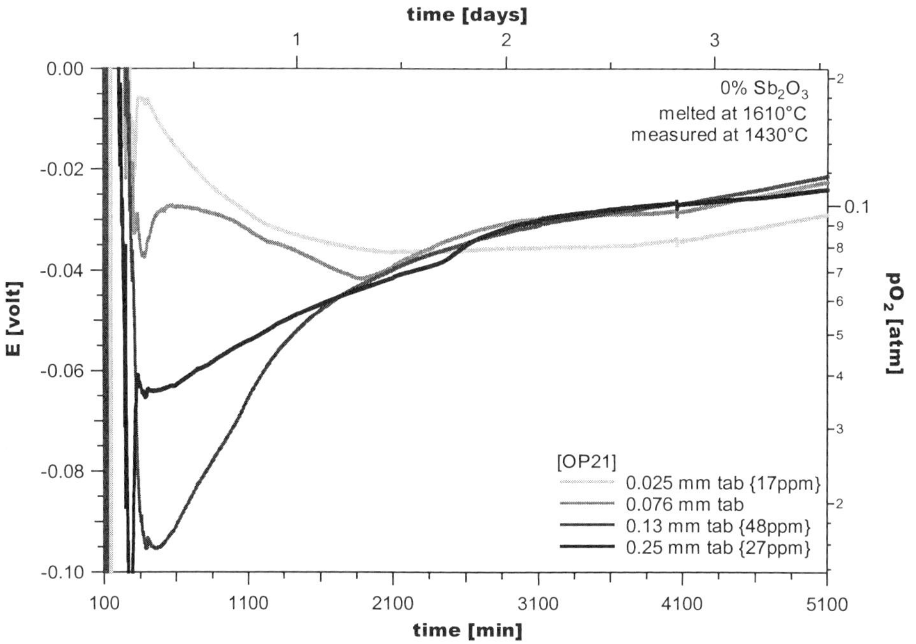

Figure 5: Effect of Pt tramp element content on $P_{O_2}^{meas}$.

As shown in Figure 6, Pt and Ir measuring electrodes immersed in the same melt reported widely different $P_{O_2}^{meas}$ at 1300°C — $P_{O_2}^{meas}$ [Ir] is 5 orders of magnitude lower than $P_{O_2}^{meas}$ [Pt]. This may result from the different alloying behavior of the two metals. For reference, $P_{O_2}^{melt}$ for this glass at 1430°C is thought to be in the range 10^{-2}-10^{-1} atm.

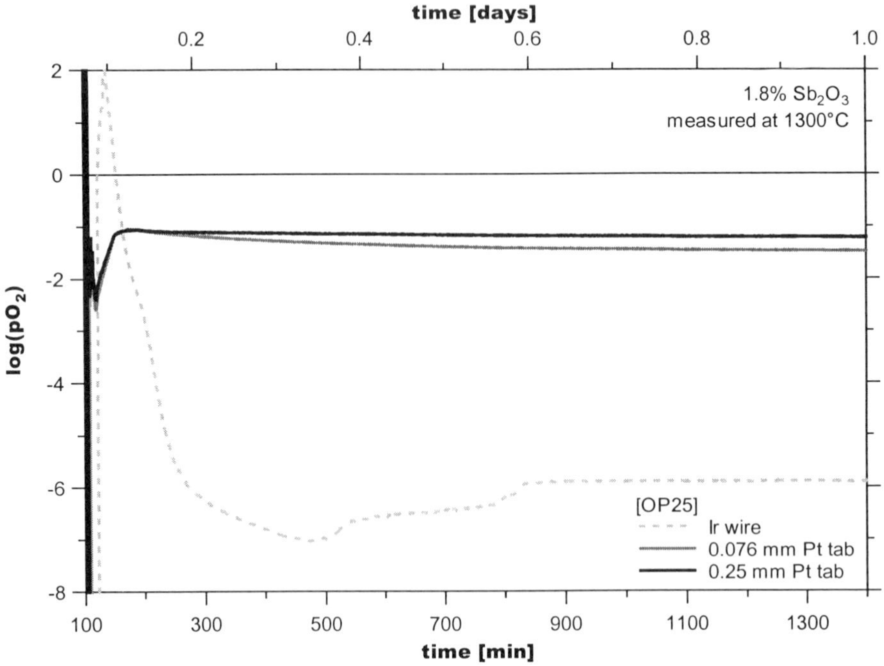

Figure 6: Effect of measuring electrode composition (Pt *vs.* Ir) on $P_{O_2}^{meas}$.

Alloying of the Pt measuring electrode with melt constituents, and transport of reaction products away from the measuring electrode, lead to initial transients and drift in $P_{O_2}^{meas}$.

It was thought that alloying and diffusion processes in glass melts were complete within several hours, but it is now clear that, at least in stagnant, viscous melts with low oxygen diffusivity, these processes are still incomplete even after several days (see Figures 2 and 3). As expected due to greater alloying, the magnitude and duration of drift in $P_{O_2}^{meas}$ increases with measuring electrode thickness and Sb_2O_3 concentration.

There is evidence that drift in $P_{O_2}^{meas}$ is due to diffusion of oxygen (produced by the alloying reactions) away from the Pt measuring electrode. It was hypothesized that the altered (*i.e.* oxidized) glass layer adjacent to the Pt measuring electrode slowed equilibration of the electrode with the melt, and that diffusion of oxygen through the glass melt was the rate-limiting step. In order to test the effect of periodically removing this altered glass from the Pt measuring electrode, two identical 0.25 mm thick Pt measuring electrodes were immersed in the same melt containing 1.8wt% Sb_2O_3. Tab #2 was periodically pulled out of the melt, quenched, cleaned of adhered glass by gentle tapping with a hammer, and reimmersed, while tab #1 was left undisturbed in the melt. As expected, the Sb concentration in tab #2 (310ppm) was significantly higher than that in #1 (212ppm), and $P_{O_2}^{meas}$ [#2] < $P_{O_2}^{meas}$ [#1] as shown in Figure 7.

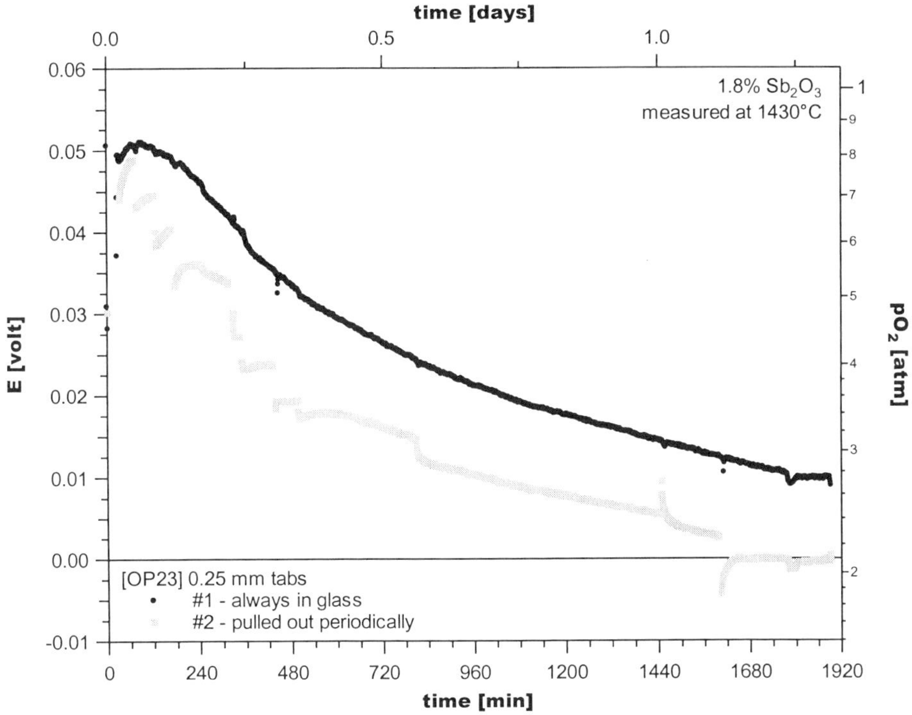

Figure 7: Effect of periodically removing glass from Pt measuring electrode.

Thermodynamic values derived from $P_{O_2}^{meas}$ vs. T curves are probably reliable.

There is strong evidence that thermodynamic values for redox reactions (*e.g.* $Sb_2O_5 \rightarrow Sb_2O_3 + O_2$) derived from $P_{O_2}^{meas}$ *vs.* T curves are reliable. Even though $P_{O_2}^{meas}$ is a function of: (i) thickness of the measuring electrode; (ii) concentration of alloyable constituents in the glass melt; and (iii) composition of the measuring electrode, the change in $P_{O_2}^{meas}$ with temperature, $\frac{d\left(\ln\left(P_{O_2}^{meas}\right)\right)}{d(1/T)}$, is not, indicating that this value and the thermodynamic values derived from it (*e.g.* $\Delta H = R\frac{d\left(\ln\left(P_{O_2}^{meas}\right)\right)}{d(1/T)}$) are accurate. However, it should be noted that $P_{O_2}^{melt}$ calculated using this ΔH for the melt under study was too high, indicating that ΔH, and therefore, $\frac{d\left(\ln\left(P_{O_2}^{meas}\right)\right)}{d(1/T)}$ may be incorrect. Specifically, $P_{O_2}^{melt}$ calculated at 1430°C assuming (i) prior equilibration with 1 atm O_2 at the melting temperature, and (ii) $\Delta H(Sb_2O_3) = 317$ kJ·mol^{-1}, was too high, as shown in Figure 2. Since alloying of the Pt electrode results in oxidation of the adjacent glass, $P_{O_2}^{meas}$ should always be greater than the calculated (Nernstian) P_{O_2}. As noted above, calculation of $P_{O_2}^{melt}$ at any temperature requires both ΔH and a single (T, $P_{O_2}^{melt}$) point, usually taken as (T^{melt}, $P_{O_2}^{melt}$ =1 atm). It is possible, therefore, that ΔH is correct, and that the error in the calculated $P_{O_2}^{melt}$ is due instead to the assumption that the melt equilibrates with 1 atm O_2 at the melting temperature.

In support of the accuracy of thermodynamic values derived from $P_{O_2}^{meas}$ *vs.* T curves, the slopes of cooling curves, $\frac{d\left(\ln\left(P_{O_2}^{meas}\right)\right)}{d(1/T)}$, were relatively independent of tab thickness. As shown in Figure 8, even though $P_{O_2}^{meas}$ reported by any two tabs at a particular time or temperature differed by as much as a factor of 5, the change in $P_{O_2}^{meas}$ with temperature, $\frac{d\left(\ln\left(P_{O_2}^{meas}\right)\right)}{d(1/T)}$, was essentially the same for all tabs.

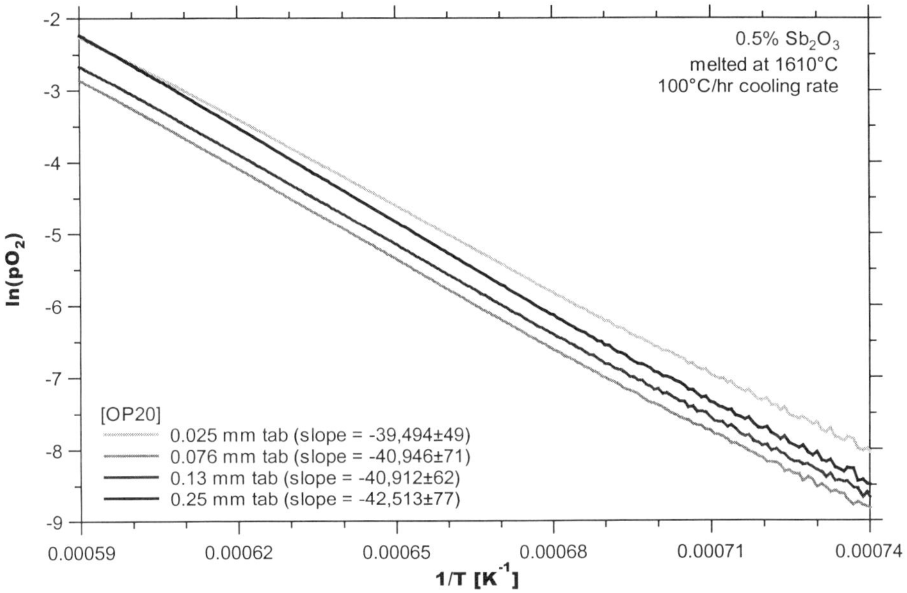

Figure 8: Effect of measuring electrode thickness on slope.

Further support for the accuracy of $\frac{d\left(\ln\left(P_{O_2}^{meas}\right)\right)}{d(1/T)}$ is shown in Figure 9. The slopes were relatively independent of the concentration of alloyable constituents in the melt. The melt with 0.5wt% Sb_2O_3 showed slightly lower slopes than the one with 2wt% Sb_2O_3, but the difference was <10%.

Final support for the accuracy of $\frac{d\left(\ln\left(P_{O_2}^{meas}\right)\right)}{d(1/T)}$ is shown in Figure 10. The slopes are independent of the base glass composition and viscosity (order of magnitude difference in viscosities) for these two different melts, each with 0.5wt% Sb_2O_3.

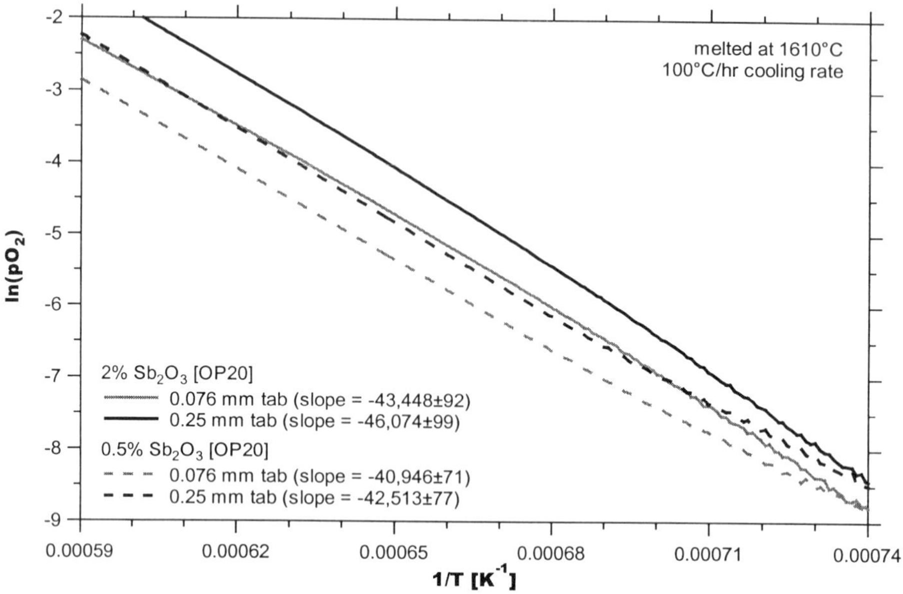

Figure 9: Effect of Sb_2O_3 concentration on slope.

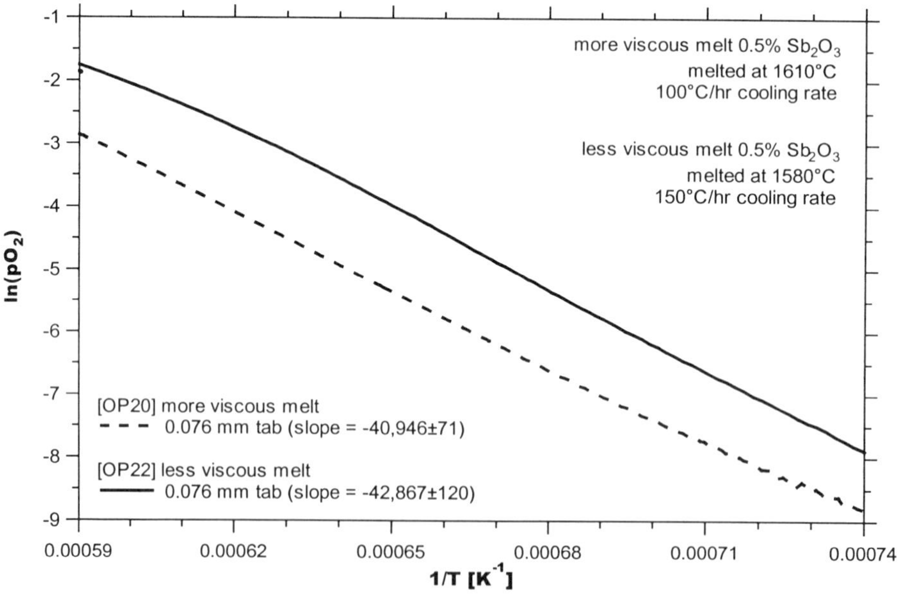

Figure 10: Effect of base glass on slope.

CONCLUSIONS

Whenever a glass melt contains reducible constituents (*e.g.* Sb_2O_3), alloying leads to oxidation of the glass melt adjacent to the Pt measuring electrode. Depending on the glass melt, the change in P_{O_2} may be very large and long lasting — an order of magnitude and several days for melts with low oxygen diffusivities. Even though the change may be smaller and shorter lived in melts with higher oxygen diffusivities, the same processes occur — the severity of the problem scales inversely with D_O^*. This result clearly has serious implications for electrochemical techniques applied to glass melts.

ACKNOWLEDGMENTS

The author gratefully acknowledges the fundamental contributions of Dr. Casey Dorfeld to the basic understanding of alloying reactions between metals and glass melts; the insight and helpful suggestions of Mr. Dave Lineman, Dr. Casey Dorfeld, Mr. Steve Currie, and Dr. Frank Woolley; and the reliable, innovative, experimental work of Mr. Steve Currie.

REFERENCES

1. A.J. Bard and L.R. Faulkner, *Electrochemical Methods. Fundamentals and Applications*, John Wiley & Sons, New York, 1980.
2. G.S. Goto, *Solid State Electrochemistry and Its Applications to Sensors and Electronic Devices,* Materials Science Monographs, 45, Elsevier, Tokyo, 1988.
3. D.R. Gaskell, *Introduction to Metallurgical Thermodynamics,* Second Edition, McGraw-Hill Book Company, New York, 1981.

Electrical Properties

A HIGH-ACCURACY, CALIBRATION-FREE TECHNIQUE FOR MEASURING THE ELECTRICAL CONDUCTIVITY OF GLASS MELTS

Susan L. Schiefelbein
Corning, Inc.
HP ME 02 38
Corning, NY 14831

ABSTRACT

The coaxial cylinders technique, a high-accuracy, calibration-free technique developed to measure the electrical conductivity of previously inaccessible conductive, corrosive liquids, has been successfully applied to a glass melt. Briefly, the new technique employs coaxial cylindrical electrodes positioned at successive immersions in the liquid. At each immersion, a.c. impedance measurements are made over a wide frequency range. The electrical conductivity of the liquid is then calculated from the change in measured conductance with immersion. The new technique has four main advantages: (i) it is a high-accuracy technique, (ii) it does not require calibration in a standard reference liquid, (iii) it permits measurement of previously inaccessible conductive, corrosive liquids, and (iv) it permits measurement of liquids obstructed from view, such as those in a furnace.

BACKGROUND

The coaxial cylinders technique was developed so that accurate electrical conductivity measurements could be made in relatively conductive, corrosive liquids like molten oxides. All published techniques suitable for use in corrosive liquids (*i.e.* techniques employing all-metal cells) are either low-accuracy (*e.g.* the crucible technique), or limited to resistive liquids (*e.g.* the interdigitated technique).

Electrical conductivity is an intensive property and cannot be measured directly; it must be calculated from a measurement of the corresponding extensive property, resistance:

$$R = \rho\left(\frac{l}{A}\right) = \left(\frac{1}{\kappa}\right)\left(\frac{l}{A}\right) = \frac{G}{\kappa} \qquad \{1\}$$

where R is resistance, ρ is electrical resistivity, κ is the electrical conductivity, l is the effective length of the current path, A is the effective cross-sectional area of the current path, and G is the cell factor (the current path length-to-area ratio). Nearly all published techniques for measuring the electrical conductivity of liquids are comparative techniques; they rely on calibration in a standard liquid. Calibration[1] is accomplished by measuring the resistance of a standard liquid, R_{std}, and calculating the cell factor, G, by the relationship, $G = \kappa_{std} R_{std}$. The conductivity of the liquid of interest, κ_{liq}, is then calculated by measuring its resistance and using the previously determined cell factor, $\kappa_{liq} = G/R_{liq}$.

This comparative procedure works well only when the cell factor is constant,[2] *i.e.*, the current path is independent of the electrical properties of the liquid under investigation, the electrodes, and the container. High-accuracy techniques satisfy these conditions; low-accuracy techniques do not. In order to evaluate the quality and accuracy of data obtained, it is necessary to know whether a technique is high- or low-accuracy.

Seven low-accuracy techniques based on five electrode designs were found in the literature, and are depicted schematically in Figure 1. The main problem with the two-wire [3-9], four-wire [10], four-wire with double immersion [11, 12], ring [13-15], and two toroid [16] techniques, is that the current path is not confined, and is, therefore, a function of electrical conductivity, κ, and dielectric constant, ε, of the liquid. Current spreads out in all directions in the two-wire, four-wire, and two toroid techniques. The four-wire with double immersion technique accounts for and eliminates the current that strays below the electrodes, but not the current that spreads out radially beyond the electrodes. The ring technique does not account for current that strays above and below the cylindrical electrodes. The main problem with the crucible [17-19] and the differential crucible [20] techniques is that the current distribution is a function of κ and ε of both the liquid and the electrodes.

---

[1] Janz and Tomkins [1] give a thorough review of calibration conventions. KCl (aq) solutions, as determined by Jones and Bradshaw [2], are often used as conductivity standards.

[2] By definition, calibration also works if the electrical properties of the liquid of interest are identical to those of the standard — accuracy decreases as these diverge.

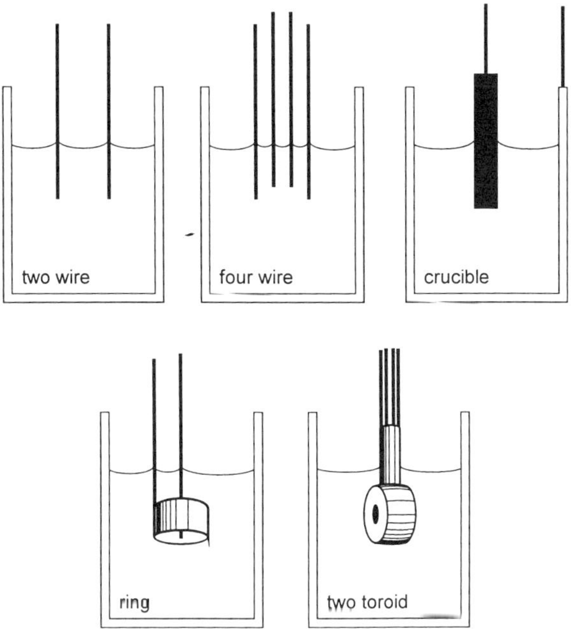

Figure 1: Five electrode designs for the seven low-accuracy techniques.

Four high-accuracy techniques based on three electrode designs were found in the literature, and are depicted schematically in Figure 2. Factors to consider in choosing one of these techniques for a specific application include (i) chemical compatibility between the liquid and the materials of construction of the cell, (ii) the range limits of the available impedance measuring instrument, and (iii) the conductance of the liquid under investigation (*i.e.*, the conductivity as seen through the cell factor of the particular electrode configuration). The interdigitated electrode design [21], featuring relatively short and wide current paths, is best suited for use in resistive liquids (*e.g.*, transformer oil), while the capillary design [2, 22-29], featuring relatively long and narrow current paths, is best suited for use in conductive liquids (*e.g.*, molten NaCl). The meandering winding technique [30], which has been used for measuring the electrical conductivity of solid metals, is, in principle, suitable for use with conductive liquids. As for the molten salts and glasses of interest, their relatively high conductivity excludes the interdigitated design, while their chemical incompatibility with dielectric materials excludes the capillary techniques and the meandering winding technique.

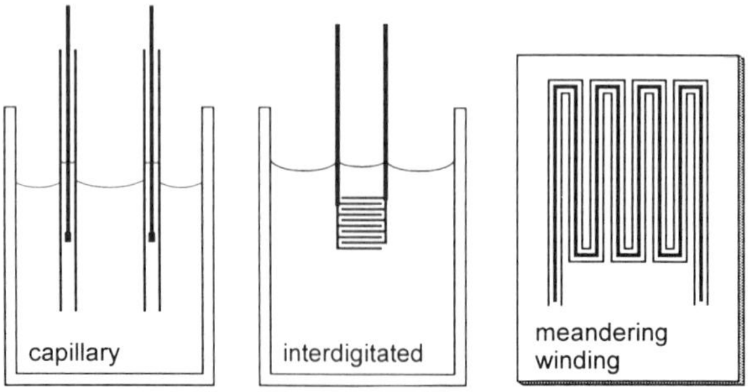

Figure 2: Three electrode designs for the four high-accuracy techniques.

The coaxial cylinders technique was developed specifically for use in conductive, highly corrosive melts [31-33]. Besides enabling the measurement of liquids inaccessible by other high-accuracy techniques, the coaxial cylinders technique has two unique advantages: (i) it does not require calibration in a standard reference liquid; and (ii) only the *relative* (not *absolute*) depth of immersion of the electrodes need be known.

The coaxial cylinders technique employs cylindrical electrodes: the inner electrode is a rod coaxially positioned inside the outer tube electrode. At the top the two are positioned by a dielectric separator that never contacts the liquid under investigation. Briefly, the technique is deployed as follows: the electrodes are immersed in the liquid to an arbitrary initial depth, and a.c. impedance is measured over a wide frequency range. The electrodes are then immersed deeper, and a.c. impedance is again measured over the same frequency range. This process is repeated at several immersions. The electrical conductivity is calculated from the change in measured conductance with immersion.

The technique has been certified in KCl (aq) reference solutions [31], and applied to $CaO{\cdot}MgO{\cdot}SiO_2$ melts [31, 32]. This paper describes successful application of the new technique to a glass melt (Corning code 7059).

THEORY

For the coaxial cylinder electrode geometry, κ is related differentially to the radial component of the resistance, R_{liq}^{radial}, as follows [31]:

$$\kappa = \left(\frac{\ln(b/a)}{2\pi}\right) \cdot \frac{d\left(1/R_{liq}^{radial}\right)}{d\xi} \qquad \{2\}$$

where b is the inner diameter of the outer electrode, a is the outer diameter of the inner electrode, and ξ is the relative depth of immersion.

An electrochemical impedance spectroscopy (EIS) measurement yields the total impedance of the measuring system, $\tilde{Z}_{meas}$ (the tilde denotes a complex number). The equivalent circuit for the measurement is shown in Figure 3.

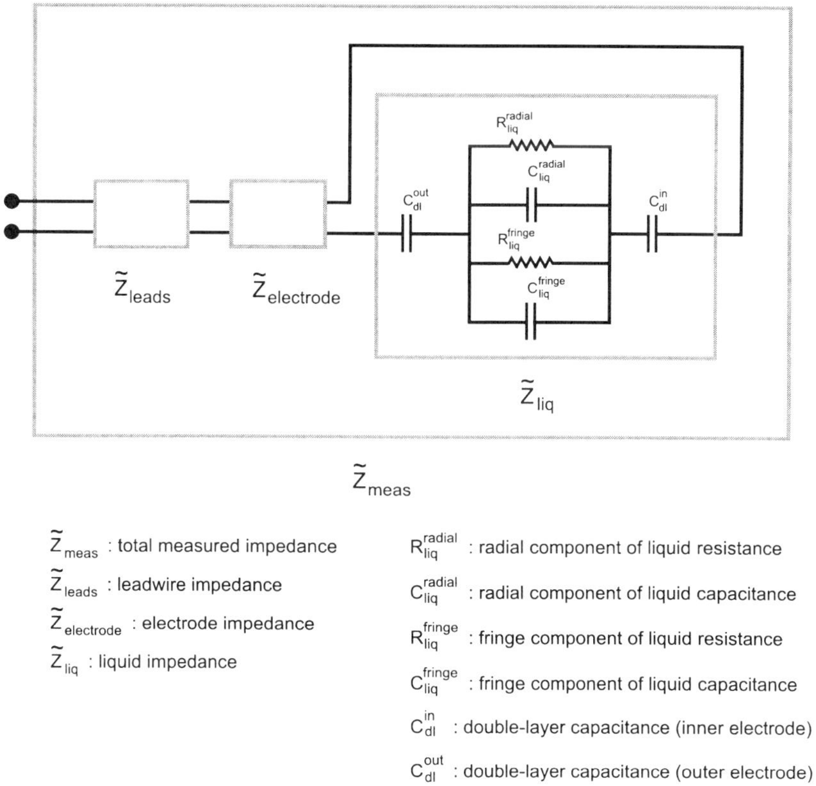

Figure 3: Equivalent circuit.

Summing series impedances gives:

$$\tilde{Z}_{meas} = \tilde{Z}_{leads} + \tilde{Z}_{electrode} + \tilde{Z}_{liq} \qquad \{3\}$$

where $\tilde{Z}_{leads}$ is the impedance of the leadwires from the impedance measuring instrument, $\tilde{Z}_{electrode}$ is the impedance of the electrodes, and $\tilde{Z}_{liq}$ is the impedance of the liquid under investigation.Since we are only interested in the properties of the liquid, $\tilde{Z}_{leads}$ and $\tilde{Z}_{electrode}$ must be eliminated from $\tilde{Z}_{meas}$. This is done by measuring the impedance of the system with the electrodes shorted, then subtracting this from $\tilde{Z}_{meas}$.[3, 4] The results are plotted on an impedance plane plot: the negative of the imaginary part of $\tilde{Z}_{liq}$, denoted $\left(-Z_{liq}^{im}\right)$, is plotted against the real part of $\tilde{Z}_{liq}$, denoted Z_{liq}^{real}. In addition, since κ is related to resistance (not capacitance or inductance), it is necessary to isolate the *purely resistive* part of $\tilde{Z}_{liq}$, denoted $\left(Z_{liq}^{real}\right)^*$. The value of Z_{liq}^{real} corresponding to the minimum in $\left(-Z_{liq}^{im}\right)$ is taken as the purely resistive part of $\tilde{Z}_{liq}$, $\left(Z_{liq}^{real}\right)^*$.[5] From the equivalent circuit we know that $\frac{1}{\left(Z_{liq}^{real}\right)^*} = \frac{1}{R_{liq}^{radial}} + \frac{1}{R_{liq}^{fringe}}$, and since R_{liq}^{fringe} is not a function of ξ,[6] (*i.e.* $\frac{d\left(1/R_{liq}^{fringe}\right)}{d\xi} = 0$), Equation {2} becomes:

$$\kappa = \left(\frac{\ln(b/a)}{2\pi}\right) \cdot \frac{d\left(1/\left(Z_{liq}^{real}\right)^*\right)}{d\xi} \qquad \{4\}$$

---

[3] $\tilde{Z}_{leads}$ is predominantly inductive and therefore extremely sensitive to placement of the leadwires. The position of the leadwires must be the same for all measurements.

[4] The impedance of the electrode is assumed to be invariant with depth of immersion. For very conductive liquids, this assumption is not valid and the electrode must be treated as a transmissionline [31].

[5] Mathematical validation of this assignment is given in reference [31].

[6] R_{liq}^{fringe} is constant so long as the bottom of the electrodes remain a distance of at least $2(b - a)$ above the floor of the container [31].

At least five measurements are made at successive immersions, and $\frac{1}{\left(Z_{liq}^{real}\right)^{*}}$ is plotted *vs.* ξ. According to Equation {4}, the slope of this plot, $\frac{d\left(1/\left(Z_{liq}^{real}\right)^{*}\right)}{d\xi}$, is multiplied by the constant $\frac{\ln(b/a)}{2\pi}$ (known from electrode geometry) to yield κ. Since this constant depends only on electrode geometry, the coaxial cylinders technique does not require calibration in a reference liquid.

In summary, the procedure for obtaining κ from $\tilde{Z}_{meas}$ consists of the following steps:

1. Impedance measurements are taken over a wide frequency range at many successive depths of immersion.

2. Using a short circuit correction, $\tilde{Z}_{liq}$ at each immersion is isolated from $\tilde{Z}_{meas}$.

3. The value of $\left(Z_{liq}^{real}\right)^{*}$ at each immersion is found by choosing the value of Z_{liq}^{real} at which $\left(-Z_{liq}^{im}\right)$ is a minimum.

4. $\frac{1}{\left(Z_{liq}^{real}\right)^{*}}$ is plotted *vs.* the corresponding relative immersion, ξ.

5. The slope of this straight line, $\frac{d\left(1/\left(Z_{liq}^{real}\right)^{*}\right)}{d\xi}$, is multiplied by the constant $\frac{\ln(b/a)}{2\pi}$ to yield the electrical conductivity of the liquid, κ.

EXPERIMENT

A Solartron 1260 FRA was used to measure impedance. The 1260 FRA settings used in this study are given in Table I. PC control of the FRA was accomplished with Z60 software (Scribner Assoc., Inc., Southern Pines, NC).

Table I: 1260 parameters

Analyzer	Experiment
auto integration type: off	sweep type: log
measurement integration time: 2 s	log increment type: steps per decade
measurement delay time: 0.1 s	log increment value: 25
mode: impedance	initial sweep frequency: 1 MHz
source: V1/i	final sweep frequency: 100 Hz
range input V1: auto	bias voltage: 0
range input i: auto	monitor: off
coupling input V1: dc	generator voltage: 50 mV
coupling input i: dc	
input type V1: differential	
outer conductor type, V1: grounded[7]	

Electrodes were designed and constructed from 80Pt-20Rh as shown in Figure 4. The outer electrode was 10.16 cm long and had a 3.175 cm i.d.; the inner electrode had a 0.635 cm i.d. A 2.54 cm long, close-fitting zircon cylinder was used for electrical isolation, and to position the inner electrode coaxially within the outer electrode. A simple two lead design was chosen for this work.[8] The outer electrode had several 0.3175 cm diameter holes, just below the zircon, to permit the gas flow required when raising and lowering the electrodes in the melt. An XYZ translator, built in-house, was used to vertically translate the electrodes in the glass melt. Several refractory heat shields (8.89 cm diameter disks) were fitted on the leads ~10 cm above the top of the outer electrode to reduce heat loss and to improve alignment and stability of the entire electrode

[7] The preferred setting, "floating," could not be used; the "grounded" setting had to be used to avoid input and generator overload errors due to noise pick up from the furnace.
[8] It should be noted that a four lead design (separate current and voltage leads) is always a better choice, but the two lead design is satisfactory when the measured impedance is in the intermediate range, ~10 to ~1,000 Ω.

assembly. A type S thermocouple in a 0.635 cm o.d., 80Pt-20Rh closed-end sheath was threaded through the heat shields and its end positioned near the bottom of the electrodes.

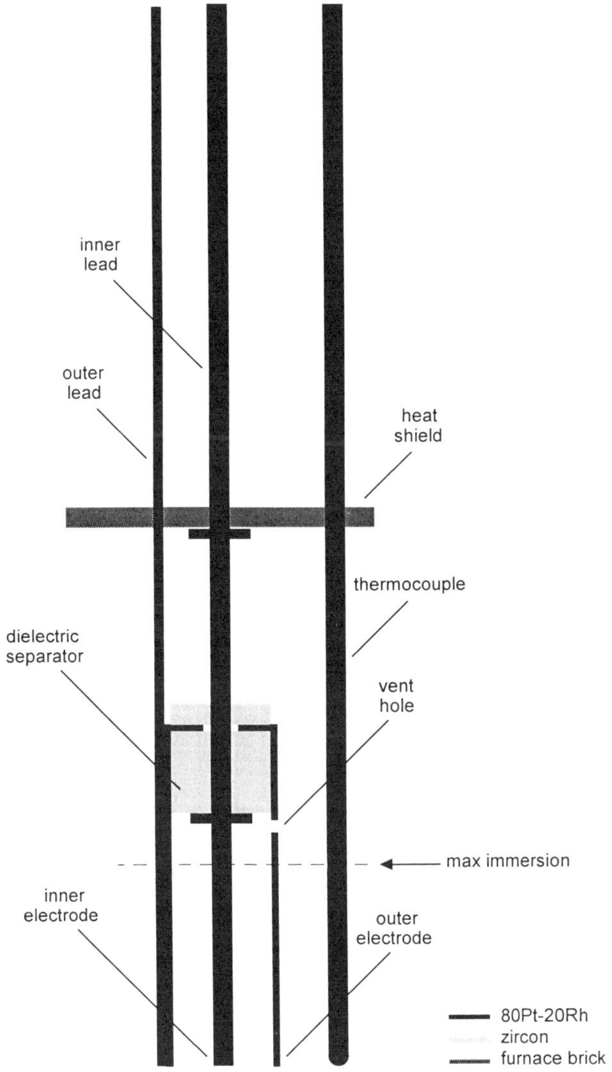

Figure 4: Pt-Rh coaxial cylinder electrodes.

Prior to use at high temperature, all 80Pt-20Rh parts were thoroughly cleaned and dried. The electrodes and zircon separator were then carefully assembled, and the coaxial position of the inner electrode verified.

A fused silica crucible, 7.62 cm i.d., 15.24 cm tall, filled with clean code 7059 cullet was placed on the furnace hearth, and the furnace was ramped to 1350°C. One hour after the furnace had attained 1350°C, the melt surface was located by slowly lowering the electrodes while making an impedance measurement — the impedance dropped precipitously when the electrodes first contacted the melt. The (approximate) vertical position of the surface was recorded. The electrodes were then immersed into the melt beyond the deepest immersion used for impedance measurements and left for one hour.

The electrodes were then raised up, and re-immersed to the first immersion, approximately 0.76 cm ($\xi = 0$). After 2 minutes (for thermal equilibrium), the temperature was recorded and impedance measured over the frequency range 1 MHz - 100 Hz (this range was chosen to include the critical point in the impedance plane plot). The electrodes were then lowered to the next immersion, the relative immersion recorded from the vertical vernier on the XYZ translator, and the process repeated. It is worth noting that the electrodes were always *lowered* to the next immersion, because *raising* the electrodes to the next immersion had caused reproducibility problems in previous molten oxide measurements [31, 32] (probably due to the hysteresis in the contact angle of liquid oxides on solid metals [34]). This process was repeated at a minimum of five successive immersions. After the last measurement, the electrodes were raised out of the melt. The entire process was then repeated at the same temperature.

In a separate experiment, the electrodes were shorted by winding a 0.1016 cm diameter Pt wire between the bottoms of the inner and outer electrodes. Impedance measurements were made at the same temperature, over exactly the same frequency range, and at the same increments as for the glass melt. The raw data for the first run (6 immersions), as well as the short, are shown in Figure 5.

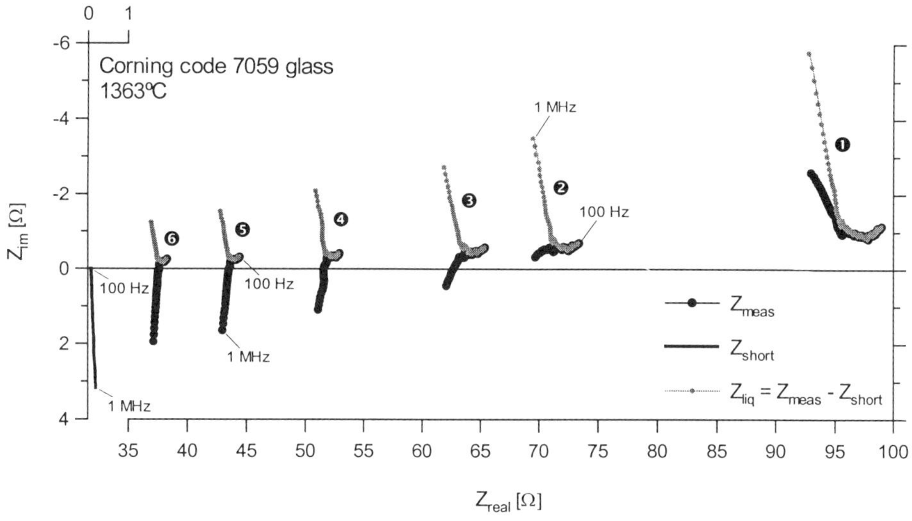

Figure 5: Measured, short, and difference (liquid) impedances at six immersions.

The impedance of the short was subtracted from that of the glass melt (point by point over the entire frequency range) at each immersion. The result is shown in Figure 5. Next, the value of $\left(Z_{liq}^{real}\right)^*$ at each immersion was found by choosing the value of Z_{liq}^{real} at which $\left(-Z_{liq}^{im}\right)$ was a minimum. $\frac{1}{\left(Z_{liq}^{real}\right)^*}$ was then plotted *vs.* the corresponding relative immersion, ξ,[9] see Figure 6.

[9] The measured relative immersion must be corrected to compensate for the liquid displaced by the electrodes (Eureka!).

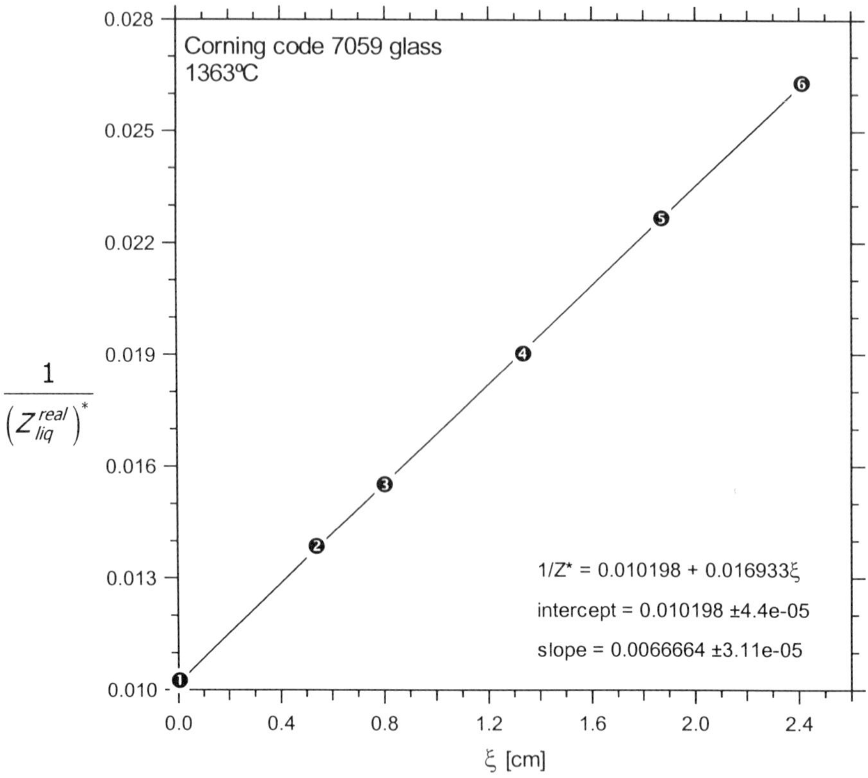

Figure 6: Plot of $\dfrac{1}{\left(Z_{liq}^{real}\right)^{*}}$ *vs.* ξ for six immersions.

Finally, the slope of the straight line, 0.00667 $\Omega^{-1}cm^{-1}$, was multiplied by the constant $\dfrac{\ln(b/a)}{2\pi} = 0.25615$ to yield the electrical conductivity of the liquid, $\kappa = 0.0017\ \Omega^{-1}\cdot cm^{-1}$. The average of two measurements at the same temperature was 0.0017 $\Omega^{-1}\cdot cm^{-1}$.

RESULTS

The electrical conductivity of code 7059 glass was measured at 1363±2°C using the coaxial cylinders technique. The average value was 0.0017 $\Omega^{-1}\cdot cm^{-1}$. The value measured by Corning's Physical Properties Lab in the same apparatus used for viscosity measurements (essentially a crucible technique), was 0.0014 $\Omega^{-1}\cdot cm^{-1}$. The difference of ~18% is surprisingly small considering the vast difference in rigor of the two techniques.

This was the first test of the technique in a viscous melt, and no particular problems were encountered. The technique is clearly applicable to viscous liquids such as glass melts.

ACKNOWLEDGMENTS

The author gratefully acknowledges the collaboration of Prof. Donald R. Sadoway, Dr. Kevin G. Rhoads, and Dr. Naomi Fried in the initial development of the coaxial cylinders technique at M.I.T.; and the technical assistance of Mr. Stephen C. Currie in the glass melt experiments at Corning, Inc.

REFERENCES

1. G.J. Janz and R.P.T. Tomkins, "Conductance Cell Calibrations: Current Practices," *J. Electrochem. Soc.*, **124**, 55C-59C, 1977.
2. G. Jones and B.C. Bradshaw, "The Measurement of the Conductance of Electrolytes. V. A Redetermination of the Conductance of Standard Potassium Chloride Solutions in Absolute Units" *J. Amer. Chem. Soc.*, **55**, 1780-1800, 1933.
3. V.P. Boiko and T.Yu. Boiko, "A Cell to Measure the Electrical Conductivity of Solutions," *Instruments and Experimental Techniques*, **35**, 725-727, 1992.
4. R.E. Tickle, "The Electrical Conductance of Molten Alkali Silicates," *Phys. Chem. Glasses*, **8**, 101-112, 1967.
5. R. Winand, A. Fontana, L. Segers, P. Hannaert, and J. Lacave, "Electrowinning Low-carbon Manganese from Molten Oxide Mixtures," in *Proceedings of Molten Salts Electrolysis in Metal Production*, London, 1977.
6. U. Pal, T. Debroy, and G. Simkovich, "Electrical Conductivity of $PbO-SiO_2$ Liquids Containing Pb Precipitates," *Can. Metall. Q.*, **23**, 295-302, 1984.
7. H. Keller, K. Schwerdtfeger, and K. Hennesen, "Tracer Diffusivity of Ca^{45} and Electrical Conductivity in $CaO-SiO_2$ Melts," *Metall. Trans. B.*, **10**, 67-70, 1979.

8. H.J. Engell and P. Vygen, "Ionen- und Elektronenleitung in CaO-FeO-Fe_2O_3-SiO_2 Schmelzen," *Ber. Bunsengesell. Phys. Chemie*, **72**, 5-12, 1968.
9. J. O'M. Bockris and G.W. Mellors, "Electric Conductance in Liquid Lead Silicates and Borates," *J. Phys. Chem.*, **60**, 1321-1328, 1956.
10. K. Ogino, H. Hashimoto, and S. Hara, "Measurement of the Electrical Conductivity of ESR Fluxes Containing Fluoride by Four Electrodes Method with Alternating Current," *Tetsu-to-Hagane*, **64**, 41-47, 1978.
11. Y. Ohta, A. Miyangana, K. Morinaga, and T. Yanagase, "Study on the Measuring Method of the Electrical Conductivity of Ionic Solutions and Melts," *J. Japan Inst. Met.*, **45**, 1036-1043, 1981.
12. A. Silny and B. Haugsdal, "Electrical Conductivity Measurements of Corrosive Liquids at High Temperatures," *Rev. Sci. Inst.*, **64**, 532-537, 1993.
13. J.C. Nichol and R.M. Fuoss, "A New Cell Design for Precision Conductimetry," *J. Phys. Chem.*, **56**, 697-699, 1954.
14. T. Yanagase, Y. Suginohara, and K. Ikeda, "The Effects of Additional Oxide on the Electrical Conductivity of Barium Silicate Melts," *J. Japan Inst. Met.*, **33**, 448-454, 1969.
15. M. Hajduk and T. El Gammal, "Schrifttumszusammenstellung von Leitfähigkeitsmessungen an CaF_2-haltigen Schlackenschmelzen," *Stahl und Eisen*, **99**, 113-116, 1979.
16. T.S. Light, E.J. McHale, and K.S. Fletcher, "Electrodeless Conductivity," *Talanta*, **36**, 235-241, 1989.
17. L. Segers, A. Fontana, and R. Winand, "Conductivité électrique des Silicates Fondus Méthode de Mesure," *Silicates Industriels*, **12**, 341-345, 1975.
18. E.A. Dancy and G.J. Derge, "The Electrical Conductivity of FeO_x-CaO Slags," *Electrochim. Acta*, **236**, 1642-1648, 1966.
19. J. Híveš, J. Thonstad, Å. Sterten, and P. Fellner, "Electrical Conductivity of the Molten Cryolite-based Ternary Mixtures Na_3AlF_6-Al_2O_3-CaF_2 and Na_3AlF_6-Al_2O_3-MgF_2," *Electrochim. Acta*, **38**, 2165-2169, 1993.
20. K. Itoh and E. Nakamura, "Electrical Conductivity and Density of Cryolite-base Melts," *Denki Kagaku*, **61**, 734-735, 1993.
21. R.B. Michnick, K.G. Rhoads, and D.R. Sadoway, "Relative Dielectric Constant Measurements in the Butyronitrile-Chloroethane System at Subambient Temperatures," *J. Electrochem. Soc.*, **144 (7)**, 2392-2398, 1997.
22. K.B. Kim and D.R. Sadoway, "Electrical Conductivity Measurements of Molten Alkaline-earth Fluorides," *J. Electrochem. Soc.*, **139**, 1027-1033, 1992.
23. J.A.A. Ketelaar and P.P.E. Maenaut, "Conductivité électrique du Sodium Fondu et son Emploi Comme sel de Reference à 1000ºC," *Electrochim. Acta*, **17**, 2195-2203, 1972.

24. F.G.K. Baucke, J. Braun, G. Röth, and R.-D. Werner, "Accurate Conductivity Cell for Molten Glasses and Salts," *Glastechnische Berichte*, **62**, 122-126, 1989.
25. F.G.K. Baucke and R.-D. Werner, "Mixed Alkali Effect of Electrical Conductivity in Glass-forming Silicate Melts," *Glastechnische Berichte*, **62**, 182-186, 1989.
26. V.S. Degtyarev, V.A. Reznichenko, S.I. Denisov, E.I. Zobov, and V.V. Nerubaschenko, "Measuring Electric Conductivity of Highly Conducting Slags," *Zavod. Lab.*, **35**, 64, 1969.
27. E.A. Pastukhov, O.A. Esin, and S.K. Chuchmarev, "The Electrical Conductivity of Molten Silicates Containing Iron Oxides," *Sov. Electrochem.*, **2**, 209-215, 1966.
28. P. Fellner, O. Kobbelvedt, Å. Sterten, and J. Thonstad, "The Electrical Conductivity of Molten Cryolite-based Mixtures Obtained with a Tube-type Cell Made of Pyrolytic Boron Nitride," *Electrochim. Acta*, **38**, 589-592, 1993.
29. Y.C. Wu and W.F. Koch, "Absolute Determination of Electrolytic Conductivity of Primary Standard KCl Solutions from 0 to 50°C," *J. Sol. Chem.*, **20(4)**, 391-401, 1991.
30. N.J. Goldfine, "Magnetometers for Improved Materials Characterization in Aerospace Applications," *Materials Evaluation*, March 1993, 396-405, 1993.
31. S.L. Schiefelbein, "A New Technique to Measure the Electrical Properties of Molten Oxides," Ph.D. Thesis, Massachusetts Institute of Technology, 1996.
32. S.L. Schiefelbein and D.R. Sadoway, "A High-accuracy, Calibration-free Technique for Measuring the Electrical Conductivity of Molten Oxides," Met. Trans. B, **28B**, 1141-1149, 1997.
33. D.R. Sadoway, K.G. Rhoads, N.A. Fried, and S.L. Schiefelbein, "High Accuracy Calibration-Free Electrical Parameter Measurements," U.S. Patent #5,489,849, 1996.
34. W.D. Kingery, H.K. Bowen, and D.R. Uhlmann, Introduction to Ceramics, John Wiley & Sons, New York, 1976.

IONIC CONDUCTIVITY and RELAXATIONS IN Li AND (Li AND K) FLUOROPHOSPHATE GLASSES

A. R. Kulkarni[†] , U. Schneider, F. Mayr, P. Lunkenheimer and A. Loidl
Experimentalphysik V, Universität Augsburg
D-86135 Augsburg, GERMANY

ABSTRACT

Fluorophosphate glasses in the system LiF-KF-$Al(PO_3)_3$ were synthesized and the glass-transition temperature, dc conductivity and electrical relaxations were studied. The conductivity spectrum of the 0.8 LiF-0.2$Al(PO_3)_3$ glass was recorded at room temperature covering 17 decades in frequency. A maximum conductivity of 38 (Ω^{-1} cm^{-1}) at 3.4 x 10^{13} Hz agreed with the theoretically predicted value. For mixed cation glasses, the frequency dependent conductivity was measured in the frequency range 20 Hz to 2 GHz and at temperatures 100 K $\leq$ T $\leq$ 523 K. While the dc conductivity showed a characteristic minimum with composition, the ac conductivity also exhibited the mixed alkali effect. The data recorded at room temperature supported the existence of two relaxation processes: the Jonscher's universal dielectric response at low frequency and a high-frequency process characterized by a power-law exponent larger than unity.

INTRODUCTION

Ionic conductivity and electrical relaxations in fast ion conducting glasses have been intensely investigated in recent years [1-3]. Ion dynamics is usually investigated from electrical conductivity relaxations represented by the ac

[†] on leave from Indian Institute of Technology-Bombay 400076, INDIA

conductivity from near zero frequency (dc) to the far or mid infrared spectrum to which relaxations of the ion and network motions contribute. Many models have been proposed in recent years to explain the data below 10^6 Hz, but only limited data are available in the range 10^6 to 10^{10} Hz. Wong and Angell [4] were the first to construct a complete spectrum for a sodium silicate glass by interpolating conductivity data recorded below 10^6 Hz and the far infrared measurements and predicted that in the GHz range the conductivity will be proportional to $\nu^{1.0}$. Cole et. al [5] measured the conductivity for the same composition and observed $\nu^{1.0}$ behavior in the GHz range. Recently, Funke and co-workers reported a complete spectrum for glasses [6-9] from 1 Hz to 10^{14} Hz and observed $\nu^{1.0}$ and $\nu^{1.3}$ behavior at higher frequencies. The former universality was explained using Jonscher's universality [10] while the latter was debated [11]. However, the experimental observations agree with both the jump relaxation model [12] and the coupling model [13,14]. Recently Ngai et al [15] reported the physical origins of these $\nu^{1.0}$ and $\nu^{1.3}$ contributions to the conductivity spectrum. Above conclusions were arrived at after analyzing data over wide frequency and temperature range [16]. In this paper, the conductivity spectrum at room temperature over 17 decades of frequency has been reported for a lithium fluorophosphate glass and discuss different contributions to hopping conductivity.

Although the mixed alkali effect in the dc conductivity has been known for a long time, there is no acceptable theory to explain the physical origins of this effect and only limited work has been reported on the dynamics of the ionic conductivity in mixed alkali glasses [17,18,19]. The major finding has been a disappearance of mixed alkali effect when temperature or frequency is raised. Our results show that the mixed alkali effect is present in the ac conductivity too [20].

In the present work, we have studied ternary LiF-KF-$Al(PO_3)_3$ glasses. We have examined the dc and ac conductivities of these glasses, as KF is progressively added to replace LiF in 80LiF-20$Al(PO_3)_3$ glass. The ac conductivity was measured over a wide range of temperature (100 - 523K) and frequency (1mHz - 150 THz). The aim was to investigate relaxations in a wide frequency range and to compare the high frequency conductivity with theoretical predictions.

EXPERIMENTAL

Starting materials, LiF, KF and Al $(PO_3)_3$ mixed in appropriate proportion (each glass weighing about 10 g) were melted in a covered platinum crucible at a temperature of 1100-1200K for 5 minutes. The melts were cast between two brass plates which were preheated to 473K to avoid thermal stress induced shattering of the samples, and the 3 cm x 1.5 cm x 0.5 cm rectangular blocks were annealed for 12 hours before slowly cooling to room temperature. The nominal compositions for the glasses are given in Table I. Glass-transition temperatures were recorded on 10 mg samples in sealed aluminum pans using a Perkin Elmer DSC-2 differential scanning calorimeter at a heating rate of 10K per minute.

For conductivity measurements, silver paste was applied on the parallel surfaces of the glass. The ac electrical conductivity was determined using computer controlled facilities. An HP4284 impedance/gain phase analyzer scanned the frequency from 20 Hz to 1 MHz at temperatures from 100K - 523K. A binary composition was investigated over a broad band of frequency from 1mHz to 150 THz employing various techniques. For the radio frequency experiments a reflectometric technique was employed. The microwave data were recorded in reflection using coaxial lines. Above 60 GHz, a quasioptical technique [21] was used with an experimental arrangement similar to a Mach-Zehnder interferometer. Reflection was used in the infrared (FIR, MIR, and NIR). The reflection data were converted to dielectric data using Kramers-Kronig relations. A complete conductivity spectrum was constructed by combining the impedance measurements at low frequencies and the dielectric data obtained at high frequencies.

RESULTS AND DISCUSSION

Figure 1 shows log σ - log ν plots for the binary composition at various temperatures and in the frequency range 1 Hz to 2 GHz. At very low temperatures, only ac conductivity was observed. As the temperature was increased, the dynamic conductivity showed a typical behavior: a frequency independent plateau and a power-law increase at high frequencies following the so-called Universal Dielectric Response (UDR) [10]

$$\sigma(\omega) = \sigma_{dc} + \sigma_0\omega^s \quad (1)$$

Where σ_{dc} is the dc (or frequency independent) conductivity, σ_0 is a temperature dependent parameter and s lies in the range $0 < s < 1$. The dc conductivity (σ_{dc}) was obtained at various temperatures by fitting eq (1) to the experimental data. The exponent s varied from 1 at very low temperatures to 0.6 at higher temperatures. This is in agreement with the new universality reported by Lee et al. [22]. The rapid fall of the conductivity at low frequencies and higher temperatures is the well-known electrode polarization phenomenon (blocking electrodes) and is observed for all compositions.

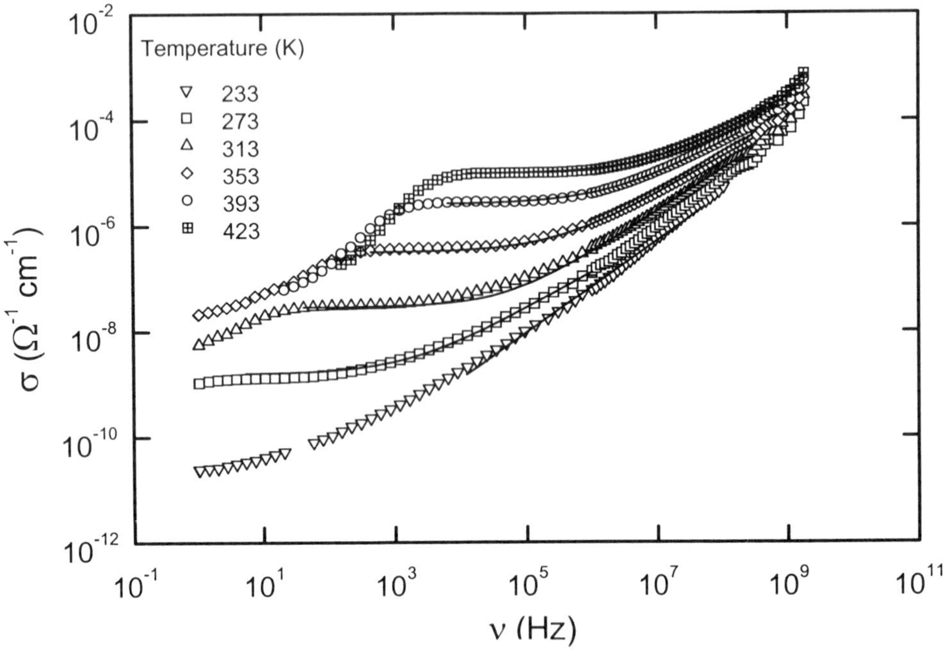

Figure 1: Conductivity spectra of the 80LiF-20Al(PO_3)$_3$ glass recorded at various temperatures shown in the legend. The dc conductivity was obtained by fitting eq (1) to the data. The fits are shown by solid lines through the data points.

Figure 2 shows the variation of the dc conductivity with inverse temperature for all glass compositions studied. The value of dc conductivity at 473K, and the activation energy obtained by fitting Arrehenius euation to the data are given in Table I. Both conductivity and T_g exhibit minima with composition variation in this system. Figure 3 shows the variation of σ_{dc} at 473K, the ac conductivities (473K) at 10^5 Hz and 10^6 Hz (after subtracting the dc conductivity) with the lithium-cation mole fraction defined by the ratio f_{Li+} = mol LiF/mol(LiF+KF). The ac conductivity shows a minimum at the same composition and in the same manner, as does the dc conductivity. This confirms that we have observed the mixed alkali effect in both the dc and hitherto unreported ac conductivity of ion conducting glasses. This is reported for the first time by us and the details are published elsewhere [20].

Table 1: Composition and the physical properties of LiF-KF-Al $(PO_3)_3$ glasses

	LiF (mole %)	KF	$Al(PO_3)_3$	Tg (K)	ΔE σ (Kcal/mol)	σ_{dc} 473K (Ω^{-1} cm^{-1})	log A (Ω^{-1} cm^{-1})	ΔE τ (Kcal/mol)
G1	0	80	20	566	19.11	1.60×10^{-7}	2.05	19.15
G2	10	70	20	536	22.83	2.25×10^{-8}	2.92	22.81
G3	20	60	20	542	24.43	4.97×10^{-9}	3.01	26.55
G4	30	50	20	540	25.96	1.46×10^{-9}	3.18	29.12
G5	40	40	20	542	23.44	4.04×10^{-9}	2.46	25.44
G6	50	30	20	558	23.24	1.99×10^{-8}	3.06	24.73
G7	60	20	20	565	20.96	2.09×10^{-7}	3.02	21.03
G8	70	10	20	578	15.89	3.33×10^{-6}	1.88	17.02
G9	80	0	20	593	13.37	4.64×10^{-5}	1.86	13.13

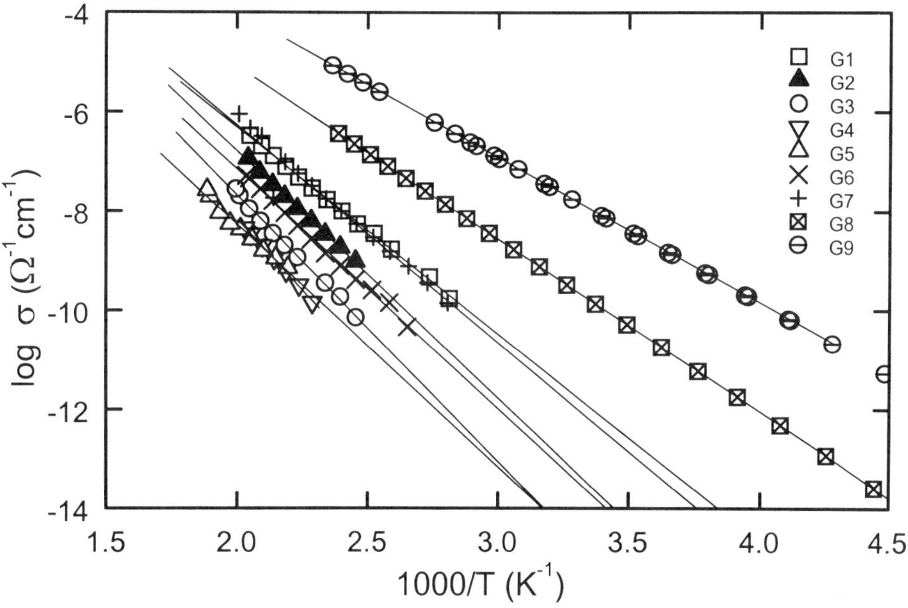

Figure 2: Arrhenius plot of the dc conductivity of LiF-KF-Al $(PO_3)_3$ glasses specified in the legend.

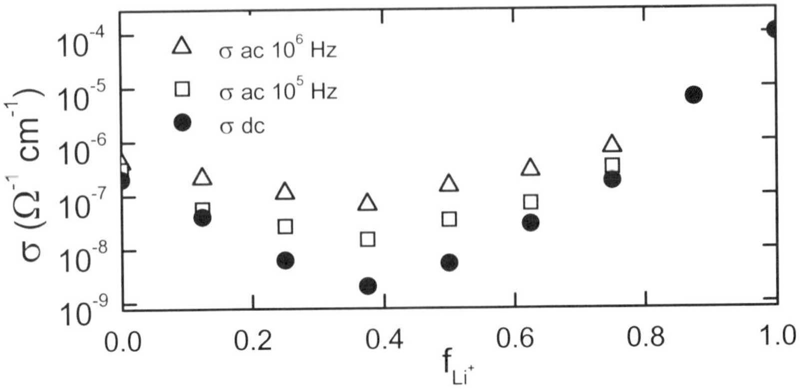

Figure 3: Dependence on lithium mole fraction of the dc and the ac conductivity at 473K. Note the minima in both the dc and the ac conductivities.

Additional dynamic characteristics of the ion motion have been studied by analyzing the ac data in the modulus formalism. Figure 4 shows the frequency dependence of the imaginary part of the complex electrical modulus at various temperatures for $60LiF\text{-}20KF\text{-}20Al(PO_3)_3$ glass. The height of the M″ curves is independent of temperature, which implies that the relative permittivity is independent of temperature. It may be noted that the electrical relaxations are asymmetric, implying that as for the most glasses the relaxation is non-exponential.

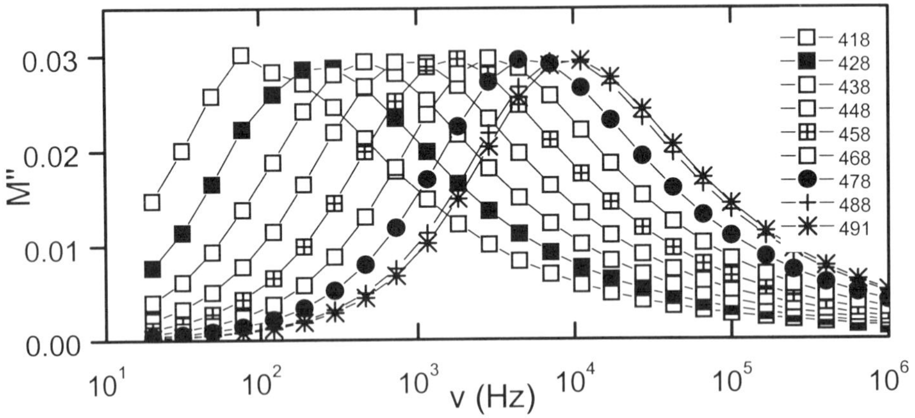

Figure 4: Imaginary part of electrical modulus, M″, versus frequency for $60LiF\text{-}20KF\text{-}20Al(PO_3)_3$ glass. The numbers in the legend indicate temperatures in K. The lines through the data points are for visual clarity.

In order to get a proper description of the relaxation , we have fitted the data with three commonly used phenomenological functions, namely, the three parameter Cole-Davidson (CD) [23] and Kohlrausch-Williams-Watts (KWW) [24] functions, and the four parameter Havriliak-Negami (HN) [25]. The fits with these functions are shown as lines in Figure 5, where data at only 453 K are shown for analysis. In addition, we show a fit using Jonscher's four-parameter expression for the complex conductivity, eq (1) and $\sigma'' = \sigma' \tan(s\pi/2) + e_\infty e_0 2\pi\nu$ [26] which was converted to the modulus data.

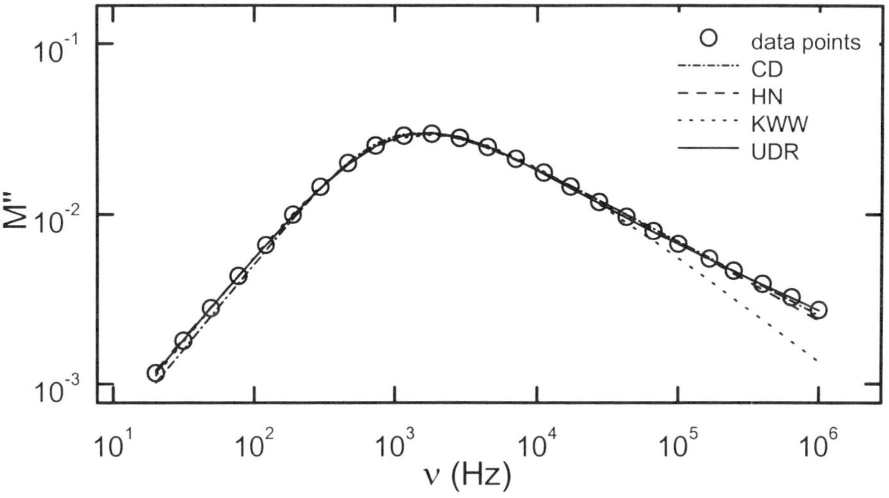

Figure 5: Modulus spectrum for a representative mixed alkali composition [60LiF-20KF-20Al$(PO_3)_3$] of the present system. Four functions, Cole-Davidson, Kohlrausch-Williams-Watt, Havriliak-Negami and UDR are fitted to the data.

It may be noted that the modulus fit obtained from the complex conductivity parameters gives the best description of the experimental data followed by the four parameter HN function. The three parameter functions, CD and KWW, fail to explain the low and high frequency deviations respectively. This does not agree with the observation of Moynihan [27,28] whose analysis shows KWW as the best function to describe relaxation phenomena in glasses in the range $10^{-2} < \omega\tau < 10^{2}$. A detailed analysis of our data using above fitting functions, over the entire frequency range, will be published elsewhere [29]. The spectra for all the compositions at various temperatures were fitted with Jonscher's four parameter function. The relaxation times at each temperature and for each composition were determined in the vicinity of the modulus peak. The relaxation times followed the Arrhenius relation. The values of the activation energies are compared with those obtained from the conductivity data in Table I. Both activation energies matched within experimental errors and suggesting that the same process is responsible for ionic conductivity and relaxation. In the present case the process is hopping of the mobile ions in the disordered solids.

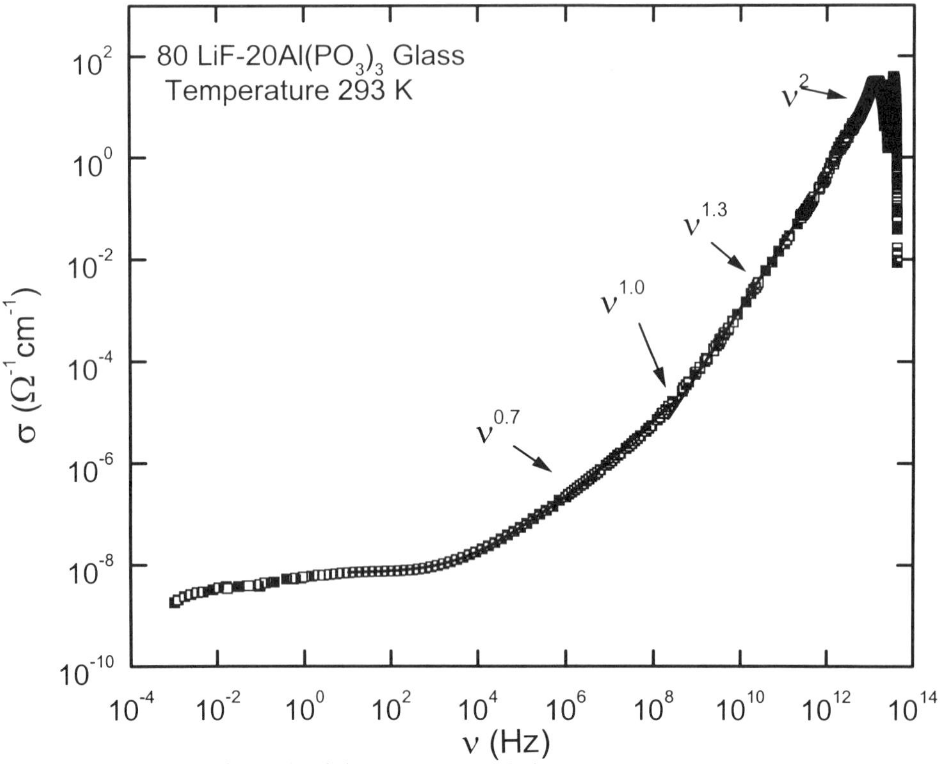

Figure 6: Broad band conductivity spectrum of the $80LiF\text{-}20Al(PO_3)_3$ glass at room temperature. The infrared reflection data were converted into dielectric data using Kramers-Kronig relations and then into conductivity data.

The broad band spectrum of $80LiF\text{-}20Al(PO_3)_3$ glass at room temperature recorded over a frequency range from 1mHz to 150 THz. The spectrum is a true representation of all the dynamical features of disordered solids. In the low frequency region three different processes are distinctly visible. The electrode polarization at very low frequencies, the frequency independent plateau (dc conductivity), and the power-law dispersion. Above this range, conductivity increases continuously up to infrared frequencies. It may be noted in Figure 6 that at room temperature we are able to see all the features. The exponents for the processes identified are depicted in Figure 6. As expected, the power-law fits up to $10^{6.5}$ Hz yield the exponent s = 0.7. At higher frequencies, and in the range from $10^{8.5}$ - 10^{12} Hz, the conductivity increases even steeper with $\nu^{\sim 1.3}$. Presumably, the superposition of these two contributions leads to the $\nu^{1.0}$ contribution being observed in a relatively narrow frequency window ($10^{6.5}$ to $10^{8.5}$). Beyond 10^{12} Hz, resonant absorption by quasi lattice modes with a stronger frequency dependence, $\sigma \propto \nu^2$ becomes significant. In this frequency range, the dispersive part of the spectrum merges into the high-frequency part of the spectrum and the conductivity is dominated by vibrational motions. The first conductivity maximum is very broad and often termed "Boson peak". In other words, at these frequencies we observe deviations of the experimental data from straightness. These are due to an additional process (vibrational motion, in this case) showing ν^2 dependence. These are usually interpreted as the low-frequency flank of the lowest-lying vibrational mode. It has also been reported that this contribution to the conductivity can be extrapolated to lower-frequencies and subtracted from the experimentally observed data. Such a procedure results in a high temperature plateau for fast ion glasses [7] in agreement with the jump relaxation model and the coupling model [13-15].

All the frequency dependencies observed for the present glass i. e. $\nu^{0.7}$, $\nu^{1.0}$, $\nu^{1.3}$, and ν^2 are consistent with the previous reports [6-9, 16]. The origins of the first three of these relaxations have been assigned to Jonscher's universality, vibrational part of the ion motion, and the frequency dependent susceptibility on the low frequency side of the Boson peak, respectively [15]. An important point to be noted is that we have been able to identify these contributions to ionic relaxations at room temperature unlike others that have studied the spectrum at higher temperatures. Also our data corroborate the arguments about the universality of the $\nu^{1.3}$ dependence in disordered solids.

Finally, we discuss the theoretical limit of conductivity achievable. While discussing the conjecture regarding $\nu^{1.0}$, Wong and Angell [4] also proposed that the far infrared conductivity will be the effective maximum value for the dc conductivity in the glass-melt system and predicted it to be 10 (Ω^{-1} cm^{-1}) at 10^{12} Hz. In the present case the conductivity, 4 (Ω^{-1} cm^{-1}) observed at room temperature and at 10^{12} Hz is in close agreement with the above-predicted value, although the highest value, 38 (Ω^{-1} cm^{-1}) is observed at 3.4 x 10^{13} Hz. The latter value agrees well with conductivity, 30 (Ω^{-1} cm^{-1}) for lithium borate systems at

3.4 x 10^{13} Hz.

CONCLUSION

We have shown that the LiF-KF-$Al(PO_3)_3$ glasses show mixed alkali effect in both the ac and the dc conductivity. The dynamic conductivity of these lithium fluorophosphate glasses shows universal features, with $\nu^{0.7}$, $\nu^{1.0}$ and $\nu^{1.3}$ behavior. The highest room temperature conductivity, (38 Ω^{-1} cm^{-1}) at 3.4 x 10^{14} Hz agrees with the theoretically predicted values.

ACKNOWLEDGMENTS

One of us (ARK) is grateful to the Alexander Von Humboldt Foundation for a fellowship grant to carry out present work.

REFERENCES

1. C. A. Angell, "Dynamic Processes in Ionic Glasses," *Chem. Rev*, **90** 523-442 (1990)
2. M. D. Ingram, "Ionic Conductivity and Glass Structure", Phil. Mag. B **60** 729-740 (1989)
3. K. L. Ngai, "A Review of Critical Experimental facts in Electrical and Ionic Diffusion in Ionically Conducting Glasses and Melts", *J. Non-Cryst. Solids*, **203** 232-245 (1996)
4. J. Wong and C. A. Angell, "Glass-Structure by Spectroscopy", Marcel Dekker, New York (1976), p. 750
5. A. Burns, G. D. Chryssikos, E. Tombari, R. H. Cole and W. M. Risen Jnr, " Dielectric Spectra of Ion Conducting Glasses to 2 GHz", *Phys. Chem. Glasses*, **30** 264 -270 (1989)
6. C. Cramer, K. Funke, T. Saatkamp, D. Wilmer, M. D. Ingram, "High Frequency Conductivity Plateau and Ionic Hopping Processes in a Ternary Lithium Borate Glass", *Z. Naturforsch* **A 50** 613-623 (1995)
7. C. Cramer, K. Funke, T. Saatkamp, "Ion Dynamics in Glass-forming Systems. I. Conductivity Spectra Below the Glass Transformation Temperature," *Phil. Mag.* **B 71** 701-711 (1995)
8. C. Cramer, K. Funke, T. Saatkamp, "Ion Dynamics in Glass-forming

Systems. II. Conductivity Spectra Above the Glass Transformation Temperature," *Phil. Mag.* **B 71** 713-719 (1995)
9. K. Funke, " Crystalline, Glassy and Molten Electrolytes: Conductivity Spectra and Model Considerations," *Mat. Res. Soc. Symp. Proc.* **455** 319-330 (1997)
10. A. K. Jonscher, "The Universal Dielectric Response", *Nature* **267** (1977) 673.
11. C. Cramer, K. Funke, C. Vortkamp-Ruckert and A. J. Dianoux, "Vibrational, Reorientational and Translational Ionic Motion in Glass," *Physica* **A 191** 358-370 (1992)
12. K. Funke, "Jump Relaxation in Solid Electrolytes (Review)," *Prog. Solid State Chemistry* **22** 111-195 (1993)
13. K. L. Ngai, " Universality of Low-frequency Fluctuation, Dissipation and Relaxation Properties of Condensed Matter I," *Comments Solid State Phys* **9** 127-140 (1979)
14. K. L. Ngai, " Universality of Low-frequency Fluctuation, Dissipation and Relaxation Properties of Condensed Matter II," *Comments Solid State Phys* **9** 141-156 (1979)
15. K. L. Ngai, H. Jain and O. Kanert, "Physical Origins of the $\omega^{1.0}$-dependent and the ω^{q}-dependent ($q \approx 1.3$) Contributions to the Conductivity Relaxations of Glassy Ionic Conductors," *J. Non-cryst. Solids* **222** 383-390 (1997)
16. C. H. Hsieh and H. Jain, " Are the Low Temperature-Low Frequency and High Temperature-High Frequency ac Conductivity of Glasses the Same Phenomenon?," *J. Non-cryst. Solids* **203** 293-299 (1996)
17. C. T. Moynihan, "Mixed Alkali Effect in Hydrate Melts," *J. Electrochem. Soc.*, **126** 2144-2149 (1979)
18. M. Tomozawa, J. M. Hyde, J. F. Cordaro and M. Yoshiyagawa, " Decay Function Kohlraush Exponent for the Electrical Modulus of Mixed Alkali Glasses", *Phys. Chem. Glasses*, **33** 69-72 (1992)
19. A. Pradel and M. Ribes, "Relaxation Processes in Ionic Conductive Chacogenide Glasses studied by Electrical and NMR Spectroscopies," *J. Non-Cryst. Solids* **131-133** 1063-1067 (1991)
20. A. R. Kulkarni, P. Lunkenheimer and A Loidl, " Mixed Alkali Effect in the AC Conductivity of Glasses," *J. Non-cryst. Solids* (submitted) (1998)
21. A.A. Volkov, Yu. G. Goncharov, G.V. Kozlov, S.P. Lebedev, and A.M. Prokhorov, "Dielectric Measurements in the Submillimeter Wavelength Region", Infrared Phys. **25**, 369 (1985).
22. W. K. Lee, J. F. Liu ans A. S. Nowick, "Limiting Behaviour of ac Conductivity in Ionically Conducting Crystals and Glasses: A New Universality", *Phys. Rev. Lett.*, **67**, 1559-1561 (1991)

23. D.W. Davidson and R.H. Cole, "Dielectric Relaxation in Glycerine", J.Chem. Phys. **18**, 1417 (1950).
24. R. Kohlrausch, "Theorie Des Elektrischen Rückstandes in der Leidener Flasche", Ann. Phys. **91**, 179 (1854); G. Williams and D.C. Watts, "Non-Symmetrical Dielectric Relaxation Behaviour Arising from a Simple Empirical Decay Function", Trans.Faraday Soc. **66**, 80 (1970).
25. S. Havriliak Jr. and S. Negami, "A Complex Plane Analysis of α-dispersion in Some Polymer Systems", J. Polym. Sci. C **14**, 99(1966).
26. A. K. Jonscher, *Dielectric Relaxation in Solids*, Chelsea Dielectric Press, London, 1983.
27. C. T. Moynihan, "Analysis of Electrical Relaxation in Glasses and Melts with Large Concentration of Mobile Ions," *J. Non-Cryst. Solids* **172-174** 1395-1407 (1994)
28. C. T. Moynihan, "Analysis of Electrical Relaxation in ionically Conducting Glasses and Melts," *J. Non-Cryst. Solids* **203** 359-363 (1996)
29. A. R. Kulkarni, P. Lunkenheimer and A Loidl, "Analysis of Electrical Relaxations in Lithium Ion Conducting Phosphate Glasses," *(to be published)*

ELECTROCHEMICAL AND SEMICONDUCTOR PROPERTIES OF POLYCRYSTALLINE COPPER(I) BROMIDE

P. Knauth* [1]; Y. Massiani[2]; M. Pasquinelli[3] and H. L. Tuller[1]
[1]Massachusetts Institute of Technology, Department of Materials Science and Engineering, Crystal Physics and Electroceramics Laboratory, Cambridge, MA 02139, USA
[2]Laboratoire de Physico-Chimie des Matériaux, Université de Provence, F-13331 Marseille Cedex 3, France.
[3]Laboratoire de Physique des Semiconducteurs, Université d'Aix-Marseille, F-13397 Marseille Cedex 20, France.

ABSTRACT

Measurements of the Hall effect and of the bias-dependence of the CuBr-electrolyte interfacial capacitance show that copper(I) bromide is a p-type semiconductor with a flat-band potential of (1.50 ± 0.05) V versus the standard hydrogen electrode. A low hole mobility of 0.4 cm^2/Vs is found, which is consistent with polaron conduction. A discrepancy is observed between the acceptor density determined from capacitance measurements ($2\ 10^{16}$ cm^{-3}) and the hole density obtained from the Hall effect ($9\ 10^{17}$ cm^{-3}). This may be due to oxygen doping prior and during the Hall experiments and/or to a redistribution of copper vacancies in the space charge field during the electrochemical experiments.

INTRODUCTION

The ionic conductivity of copper(I) bromide at moderately elevated temperatures has been studied for some time [1,2]. In its cubic β- and α-phases, this compound presents the highest ionic conductivity of all simple copper(I) halides. At reduced temperatures, a mixed ionic-electronic conductivity is observed [1], which makes copper(I) bromide an attractive candidate as the active component of gas sensors, particularly in thin-film form [3]. The high ionic

---

* On leave from: Laboratoire EDIFIS, CNRS-Université d'Aix-Marseille, F-13397 Marseille Cedex 20, France.

conductivity permits a fast response to changes of gas composition, while the electronic conductivity enables a ready integration into miniaturized electronic devices. The partial electronic conductivity of CuBr is therefore an interesting subject of investigation, especially near room temperature.

Some controversy remains regarding the type of the majority electronic charge carriers. Different techniques are, in principle, suitable to study the partial electronic conductivity of a mixed conductor, the most popular being the so-called Hebb-Wagner polarization method [1]. Here, the ionic or the electronic current is suppressed by use of blocking electrodes, e.g. a pure electronic conducting electrode to block the ionic current or a purely ionic conducting electrode to block the electronic current. Wagner [1] used a blocking graphite electrode to suppress the Cu^+ ion current in CuBr. The steady-state I-V characteristics indicated that electron holes were the majority electronic carriers, although CuBr was in equilibrium with metallic copper in these experiments, i.e. at the lowest possible chemical potential of bromine.

A p-type semiconductivity under these conditions implies either substantial acceptor impurity doping or, if the conductivity is intrinsic, that CuBr exists only as a copper-deficient non-stoichiometric compound $Cu_{1-x}Br$. Some Cu^{2+} ions would exist, in this hypothesis, for charge compensation. Given the predominant di-valent state of copper as Cu^{2+} in aqueous solution, this hypothesis is not unrealistic. However, more recently, four-point Hebb-Wagner experiments and a study of the oxygen pressure dependence of the conductivity of CuBr were not in agreement with a p-type semiconductivity of the material [4-6]. The interpretation of these experiments was that the p-type semiconductivity observed by Wagner was in fact due to decomposition of the sample resulting from electrolysis and/or oxygen doping, the pure material being n-type.

Altogether, the situation appears not yet clear, and a key objective of this study was to determine the type, concentration and mobility of the electronic charge carriers in polycrystalline CuBr by using two different complimentary techniques: Hall effect measurements and determination of the bias dependence of the CuBr-electrolyte interfacial capacitance. These experiments were done at room temperature, where Hebb-Wagner polarization experiments are practically impossible.

Preliminary experiments on the electrochemical behaviour of a polycrystalline CuBr electrode were also made to study the open-circuit voltage and the range of redox stability of the sample. Furthermore, the use of CuBr electrodes as copper sensors in aqueous solutions was also studied. In principle, the region of thermodynamic stability of the electrode is situated between its oxidation potential (formation of Cu^{2+} ions), corresponding to the anodic reaction:

$$CuBr \rightarrow Cu^{2+} + Br^- + e^- \qquad E_1 = E°_1 + (RT/F) \ln [a(Cu^{2+})\, a(Br^-)] \quad (1)$$

and its reduction potential (formation of metallic copper), corresponding to the cathodic reaction:

$$CuBr + e^- \rightarrow Cu + Br^- \qquad E_2 = E°_2 - (RT/F) \ln a(Br^-) \quad (2)$$

The standard electrode potentials versus the standard hydrogen electrode (SHE) are $E°_1 = 0.654$ V and $E°_2 = 0.033$ V [7].

EXPERIMENTS

The ceramic samples were prepared from pure CuBr powder (99.999%, Aldrich, trace analysis: 0.1 ppm Mg) by uniaxial compression under 750 MPa at room temperature. The pellets were then annealed several hours at 350 °C under argon. The relative density of the pellets was more than 98%. The residual porosity was sufficiently low, so that the ceramics were impermeable to the electrolytic solutions.

The CuBr ceramics were connected to copper wires using silver paint. The ohmic character of the contacts was checked previously. The electrode was then embedded in an epoxy resin and its surface was carefully polished with diamond paste (6 μm). The electrode-electrolyte contact area was 1.3 cm^2. The experiments were performed at 25 °C in dim light and in electrolytes de-aerated by argon bubbling. A gold foil served as counter electrode and a saturated mercurous sulphate reference electrode (E = 0.642 V/(SHE)) was used.

The electrolytic solutions were prepared with high-purity water (18.2 MΩ cm). $CuSO_4$ solutions of different concentrations were used as electrolytes. The supporting electrolyte was a 0.5 M Na_2SO_4 solution. Impedance measurements were performed with an EG&G 273 potentiostat and an EG&G 5208 lock-in amplifier. Prior to the impedance measurements, the electrodes were polarized to a d.c. potential in the domain of redox stability of CuBr (0.1-1.1 V/(SHE)). The amplitude of the a.c. signal was 10 mV at frequencies generally between 1 and 100 kHz. Measurements at low frequencies (> 10^{-3} Hz) were made in some experiments to confirm our conclusions.

The total copper concentration in solution was measured by ICP atomic emission spectroscopy (Perkin-Elmer ICP 6500) after immersion of the CuBr electrode in a 0.5 M Na_2SO_4 solution for 16 h at i) its abandon potential, ii) an anodic potential of 1.0 V/(SHE). The ICP analysis confirms that even after polarisation of the electrode during 16 h at 1.0 V/(SHE), where the anodic reaction (1) is thermodynamically possible, the copper concentration in the solution is the same as that measured at open-circuit potential. The anodic dissolution of the electrode is thus very slow.

Four-point Hall effect measurements by the Van der Pauw technique with sputtered gold contacts of 1 mm diameter were performed using a Bio-Rad HL5500 device. Currents were kept below 2 mA, because at higher currents erratic results were observed : this confirms the problems encountered by other authors when too large currents were applied to CuBr samples. The magnetic field was 0.32 T in all experiments. Current and magnetic field were reversed in each measurement and the voltage probes were automatically permutated to check for possible asymmetries and contact resistances. The experiments were performed at room temperature in air.

An X-ray diffraction analysis of the electrodes was performed after the electrochemical experiments to check the absence of any structural modification or reaction with the electrolytic solutions (Siemens D 5000 diffractometer, CuKα radiation).

RESULTS

Fig. 1 shows the open-circuit potential E_{oc} of the cell Cu | CuBr | EL | Au (EL = $CuSO_4$ or Na_2SO_4). CuBr electrodes are immersed in a 0.001 M $CuSO_4$ solution for more than 8 h or in Na_2SO_4 solution in the presence of oxygen for 1 h.

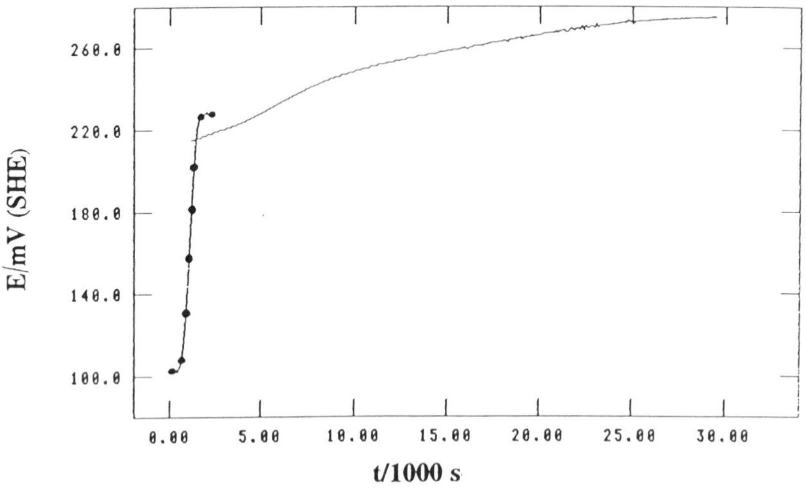

Fig. 1. Open-circuit voltage of a CuBr electrode in 0.001 M $CuSO_4$ solution or in pure supporting electrolyte (Na_2SO_4) in presence of oxygen (●).

The open-circuit voltage increases slowly and reaches a value of about 0.27 V/(SHE) at the end of the measurement. Assuming that this value is the equilibrium potential of the electrode, it can, in principle, be related to Nernst's equation for the appropriate electrode reaction. For the exchange reaction:

$$Cu^+ \rightarrow Cu^{2+} + e^- \qquad (3)$$

Nernst's equation reads : $E_3 = E°_3 + (RT/F) \ln [a(Cu^{2+}) / a(Cu^+)]$.

In a 0.001 M $CuSO_4$ solution, the mean activity coefficient is tabulated : $\gamma\pm = 0.74$; it corresponds well to the Debye-Hückel limiting law [8]. Using the experimental open-circuit potential and the Cu^{2+} activity, the activity of Cu^+ ions can be deduced : $a(Cu^+) = 1.3\ 10^{-5}$ mol/l. This is near the copper concentration experimentally measured by ICP analysis after 16 h immersion of a CuBr electrode in pure supporting electrolyte at open-circuit potential : (1.65 ± 0.09) mg/l, equivalent to $2.6\ 10^{-5}$ mol Cu / l.

Consequently, the equilibrium potential seems to be established in the range of redox stability of the CuBr electrode. Cu^+ ions are due to a dissolution of the electrode ; Cu^{2+} ions are formed by oxidation of Cu^+ ions by oxygen in the solution or can be added intentionally. However, the kinetics are too slow for purposes of copper sensing.

Fig. 2 shows impedance spectra (complex plane representation) taken at different potentials between 0.14 and 1.04 V/(SHE) in a 0.1 M $CuSO_4$ solution. The measured data can be fitted by an equivalent circuit of a resistance R in series with a parallel (R_i Q_i) element using a standard non-linear least squares algorithm [9]. R is essentially the resistance of the electrode bulk, because the resistance of the electrolytic solution containing supporting electrolyte is small. From the impedance spectra, we extrapolate a resistance of the CuBr electrode: R = (10 ± 1) $k\Omega\ cm^2$; taking the CuBr electrode thickness (≈ 0.1 cm) into account, we obtain a conductivity $\sigma \approx 10^{-5}\ \Omega^{-1}\ cm^{-1}$. The discrepancy with previous «all solid» measurements with copper electrodes (Cu | CuBr | Cu [2]) may be due to an enhanced conductivity by protons in CuBr immersed in an aqueous electrolyte, but more experiments are necessary to conclude.

The parallel R_iQ_i element describes the impedance response corresponding to the CuBr electrode-electrolytic solution interface.[1] In an elementary model, R_i represents the charge transfer resistance. A mean value in the blocking region

---

[1] An additional small contribution to the impedance at intermediate frequencies may exist due to grain boundary effects in polycrystalline CuBr. This response is not treated further in this study.

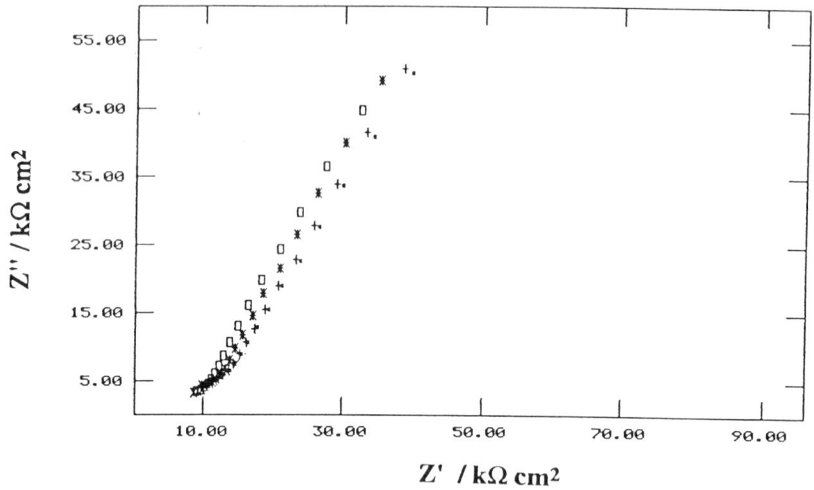

Fig. 2. Impedance spectra of a CuBr electrode in a 0.1 M $CuSO_4$ solution at different electrode potentials E/V(SHE): ■ 1.04, + 0.74, * 0.44, ❑ 0.14).

($R_i \approx 750$ kΩ cm^2, fig. 2) is not in contradiction with the value derived from the exchange current density [10]. A frequency dependent capacitance was observed between 1 and 100 kHz and described by introduction of a constant phase element (CPE), which permitted an excellent fit of the impedance spectra over the whole investigated range of frequency. The impedance of the CPE Q_i can be written $Z_Q = Q^{-1} (i\omega)^{-n}$ with a constant prefactor Q, the angular frequency ω and an exponent n. The value of n gives the nature of the element : n = 0 corresponds to a pure resistance, n = 0.5 to a Warburg impedance associated with a diffusional process and n = 1 to a pure capacitance. Experimentally, we found exponents n between 0.68 and 0.72; we thus attributed the CPE to non-ideal capacitances. As discussed in the literature [11], capacitances of polarizable electrodes exhibit a CPE behaviour, essentially because of their surface roughness.

In the case of depletion of majority electronic carriers near the interface with the electrolytic solution, the interfacial capacitances are largely related to the space charge regions in the CuBr electrode. One may apply the Mott-Schottky analysis of the semiconductor properties [12] based on the fundamental equation:

$$C^{-2} = 2\,(E - E_{fb} - RT/F) / (F\,\varepsilon\,\varepsilon_0\,N) \qquad (4)$$

In this equation, C is the space charge capacitance (identified with Q), E_{fb} the flat band potential and N the effective donor or acceptor density of the material ($\varepsilon\,\varepsilon_0$: dielectric permittivity, E: electrode potential, T: absolute temperature, R: gas constant, F: Faraday's constant). A Mott-Schottky plot of C^{-2} against potential E is presented in fig. 3. A straight line with negative slope is observed between 0.2 and 1.0 V/(SHE). This corresponds to a p-type behaviour of polycrystalline copper(I) bromide. The effective acceptor concentration N_A can be calculated from the slope using the dielectric constant of CuBr of 8 : $N_A = (2 \pm 1)\ 10^{16}$ cm^{-3}. The flat-band potential of CuBr was found to be $E_{fb} = (1.50 \pm 0.05)$ V versus the standard hydrogen electrode. Since the SHE lies 4.5 eV below the vacuum level [12], this places the top of the valence band of CuBr approximately 6 eV below the vacuum level.

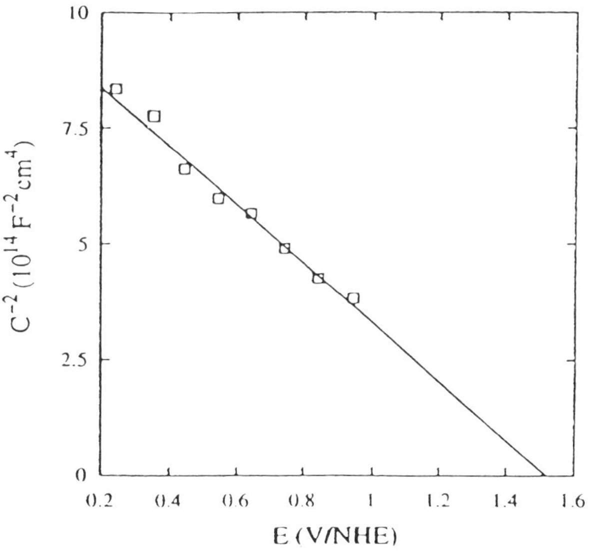

Fig. 3. Mott-Schottky plot for a polycrystalline CuBr electrode.

The sign of the Hall voltage was always positive, in accordance with electron holes as the majority electronic charge carrier. The Hall voltage U_H is related to current I and magnetic field B by the relation (w: sample thickness):

$$U_H = R_H I B / w \tag{5}$$

The average Hall coefficient was $R_H = (7 \pm 2)\ cm^3/C$. The density p of electron holes was thus $p = 1/(e R_H) = (9 \pm 2)\ 10^{17}\ cm^{-3}$ (e: elementary charge). This is nearly two orders of magnitude higher than the value deduced from Mott-Schottky analysis. The mobility μ of the electron holes could be determined from the conductivity σ using the equation:

$$\sigma = 1/\rho = e \mu p \tag{6}$$

The average mobility was $\mu = (0.4 \pm 0.1)\ cm^2/(Vs)$. This small value is compatible with a polaron mechanism of conduction.

DISCUSSION

The p-type semiconductivity of polycrystalline CuBr samples was confirmed by both Hall effect and Mott-Schottky studies. The discrepancy between the results of Hall effect and Mott-Schottky analysis relative to carrier density points to possible experimental problems. The Hall data may be biased by the presence of oxygen in near surface regions. In the electrochemical measurements, on the other hand, adsorbed oxygen is expected to be reduced by cathodic polarization. However, the Mott-Schottky experiments can be biased by a possible redistribution of mobile copper vacancies, which are the majority ionic defects, in the space charge field near the electrode-electrolyte phase boundary [13]. The drift of copper vacancies under the built-in field in the space charge region would lead to a flattening of the space charge potential and thus to a decrease of the measured acceptor density. Altogether, we assume that the acceptor density is intermediate between the two results, about $10^{17}\ cm^{-3}$.

The p-type semiconductivity can have different origins: i) a hole mobility significantly higher than the electron mobility, ii) substantial "bulk" acceptor doping, iii) intrinsic copper deficiency.

A substantial acceptor doping by oxygen was previously assumed to play a decisive role [4-6]. For electrostatic and geometrical reasons, oxygen atoms should occupy substitutional bromine sites and the doping reaction can be written in Kroger-Vink nomenclature:

$$\frac{1}{2} O_{2\,(g)} + V_{Br}^{\bullet} \rightarrow O_{Br}' + 2\,h^{\bullet} \qquad (7)$$

Assuming an acceptor density of 10^{17} cm^{-3}, this would correspond to a concentration of substitutional oxygen of 10 ppm.

Finally, if the electron hole density is nonstoichiometry related, it would correspond to a concentration of Cu^{2+} ions of 10^{17} cm^{-3}, or, equivalently, to a copper deficiency of about 10^{-5}.

CONCLUSION

The electrochemical characteristics of CuBr are intermediate between those of typical metallic electrodes and those of insulators. While the equilibrium potential of the Cu^{+}/Cu^{2+} redox couple is established, kinetics are too sluggish to permit the use of a CuBr electrode as a copper sensor in aqueous solutions. Microcrystalline copper(I) bromide is a p-type semiconductor with a low hole mobility, compatible with polaron conduction. The acceptor density can be explained by oxygen doping or an intrinsic copper deficiency.

REFERENCES

1. J. B. Wagner, C. Wagner, "Electrical Conductivity Measurements on Cuprous Halides", *J. Chem. Phys.*, 26, 1597 (1957).
2. S. Villain, M.-A. Desvals, G. Clugnet, P. Knauth, "Study of Polycrystalline CuBr and the Interface Cu / CuBr by Impedance Spectroscopy", *Solid State Ionics*, 83, 191 (1996).
3. J.-L. Seguin, M. Bendahan, G. Lollmun, M. Pasquinelli, P. Knauth, "Preparation of Thin Films of Copper(I) Bromide by r.f. Sputtering: Morphology and Electrical Properties", *Thin Solid Films*, in press.
4. I. Riess, R . Safadi, "Failure of Hebb-Wagner Polarization Measurements Due to Decomposition of the Sample", *Solid State Ionics*, 59, 99 (1993).
5. I. Riess, R. Safadi, H. L. Tuller, " Problems with Hebb-Wagner Polarization Measurements Due to Overpotentials and Decomposition of the Sample", *Solid State Ionics*, 72, 3 (1994).
6. I. Riess, R. Safadi, H. L. Tuller, "Electrical Conductivity Measurements on Cuprous Bromide, CuBr, in the Presence of Oxygen", *Solid State Ionics*, 59, 279 (1993).
7. A. J. Bard, R. Parsons, J. Jordan, *Standard Potentials in Aqueous Solutions*, Marcel Dekker, New York (1985).
8. C. H. Hamann, W. Vielstich, *Elektrochemie,* 2nd ed., VCH, Weinheim (1985), p. 64.

9. B. A. Boukamp, *«Equivalent circuit»*, 2nd edition, Univ. Twente (1989).
10. P. Knauth, Y. Massiani, "Mott-Schottky Analysis of Polycrystalline Copper(I) Bromide in Aqueous Electrolytes", *J. Electroanal. Chem.* 442, 229 (1998).
11. K. Jüttner, "Electrochemical Impedance Spectroscopy of Corrosion Processes on Inhomogeneous Surfaces", *Electrochim. Acta*, 35, 1501 (1990).
12. A. J. Bard, L. R. Faulkner, Electrochemical methods - Fundamentals and Applications, John Wiley, New York (1980).
13. M. A. Butler, "Aging Effects in Defect-Doped Semiconductor Electrodes", *J. Electrochem. Soc.*, 126, 338 (1979).

MICROSTRUCTURE AND ELECTRICAL CONDUCTIVITY OF RARE EARTH-DOPED CERIA

Yoshihiro Hirata, Hiroshi Ono, Kenji Higashi, Kazutoshi Sonoda, Soichiro Sameshima and Yasuro Ikuma*
Department of Applied Chemistry and Chemical Engineering
Kagoshima University, 1-21-40 Korimoto, Kagoshima 890-0065, Japan
*Department of Chemical Technology, Kanagawa Institute of Technology, 1030 Shimoogino, Atsugi, Kanagawa 243-02, Japan

ABSTRACT

Rare earth-doped ceria powder ($Ce_{0.8}R_{0.2}O_{1.9}$, R=Yb,Y,Gd,Sm,Nd and La) applied to the oxygen ion electrolyte for solid oxide fuel cell was prepared by the oxalate coprecipitation method and calcined at 600 °C in air. The powder produced consisted of thin particles of 1.6-2.8 μm in length and compacted isostatically to 50-60 % theoretical density under a pressure of 294 MPa. The sintering at 1600 °C for 4 h gave the dense body (relative density >98.0%) with grain sizes of 3-5 μm. The specific conductivities of rare earth-doped ceria at 800 °C were 6.7×10^{-2} - 1.4×10^{-1} $S\cdot cm^{-1}$ and dominated by the resistance of oxygen ion diffusion through bulk. The activation energies (75-93 kJ/mol) for oxygen ion diffusion in bulk was lower than those across grain boundary (101-123 kJ/mol).

INTRODUCTION

The following three types of fuel cells are widely studied to generate high efficient electric power : phosphoric acid fuel cell using proton conductor (PAFC, operation temperature 190° - 210 °C), molten carbonate fuel cell using carbonate ion conductor (MCFC, operation temperature 600° - 650 °C) and solid oxide fuel cell using oxygen ion conductor (SOFC, operation temperature 1000 °C) [1]. In the above fuel cells, SOFC based on the chemical reaction $2H_2+O_2 \rightarrow 2H_2O$ has a high potential as large scale electric power generation system without environmental pollution. Y_2O_3-stabilized ZrO_2 (YSZ) is an excellent oxygen ion electrolyte which can be used at low oxygen pressures in SOFC. The operation temperature of SOFC depends on the ionic conductivity of YSZ. Reducing the operation temperature is required to increase the life time and the choice of the constituent materials such as electrodes and metal gas separator of SOFC. It is reported that rare earth-doped ceria has a higher oxygen ion conductivity than YSZ and may be a candidate oxide to operate SOFC at lower temperatures [2-7]. This paper describes the sinterability of rare earth-doped ceria powder by oxalate coprecipitation method, microstructural development and the oxygen ion conductivities of sintered dense samples.

EXPERIMENTAL PROCEDURE

Preparation and sintering of rare earth-doped ceria powders

Rare earth-doped ceria powders with the composition of $Ce_{0.8}R_{0.2}O_{1.9}$ (R : rare earth) were prepared by heating the oxalate coprecipitate. The starting materials were $Ce(NO_3)_3 \cdot 6H_2O$ (> 98 wt%) and $R(NO_3)_3 \cdot 6H_2O$ (> 99.5 wt%) and $H_2C_2O_4 \cdot 2H_2O$ (> 99 wt%). The cerium and rare earth nitrate solutions of 0.2 M were mixed at the molar ratio of Ce^{3+} / R^{3+} = 4 / 1. The mixed solution was dropped into a stirred oxalic acid solution of 0.4 M to produce oxalate coprecipitate. The precipitate was vacuum-filtrated, washed with distilled water and dried at 40 °C overnight. The dried oxalate coprecipitate was heated at 600 °C in air to form $Ce_{0.8}R_{0.2}O_{1.9}$. The rare earth-doped ceria powders were compacted isostatically under a pressure of 294 MPa after uniaxial dry pressing at 49 MPa. The pellet of 10 mm diameter and 2 mm thickness was heated at a rate of 10 °C/min to 1100° - 1600°C and sintered for 4 h in air (SPM 6512 electric furnace, Marusho Electro-Heat Co., Ltd., Hyogo, Japan). The phases produced in sintered samples were identified by X-ray diffraction (CuK α , Model No.2013, Rigaku Co., Ltd., Tokyo, Japan). The bulk density of rare earth-doped ceria was measured by the Archimedes method using distilled water. The theoretical density was calculated from the lattice parameters of rare earth-doped ceria. The chemical compositions of bulk and grain boundaries of samples sintered at 1600 °C were examined by analytical transmission electron microscopy (JEM-2010 type, JEOL Co. Ltd., Tokyo, Japan) with energy dispersive X-ray spectroscopy (EDS). The probe diameters were 65 and 10-100 nm for selected area electron diffraction analysis and EDS analysis, respectively. After mechanical polishing, samples were sputtered by Ar ion with ion current of 0.3 mA at 4.0 kV (M 600 Ar Ion Mill, Gatan, U.S.A).

Measurement of electrical conductivity

Two planes (0.50×0.45 cm^2) of a rectangular $Ce_{0.8}R_{0.2}O_{1.9}$ (3cm length) sintered at 1600 °C were polished with diamond paste of 0.25 μm to attach Pt electrodes. The other planes were polished with abrasive paper No.600. Platinum paste was spread homogeneously on the electrode planes and heated to

300°C at a rate of 5 °C/min to prevent the formation of porous Pt membrane and further to 1300°C at 8 °C/min in air. Electrical contact was made via Pt plates in pressure contact with the Pt electrodes. The complex impedance of sample in the temperature range of 300°-1000 °C, was measured every 25 °C by a two-terminal ac bridge circuit using LCZ meter (model 4276A, 4277A and 4193A, Yokogawa Hewlett Packard Co., Tokyo, Japan) at 100 Hz - 110 MHz in dried air.

Measurement of diffusion coefficient of oxygen ion

The oxygen self-diffusion coefficients of rare earth-doped ceria sintered at 1600 °C were determined by the gas - solid isotope exchange technique using ^{18}O as tracer. The samples were annealed in the atmoshere of $P_{O_2}=6.6x10^3$ Pa (^{18}O = 60 vol%) at 800 °C. The ^{18}O concentrations in the gas phase were determined by mass spectrometer. The diffusion coefficients of oxygen ions were calculated from the ^{18}O concentrations using a diffusion equation with the boundary condition where the surface exchange rate of oxygen controlled the amount of ^{18}O diffused in spherical samples.

RESULTS AND DISCUSSION

Lattice parameter and density of rare earth-doped ceria

The concentration of oxalic acid influenced the particle size of coprecipitated platelike oxalate and derived yttria-doped ceria. The median size of $Ce_{0.8}Y_{0.2}O_{1.9}$ powder decreased from 2.9 μm for 0.1 M to 1.8 μm for 0.4 M oxalic acid, because the nucleation rate of Ce/Y oxalate increased at a higher concentration of supersaturated solution. The decrease of particle size of $Ce_{0.8}Y_{0.2}O_{1.9}$ powder increased the green density from 55.6 % theoretical density for the case of 0.1 M oxalic acid to 60.9 % for the case of 0.4 M oxalic acid. The sinterability of the powder prepared with 0.4 M oxalic acid was higher than that of the powder prepared with 0.1 M oxalic acid. The former powder reached 95 % theoretical density at 1200 °C and latter powder needed 1500 °C for the achievement of same relative density. The samples for electrical conductivity measurement were sintered at 1600 °C for 4 h from the powders produced with 0.4 M oxalic acid. The relative densities of these samples were higher than 99 % except for the Yb-doped ceria of 97.9 % relative density. The rare earth-doped CeO_2 showed XRD patterns of cubic CeO_2, indicating the formation of solid solutions of 20 mol% $RO_{1.5}$ - 80 mol% CeO_2 systems. Figure 1(a) represents the lattice parameters of cubic solid solutions. The lattice parameters increased linearly with increasing ionic radius of rare earth element. The rare earth elements were substituted for Ce ions with formation of oxygen vacancy ($Ce_{1-x}R_xO_{2-x/2}$) or occupied the interstitial sites ($Ce_{1-x}R_{4x/3}O_2$). Figure 1(b) shows the comparison of the calculated and measured

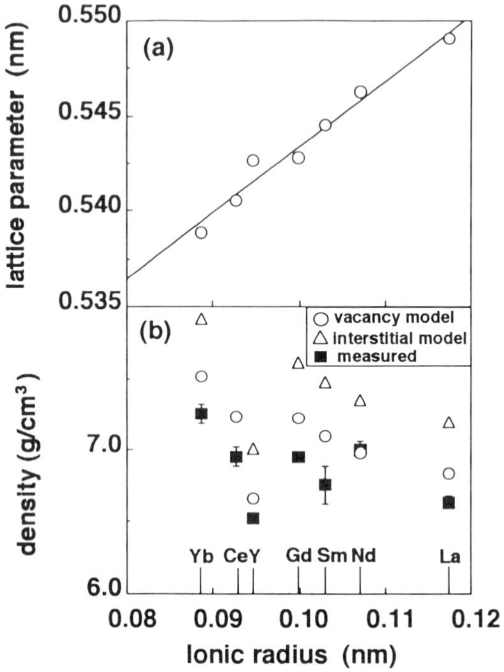

Fig. 1 Lattice parameters (a) and calculated and measured density (b) of rare earth-doped ceria as a function of ionic radius of doped cation.

densities. The measured density fitted more the calculated density for the vacancy model than the interstitial model.

Microstructure of rare earth-doped ceria

The microstructures of $Ce_{0.8}R_{0.2}O_{1.9}$ sintered at 1600 °C for 4 h and thermally etched at 1500 °C for 2 h consisted of equiaxed grains of 3-5 μm. The grain size showed a tendency to increase as the size of doped rare earth element became larger. TEM photographs showing the bulk structures and grain boundary structures of Gd-doped ceria sintered at 1600°C are shown in Fig.2. A high magnification transmission electron micrograph showing the grain boundary structure and SAD patterns for the adjacent grains are shown in Fig.3. As seen in Figs.2 and 3, no crystalline or amorphous second phase existed at the grain boundary between adjacent single crystal grains with different crystal orientation. The analysis by EDS was carried out on the elements included in the region of 100 nm diameter of different three grains (spots 1-3) and the elements near a grain

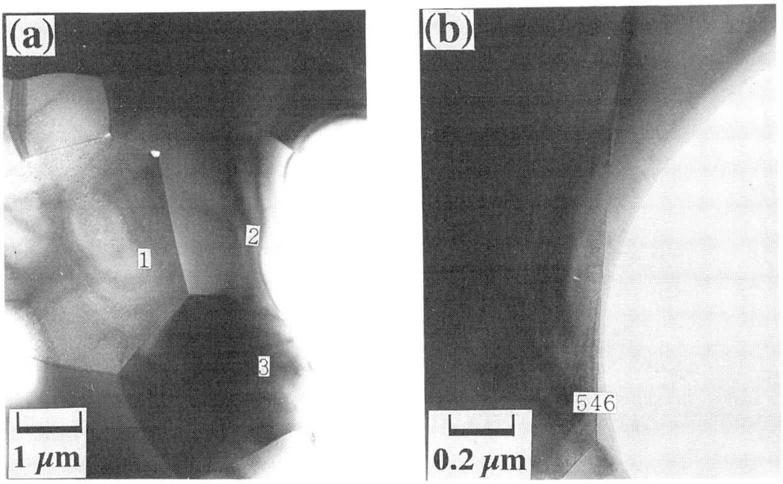

Fig.2 Transmission electron micrographs showing the shape of grain (a) and grain boundary structure (b) of the gadolinia-doped ceria.

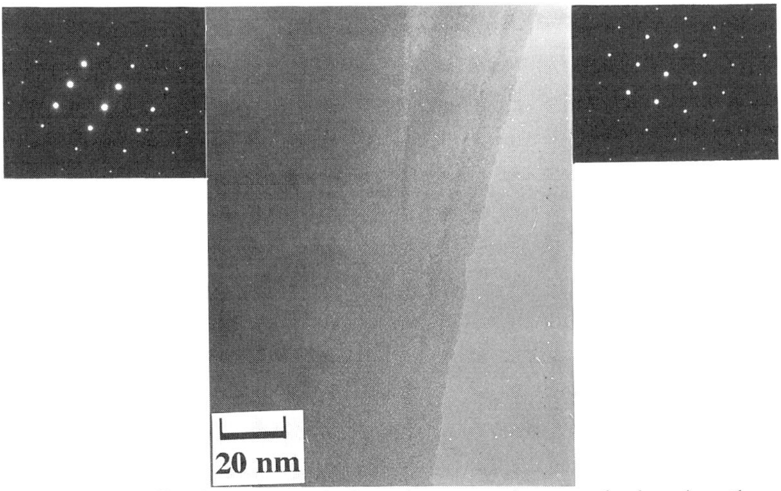

Fig.3 A high magnification transmission electron micrograph showing the grain boundary structure and selective area diffraction patterns of the gadolinia-doped ceria.

Table I Semi-quantitaive compositions (atomic%) of rare earth doped-ceria at similar points 1 to 6 in Fig.2.

element	1	2	3	4	5	6	ave.
Ce	81.71	81.22	81.01	81.42	82.83	81.19	81.45
Yb	18.29	18.78	18.99	18.59	17.17	18.81	18.55
Ce	78.01	78.02	78.77	78.97	78.46	79.98	79.00
Y	21.99	21.98	21.23	21.03	21.54	20.02	21.01
Ce	81.04	81.99	81.55	80.46	81.69	80.04	80.54
Gd	18.96	18.01	18.45	19.34	18.31	19.96	19.46
Ce	79.70	80.30	79.54	78.29	79.52	80.52	80.11
Sm	20.30	19.70	20.46	21.71	20.48	19.48	19.89
Ce	77.58	78.18	77.18	77.12	78.20	76.63	77.10
Nd	22.42	21.82	22.82	22.88	21.80	23.37	22.90
Ce	77.37	77.20	75.33	73.97	78.14	77.24	77.31
La	22.63	22.80	24.67	26.03	21.86	22.76	22.70

boundary (spots 4-6) which were measured with the probe of 10 nm diameter of EDS. Most of spectra were ascribed to cerium and doped rare earth elements. Table I summarizes the composition ratios at spots 1 to 6 by the semi-quantitative method[8]. The atomic ratios of Ce to rare earth element were almost constant through the spots of 1 to 6 in each sample, indicating that no segregation of rare earth element occurred during the sintering at 1600 °C.

Electrical conductivity of rare earth-doped ceria

Figure 4 shows the complex impedance plots at 300°-550 °C for $Ce_{0.8}Gd_{0.2}O_{1.9}$ sintered at 1600 °C. They consisted of a smaller semicircle at lower frequency range and a larger semicircle at higher frequency range. The larger and smaller semicircles were interpreted to present the oxygen ion migration in bulk and across grain boundaries, respectively. This model gives the following complex impedance,

$$Z = \left(\frac{R_1}{1+\omega^2 C_1^2 R_1^2} + \frac{R_2}{1+\omega^2 C_2^2 R_2^2}\right) - j\left(\frac{\omega C_1 R_1^2}{1+\omega^2 C_1^2 R_1^2} + \frac{\omega C_2 R_2^2}{1+\omega^2 C_2^2 R_2^2}\right) \quad (1)$$

where R and C are the resistance and capacitance associated with oxygen ion migration through bulk (subscript 1) or across grain boundary (subscript 2). The relaxation time (=1/RC) for grain boundary was 10-200 times as long as that for

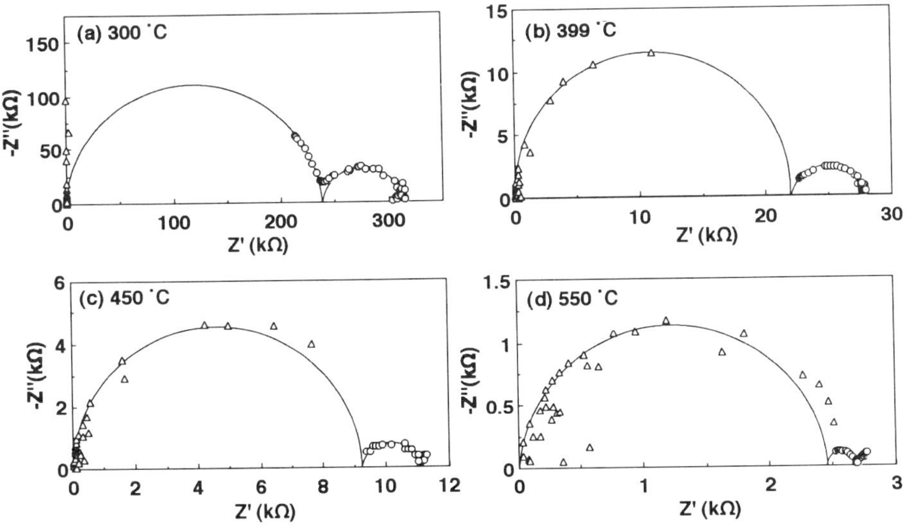

Fig. 4 Impedance diagrams of $Ce_{0.8}Gd_{0.2}O_{1.9}$ at 300 °- 550 °C.
○: equipment A, 100 Hz - 1 MHz, △: equipment B, 400 kHz - 110 MHz

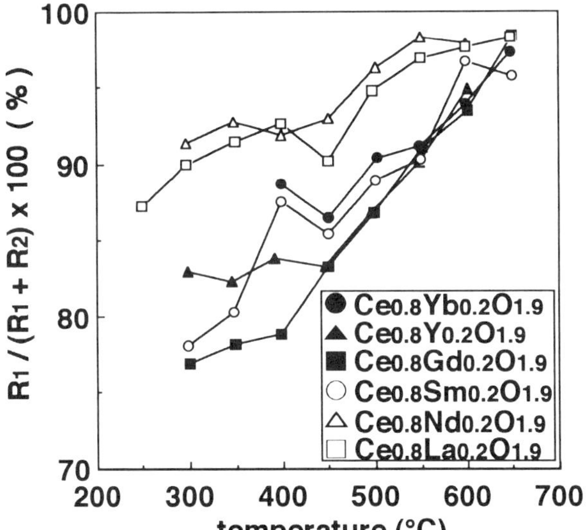

Fig.5 Ratio of bulk resistance (R_1) to grain boundary resistance (R_2) as a function of temperature for $Ce_{0.8}R_{0.2}O_{1.9}$ sintered at 1600 °C for 4 h.

bulk. Figure 5 shows the ratio of bulk resistance to total sample resistance as a function of measuring temperature. The total sample resistance was significantly dominated by the bulk resistance. This tendency became larger at higher temperatures and also in the samples with larger rare earth.

Figure 6 shows the Arrhenius plots of electrical conductivity determined from the bulk resistance, grain boundary resistance and total resistance for $Ce_{0.8}R_{0.2}O_{1.9}$. The bulk and sample resistance was converted to conductivity using the length of sample and surface area of electrodes. On the other hand, the electrical conductivity for grain boundary was calculated from the numbers of grain boundaries depending on the average grain size and the assumption of 1 nm for the width of each grain boundary. Figure 6 suggests that (1) the conductivity for bulk was higher than that for grain boundary, (2) the activation energy was lower for bulk than for grain boundary and (3) the total conductivity was close to the conductivity for bulk. The relation between the results shown in Figs.5 and 6 was interpreted as follows: Although the conductivity of oxygen ions across grain boundary was lower than that of bulk, the summation of the width of grain boundary between the two electrodes was still thin as compared with the length of bulk. Figure 7 shows (a) the electrical conductivity at 800°C and (b) the activation energy as a function of the ionic radius of doped cations. The conductivity was almost independent of size of doped rare earth element and 10^{-1} and 10^{-3} $S \cdot cm^{-1}$ for bulk and grain boundary, respectively. The total conductivity was comparable to the following reported values: 7.7×10^{-2} $S \cdot cm^{-1}$ for $Ce_{0.8}Y_{0.2}O_{1.9}$ [9], 8.8×10^{-2} $S \cdot cm^{-1}$ for $Ce_{0.8}Sm_{0.2}O_{1.9}$ [9], 9.0×10^{-2} $S \cdot cm^{-1}$ for $Ce_{0.8}Gd_{0.2}O_{1.9}$ [10]. Figure 7(b) indicates that both the activation energies for bulk and grain boundary decreased as the ionic radius increased from 0.086 to 0.10 nm.

Diffusion coefficient of oxygen ion

Figure 8 shows that the diffusion coefficients of oxygen ions for $Ce_{0.8}R_{0.2}O_{1.9}$ and CeO_2 sintered at 1600 °C for 4 h. These data were obtained by putting the measured bulk conductivity into the Nernst-Einstein equation (Eq.(2)),

$$\sigma = \frac{e^2 ZiCiDiF}{kT} \tag{2}$$

where σ is the elecrical conductivity ($S \cdot cm^{-1}$), e the electronic charge 1.602×10^{-19} C, Zi the valence of carrier (2), Ci the number of oxygen ions per 1 cm^3, Di the diffusion coefficient ($cm^2 \cdot s^{-1}$), F the correlation factor (0.652 for fluorite structure), k the Boltzmann constant and T temperatue (K). The diffusion

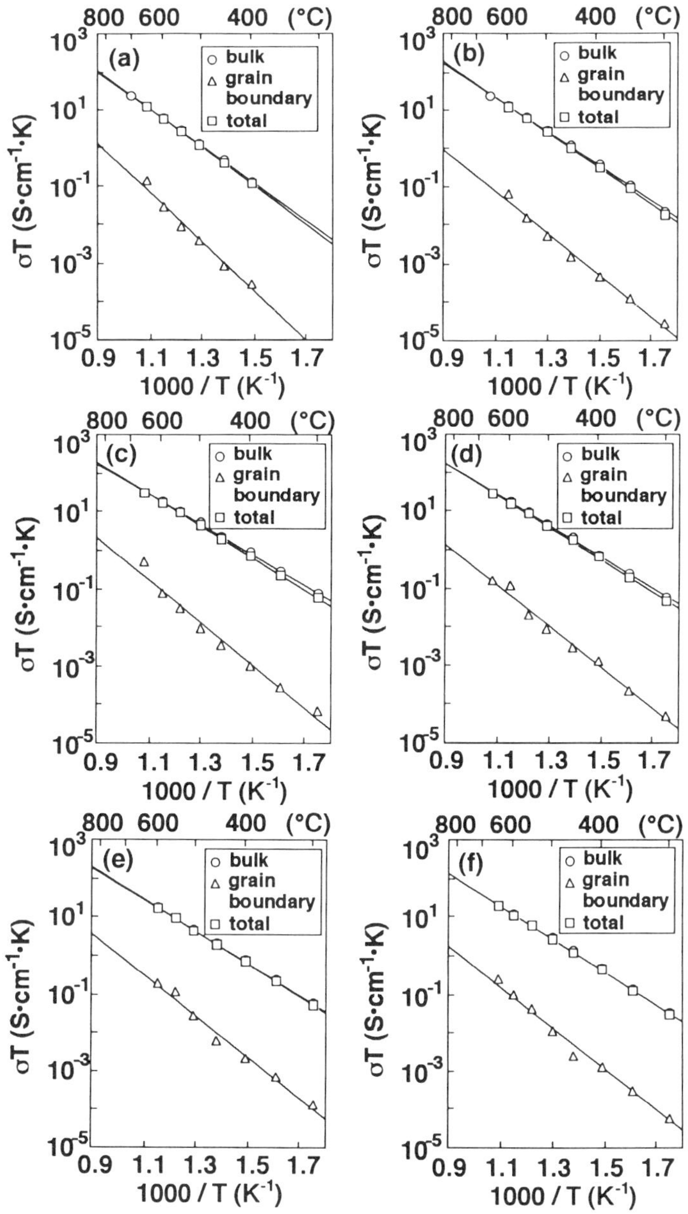

Fig. 6 log σT vs 1000 / T of $Ce_{0.8}Yb_{0.2}O_{1.9}$ (a), $Ce_{0.8}Y_{0.2}O_{1.9}$ (b), $Ce_{0.8}Gd_{0.2}O_{1.9}$ (c), $Ce_{0.8}Sm_{0.2}O_{1.9}$ (d), $Ce_{0.8}Nd_{0.2}O_{1.9}$ (e) and $Ce_{0.8}La_{0.2}O_{1.9}$ (f).

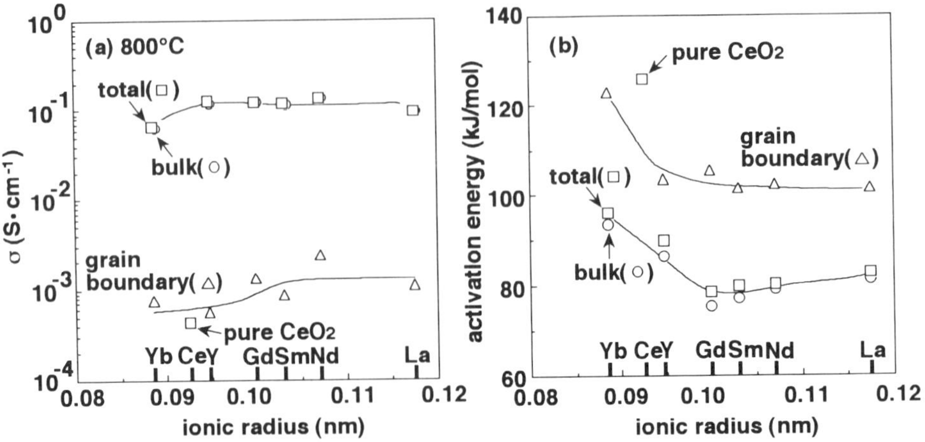

Fig.7 Ionic radius dependence of (a) electrical conductivity and (b) activation energy for $Ce_{0.8}R_{0.2}O_{1.9}$ sintered at 1600 ℃ for 4 h.

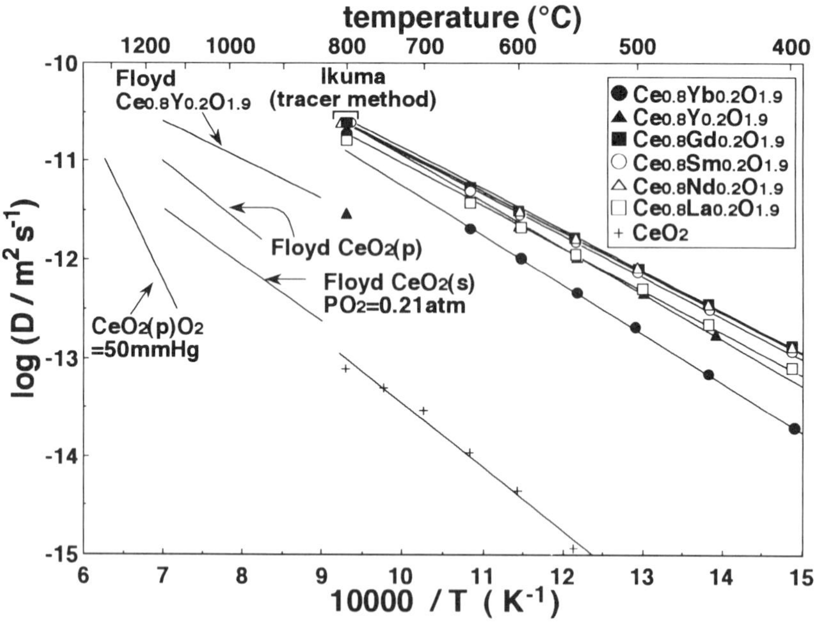

Fig.8 Diffusion coefficient calculated from electrical conductivity for $Ce_{0.8}R_{0.2}O_{1.9}$ and CeO_2 sintered at 1600℃ for 4 h.

coefficients of oxygen ions at 800°C were in the range of 1.26×10^{-11} to 2.69×10^{-11} $m^2 \cdot s^{-1}$ and the influence of doped rare earth elements was small. These values except for $Ce_{0.8}Y_{0.2}O_{1.9}$ were in good agreement with the diffusion coefficients of oxygen ions measured by the tracer method, indicating that the ionic transport number of oxygen was to close to 1. On the other hand, the diffusion coefficient of oxygen ion for $Ce_{0.8}Y_{0.2}O_{1.9}$ measured by tracer method was close to the reported value [11]. Therefore, the higher diffusion coefficient from Eq.(2) compared with the tracer method for $Ce_{0.8}Y_{0.2}O_{1.9}$ may be due to the electronic conductivity. As seen in Fig.8, the diffusion coefficients of oxygen ions between electrical conductivity and tracer method for pure CeO_2 showed a good agreement. However, no oxygen vacancy for CeO_2 decreased the magnitude of diffusion coefficient by 1/100 compared with rare earth-doped ceria.

CONCLUSIONS

Dense rare earth-doped ceria ($Ce_{0.8}R_{0.2}O_{1.9}$), which was prepared from the oxalate coprecipitate by sintering at 1600 °C, contained no second phase in bulk or at grain boundary. The oxygen ion conductivity of $Ce_{0.8}R_{0.2}O_{1.9}$ (R=Yb, Y, Gd, Sm, Nd and La) at 800 °C was 6.7×10^{-2}-1.4×10^{-1} $S \cdot cm^{-1}$ and dominated by the bulk resistance for diffusion of oxygen ions. The activation energy for diffusion of oxygen ions was lower for bulk (75.3 - 93.3 kJ/mol) than for grain boundry (101 - 123 kJ / mol). Both the activation energies decreased as the ionic radius of doped rare earth element increased from 0.086 to 0.10 nm.

ACKNOWLEDGEMENTS

The authors are grateful to Dr. Shibuya of Ube Industries, Ltd. and Dr. Shoda of Ube Scientific Analysis Laboratory for the TEM observation. A part of this work was supported by a Grant-in-Aid for Scientific Researach (C, No.09650926) of the Ministry of Education, Science and Culture and by a research project of the Japan Petroleum Institute commissioned by the Petroleum Energy Center with the subsidy of the Ministry of International Trade and Industry , Japan.

REFERENCES

[1] Y.Miyazaki, "Fuel Cell Power Generation," *Bull.Ceram. Soc. Jpn.*, **30** (4), 322-328 (1995).

[2] H.Inaba and H.Tagawa, " Ceria-based Solid Electrolytes," *Solis State Ionics*, **83** (1), 1-16(1996).

[3] B.C.H.Steele, K.Zheg, R.A.Rudkin, N.Kiratzis and M.Christie, "Properties and Applications of Ce(Gd)O_{2-x} Electolytes in the Temperature Range 500-700 °C, " pp.1028-1038 in *Proceedings of the Fourth International Symposium on Solid*

Oxide Fuel Cells (SOFC-IV), Edited by M.Dokiya, O.Yamamoto, H.Tagawa and S.C.Singhal, The Electrochemical Society, Inc., New Jersey, 1995.

[4] T.Kudo and H.Obayashi, "Oxygen Ion Conduction of the Fluorite-Type $Ce_{1-x}Ln_xO_{2-x/2}$," *J. Electrochem. Soc.*, **122** (1), 142-147 (1975).

[5] H.L.Tuller and A.S.Nowick, "Doped Ceria as a Solid Oxide Electrolye," *J. Electrochem. Soc.*, **122** (2), 255-259 (1975).

[6] R.T.Dirstine, R.N.Blumenthal and T.F.Kuech, "Ionic Conductivity of Ceria, Yttria, and Rare Earth-Doped Cerium Dioxide," *J. Electrochem. Soc.*, **126** (2), 264-269 (1979) .

[7] H.Yahiro, T.Ohuchi, K.Eguchi and H.Arai, "Electrical Properties and Microstructure in the System Ceria-Alkaline Earth Oxide," *J. Mater. Sci*, **23**, 1036-1041 (1988).

[8] G.Cliff and W.Lorimer, "The Quantitative Analysis of Thin Specimens," *J.Microscopy*, **103**, 203-207 (1974).

[9] J. V. Herle, T. Horita, T. Kawada, N. Sakai, H. Yokokawa and M. Dokiya, "Sintering Behaviour and Ionic Conductivity of Yttria-Doped Ceria," *J. Europ. Ceram. Soc.*, **16**, 961-973 (1996).

[10] F.P.F.V. Berkel, G.M. Christie, F.H.V. Heuveln and J.P.P. Huijsmans, "Electrolytes for Low Temperature Solid Oxide Fuel Cells" ; pp.1062-1071 in *Proceedings of the Fourth International Symposium on Solid Oxide Fuel Cells (SOFC-IV)*, Edited by. M.Dokiya, O.Yamamoto, H.Tagawa and S.C.Singhal, The Electrochemical Society, Inc., New Jersey, 1995.

[11] J.M.Floyd, "Interpretaion of Transport Phenomena in Non-stoichiometric Ceria," *Ind. J. Tech.*, **11** (11), 589-594 (1972).

STRUCTURE AND PROPERTY RELATIONSHIP IN THE MIXED-CONDUCTING Sr-Fe-Co-O SYSTEM

B. Ma, U. Balachandran, J.P. Hodges, J.D. Jorgensen, D.J. Miller, and J.W. Richardson, Jr.
Argonne National Laboratory, Argonne, IL 60439

ABSTRACT

Mixed-conducting ceramic oxides have potential uses in high-temperature electrochemical applications such as solid oxide fuel cells, advanced batteries, sensors, and oxygen-permeable membranes. The Sr-Fe-Co-O system combines high electronic/ionic conductivity with appreciable oxygen permeability at elevated temperatures. Dense ceramic membranes made of this material can be used to separate high-purity oxygen from air without the need for external electrical circuitry, or to partially oxidize methane to produce syngas. Samples of $Sr_2Fe_{3-x}Co_xO_y$ (with x = 0, 0.6, 1.0, and 1.4) were prepared by solid-state reaction in atmospheres with various oxygen partial pressures (pO_2) and were characterized by X-ray diffraction, scanning electron microscopy, and electrical conductivity measurements. Phase components of the samples are dependent on cobalt concentration and synthesis pO_2. Total conductivity increases with increasing temperature and cobalt content in the material. Higher ionic transference numbers have been observed in samples with lower cobalt contents. Current-voltage characteristics determined in a gas-tight cell indicate that a bulk effect, rather than a surface exchange effect, is the main limiting factor for oxygen permeation through membranes made of $Sr_2Fe_2CoO_y$. Oxygen permeability measurements at various temperatures showed that oxygen permeability increases with increasing temperature, as expected. At 900°C, an oxygen permeation flux of 2.5 $scc\cdot cm^{-2}\cdot min^{-1}$ was obtained for a $Sr_2Fe_2CoO_y$ disk membrane of 2.9 mm thickness.

INTRODUCTION

Oxides with mixed electronic and ionic conductivities are receiving increased attention because of their applications in high-temperature electrochemical processes [1-5]. For example, they are used as electrodes in solid oxide fuel cells, and if their oxygen permeability is high enough, they can be used as oxygen-separation membranes. Sr-Fe-Co-O has not only high combined electronic and oxygen ionic conductivity but also oxygen permeability that is superior to that of other reported ceramic oxides [6-11]. It is a good candidate material as a dense membrane for the production of high-purity oxygen by gas separation. In an oxygen separation reactor, oxygen can be transported from the high-oxygen-partial-pressure (high-pO_2) side to the low-pO_2 side under the driving force of the pO_2 difference, $\Delta\log(pO_2)$, without the need for external electric circuitry. Balachandran et al. [7] demonstrated that extruded membrane tubes made of Sr-Fe-Co oxide can be used for partial oxidation of methane to produce syngas ($CO + H_2$) in a methane conversion reactor that operates at ≈850°C. In this type of reactor, oxygen on one side of the membrane was separated from air on the other side of the membrane. Methane conversion efficiencies of >98% were observed and the reactor tubes have operated for >1000 h. Moreover, the oxygen flux obtained from the separation of air in this type of conversion reactor is considered commercially feasible. The use of this technology can significantly reduce the cost of oxygen separation [12].

To obtain sufficient oxygen flux, the gas separation reactor must be operated at high temperatures and substantial pO_2 differences. Therefore, stability of the ceramic membranes is an important issue. Certain compounds of $SrFe_{1-x}Co_xO_{3-\delta}$ were found unsuitable for use in gas separation because they lack stability in highly reducing environments [13]. Thermodynamic calculations can normally provide accurate data on the stability of a system. However, specific examples of discrepancies between the calculated thermodynamic stability and experimental results indicate the importance of obtaining supporting data from real systems [14].

In this paper, we report our studies of the crystal structures and transport properties of the $Sr_2Fe_{3-x}Co_xO_y$ materials. Electrical conductivity was determined in atmospheres with various pO_2 levels at elevated temperatures. Oxygen permeability was measured with a gas-tight electrochemical cell. Room-temperature X-ray diffraction (XRD) and scanning electron microscopy (SEM) experiments were also conducted to characterize the material.

SYNTHESIS AND CHARACTERIZATION

The $Sr_2Fe_{3-x}Co_xO_y$ (with x = 0, 0.6, 1.0, and 1.4) powders were made by a solid-state reaction method, with appropriate amounts of $SrCO_3$, $Co(NO_3)_2 \cdot 6H_2O$, and Fe_2O_3. Mixing and grinding were conducted in isopropanol with zirconia media for 10 h. After drying, the mixture was calcined in air at 850°C for 16 h, with intermittent grinding. After final calcination, the powder was ground with an agate mortar and pestle to an average particle size of ~7 μm. The resulting powder was pressed uniaxially with a 1200-MPa load into pellets measuring 21.5 mm in diameter and ~2-5 mm thick. The pellets were then sintered in air at ≈1200°C for 5 h. The true density was measured using a powder sample with an AccuPyc 1330 pycnometer. The bulk density was determined by the Archimedean method, with isopropanol as the liquid medium. The measured density of air-sintered $Sr_2Fe_2CoO_y$ samples was ~95% of the theoretical value. Thin bars ≈1 x 5 x 20 mm^3 were cut from the samples and subjected to post-annealing heat treatments in various low-pO_2 environments at 900-1100°C. The sintered pellets were polished on both sides and used for permeation tests.

Room-temperature X-ray powder diffraction analysis was carried out with a Scintag XDS-2000 diffractometer. Data were taken with Cu K_α radiation. A high-purity intrinsic Ge energy-dispersive detector was used to minimize background due to sample fluorescence. A continuous scan with a 2θ scanning rate of 1°/min and chopping step of 0.03° was used to collect data. SEM observations were conducted with a JEOL JSM-5400 scanning microscope at an accelerating voltage of 20 KeV.

The total conductivity of the sample was determined by the conventional four-probe method, and ionic conductivity was measured with the electron blocking method. Platinum wires were wound onto the specimen bar to serve as current and voltage leads. The resistance of the specimen was measured with an HP 4192A LF impedance analyzer at 23 Hz. Details of the experimental configuration for measuring conductivity were reported in Ref. 10. A K-type thermocouple was attached to an yttria-stabilized zirconia (YSZ) plate on which the sample bars were placed for the heat treatments. The thermocouple was used for both controlling and measuring the temperature of the programmable electric furnace. Temperature uncertainty within the uniform hot zone of the furnace was ±1°C. The desired gaseous environments were obtained by flowing premixed gases through the system.

A schematic diagram of the gas-tight electrochemical cell used in the oxygen permeation experiments is shown in Fig. 1. A sintered pellet of $Sr_2Fe_2CoO_y$ was sealed to a YSZ crucible with Pyrex glass seal. Electrical leads (10-mil Pt wires) were separated from the YSZ crucible and $Sr_2Fe_2CoO_y$ disk membrane by two alumina rings. Electrodes were placed on the bottom and the sidewall of the YSZ crucible. The bottom electrodes were used as pumping electrodes to pump oxygen from the gas-tight cell chamber, whereas the other electrodes were used to determine the pO_2 inside the cell. The pO_2 inside the cell can be determined from the measured electromotive force (EMF), E, generated on the side-wall of the YSZ crucible by

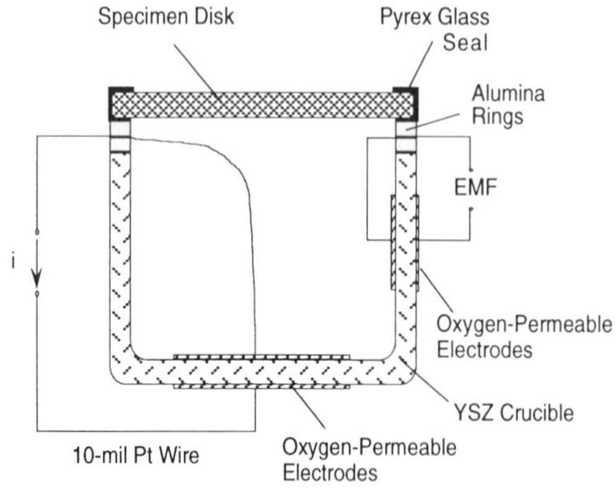

Fig. 1. Schematic drawing of cross-sectional view of gas-tight electrochemical cell used to measure oxygen permeability.

$$pO_2^{II} = pO_2^{I} \exp\left(\frac{4FE}{RT}\right) \qquad (1)$$

where pO_2^{II} and pO_2^{I} are the oxygen partial pressure values inside and outside the gas-tight cell, respectively. Other variables are: F, Faraday's constant; R, gas constant; and T, absolute temperature. Reducing oxygen environments were achieved by pumping oxygen out of the gas-tight cell by using the pumping electrodes on the YSZ crucible. Oxygen permeates the $Sr_2Fe_2CoO_y$ disk membrane because of the pO_2 difference on the two sides of the membrane. Under steady-state conditions, the amount of oxygen that enters the cell (by permeating the specimen disk) is equal to that pumped out by the YSZ oxygen pump. Therefore, the flow of oxygen through the specimen can be determined from the current applied to the YSZ oxygen pump. The oxygen permeation flux j_{O_2} (in $mol\cdot cm^{-2}\cdot s^{-1}$) is related to the applied current i (in A) by

$$j_{O_2} = \frac{i}{4FS} \quad (2)$$

where S is the effective cross-sectional area of the specimen.

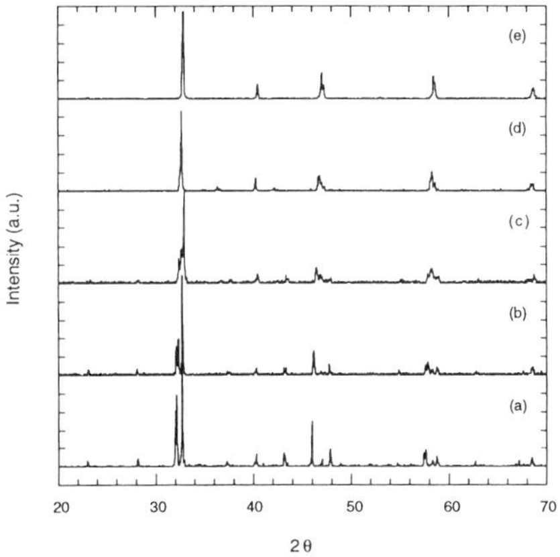

Fig. 2. Room-temperature X-ray diffraction patterns of air-sintered samples:
(a) $Sr_2Fe_3O_y$, (b) $Sr_2Fe_{2.4}Co_{0.6}O_y$,
(c) $Sr_2Fe_2CoO_y$, (d) $Sr_2Fe_{1.6}Co_{1.4}O_y$,
and (e) $SrFe_{0.8}Co_{0.2}O_y$.

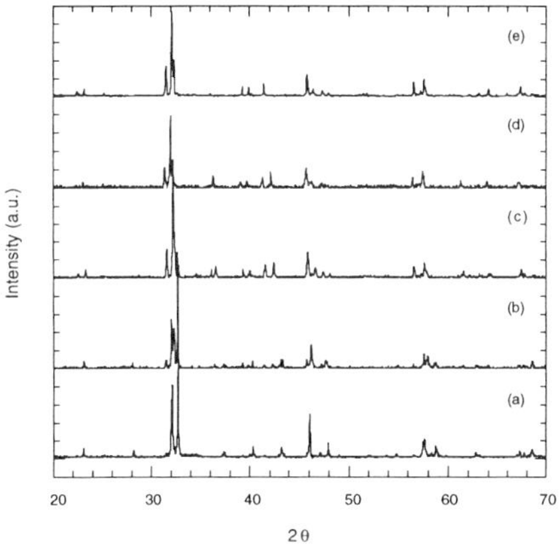

Fig. 3. Room-temperature X-ray diffraction patterns of argon-sintered samples:
(a) $Sr_2Fe_3O_y$, (b) $Sr_2Fe_{2.4}Co_{0.6}O_y$,
(c) $Sr_2Fe_2CoO_y$, (d) $Sr_2Fe_{1.6}Co_{1.4}O_y$,
and (e) $SrFe_{0.8}Co_{0.2}O_y$.

RESULTS AND DISCUSSION

Room-temperature X-ray diffraction (XRD) patterns of air-sintered $Sr_2Fe_{3-x}Co_xO_y$ are plotted in Fig. 2 for different x values. The powder XRD pattern of the x = 0 composition (Fig. 2a) agrees with the result reported for a single-crystal $Sr_4Fe_6O_{13}$ sample [15]. For comparison, the XRD pattern of the perovskite phase $SrFe_{0.8}Co_{0.2}O_{3-\delta}$ is plotted in Fig. 2e. Air-sintered $Sr_2Fe_2CoO_y$ (x = 1, Fig. 2c) is a multiple phase material containing ~25% perovskite phase, as obtained from Rietveld analysis [16]. Air-sintered $Sr_2Fe_{2.4}Co_{0.6}O_y$ (x = 0.6, Fig. 2b) contains less than 10% perovskite phase, while air-sintered $Sr_2Fe_{1.6}Co_{1.4}O_y$ (x = 1.4, Fig. 2d) has more than 90% perovskite phase. Traces of cobalt oxide impurity phase were also observed in the $Sr_2Fe_{1.6}Co_{1.4}O_y$ sample. XRD patterns of the argon-annealed samples are shown in Fig. 3.

At high temperatures, the perovskite phase can be reduced to the brownmillerite structure [17] in a reducing argon environment ($pO_2 \approx 10^{-6}$ atm). Conversely, argon-annealed $Sr_2Fe_3O_y$ (x = 0, Fig. 3a) has the same structure as that of the air-sintered material and contains no brownmillerite phase, although other argon-annealed cobalt-containing samples all contain an amount brownmillerite phase. The content of the brownmillerite phase in annealed $Sr_2Fe_{3-x}Co_xO_y$ increases with increasing x, as shown in Fig. 3b, 3c, and 3d. The cobalt oxide phase can be seen clearly in the cobalt-containing samples. An XRD of a pure brownmillerite phase of the argon-annealed $SrFe_{0.8}Co_{0.2}O_{3-\delta}$ sample is shown in Fig. 3e for comparison. Phase components and lattice parameters of the air-sintered materials are listed in Table 1. Where "O" and "P" are standing for orthorhombic- and perovskite-phase, respectively.

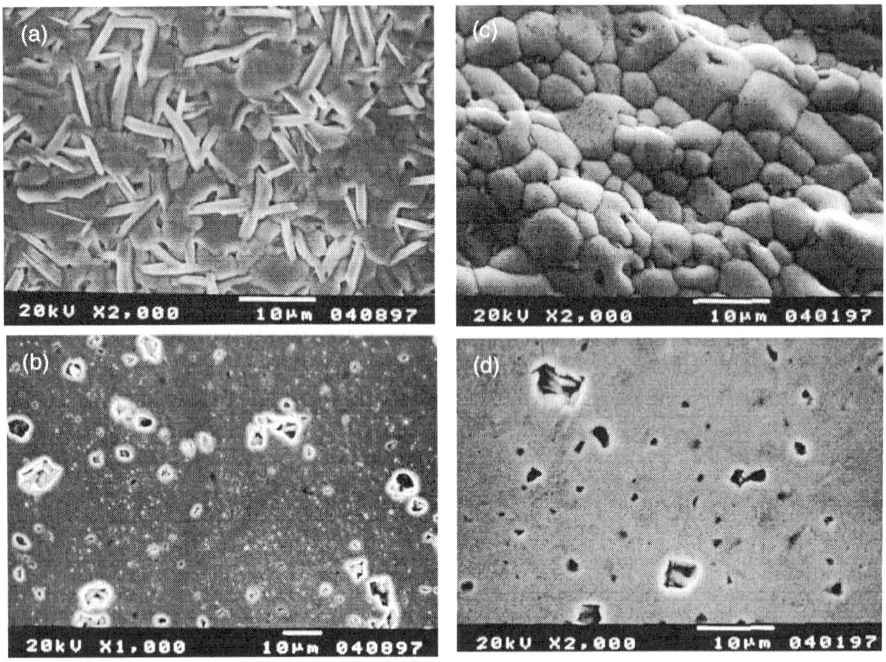

Fig. 4. SEM morphology of $Sr_2Fe_2CoO_x$ sample: (a) original surface of air-sintered sample, (b) polished surface of air-sintered sample, (c) original surface of argon-annealed sample, (d) polished surface of argon-annealed sample.

SEM images of the air-sintered and argon-annealed $Sr_2Fe_2CoO_y$ samples are shown in Fig. 4. They reveal a dense homogeneous structure for both samples. Energy-dispersive X-ray (EDX) elemental analysis showed that the overall atomic ratio of the metal elements is consistent with that in the chemical formula. The polished-surface SEM images showed that the argon-annealed sample is denser and contains fewer pores than the air-sintered sample. A spherical cobalt-rich phase was identified in the argon-annealed sample. By correlating SEM observation with XRD results, we identified the spherical phase as cobalt oxide (CoO, JCPD No. 9-402).

Table 1. Lattice parameters of air-sintered $Sr_2Fe_{3-x}Co_xO_y$ samples

Co content	Phase Components	a (Å)	b (Å)	c (Å)	V (Å^3)
0.0	O (>98%)	11.126	18.978	5.5864	1179.6
0.3	O (>95%) + P (<5%)	11.085	19.016	5.5670	1173.5
0.6	O (>90%) + P (<10%)	11.065	19.008	5.5618	1169.8
1.0*	O (>70%) + P (<25%)	11.017	19.030	5.5463	1162.8

*Contains ~5% CoO phase.

The total conductivity of the $Sr_2Fe_{3-x}Co_xO_y$ sample as a function of temperature is shown in Fig. 5. The conductivity increases as temperature and cobalt content increase. At 900°C, the conductivity of $Sr_2Fe_2CoO_y$ is ~20 and ~6 S·cm^{-1} in flowing air and argon environments, respectively. Conductivities of $Sr_2Fe_{2.4}Co_{0.6}O_y$ and $Sr_2Fe_3O_y$ are ≈10 and ≈0.6 S·cm^{-1}, respectively, at 900°C in air. The ionic transference numbers ($t_{ion} = \sigma_{ion} / \sigma_{total}$) of $Sr_2Fe_2CoO_y$ and $Sr_2Fe_{2.4}Co_{0.6}O_y$ as a function of temperature in air are shown in Fig. 6. The ionic transference numbers for both samples are not strongly dependent on temperature. A higher ionic transference number was observed for the sample with lower cobalt content.

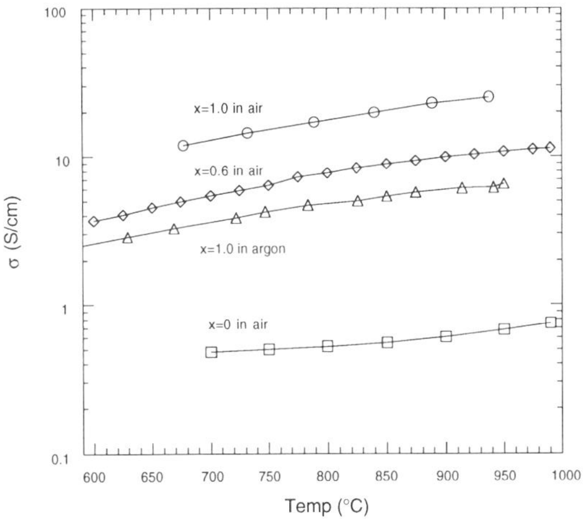

Fig. 5. Temperature dependence of $Sr_2Fe_{3-x}Co_xO_y$ samples.

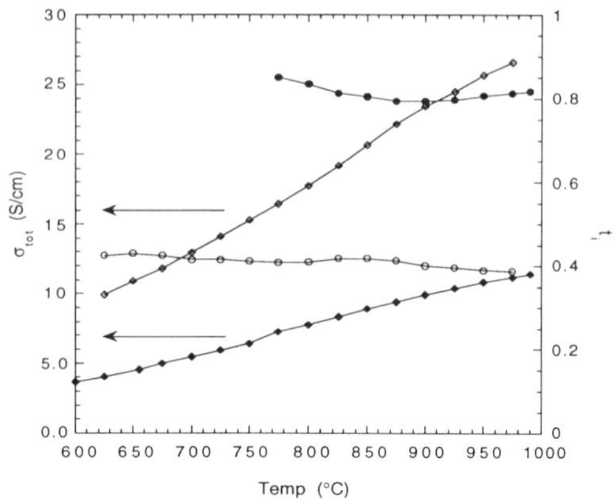

Fig. 6. Temperature dependence of the total conductivity and the ionic transference number of $Sr_2Fe_2CoO_y$ (circles) and $Sr_2Fe_{2.4}Co_{0.6}O_y$ (diamonds).

For the oxygen permeation experiments, the EMF measured at the sensor electrodes as a function of pumping current at various temperature is plotted in Fig. 7. At a given temperature, the EMF at the sensor electrodes increases with increasing pumping current. EMF increases with decreasing temperature at a given pumping current, and the pumping current increases with increasing temperature to achieve a specific EMF value because the oxygen permeability increases at higher temperatures.

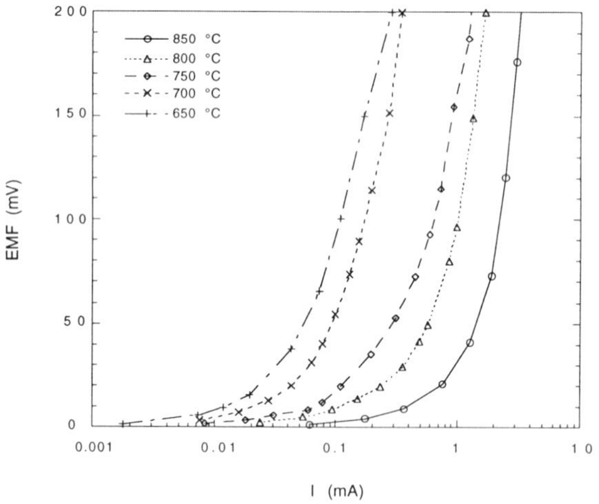

Fig. 7. Electromotive force as a function of pumping current at various temperatures.

During our experiment, the gas-tight cell was surrounded by flowing air. The pO_2 inside the cell can be determined by solving Eq. 1. The oxygen permeation flux, j_{O_2}, has been calculated from the steady-state pumping current by solving Eq. 2. To examine the contributions of bulk and interface effects to the oxygen permeation mechanism of the $Sr_2Fe_2CoO_y$ sample, we fit the permeation data with the equation [18]:

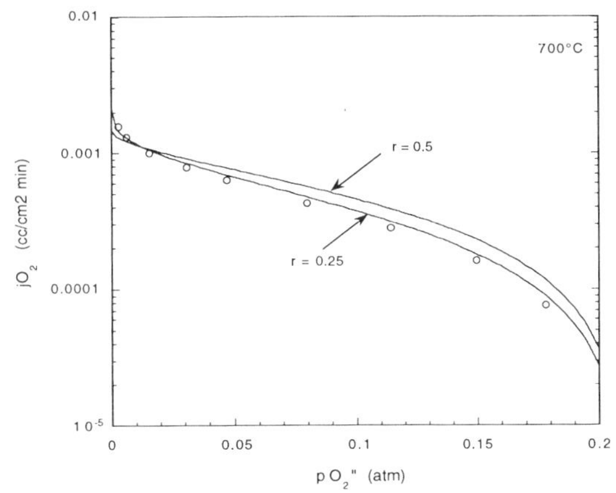

Fig. 8. Steady-state oxygen permeation flux as a function of pO_2 inside the gas-tight cell, and its fitting to Eqs. 3 (r = 0.5) and 4 (r = 0.25).

$$j_{O_2} = \alpha\left[(pO_2^{I})^r - (pO_2^{II})^r\right] \quad (3)$$

where r is the kinetic order in oxygen of the surface reaction, and a is the exchangerate at pO_2 = 1 atm. For a permeation process where the surface reaction (exchange) effect is the controlling factor, we would have r = 1/2. The surface exchange constant a (in $cm\cdot min^{-1}\cdot atm^{-1/2}$) can be determined by a

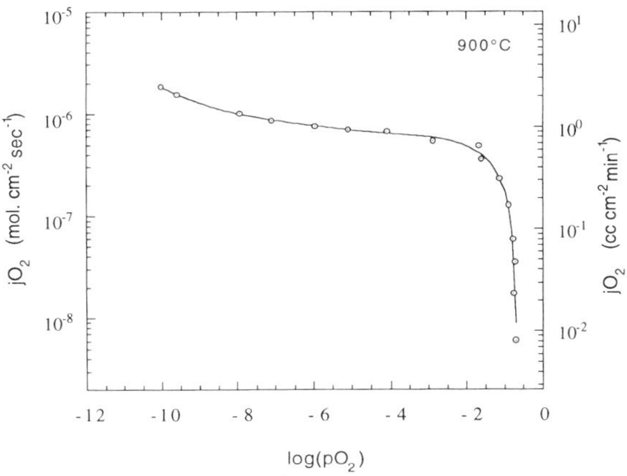

Fig. 9. The pO_2 dependence of steady-state oxygen permeation flux for a 2.9-mm-thick disk membrane.

least-squares fitting of the experimental data. Where the bulk effect is the controlling factor or, in another words, the surface exchange rate is sufficiently large, the oxygen permeation flux can be defined by [19]:

$$j_{O_2} = \frac{\beta}{L}\left[(pO_2^{I})^{1/4} - (pO_2^{II})^{1/4}\right] \tag{4}$$

where L is the thickness of the membrane (in cm) and β is a bulk diffusion parameter (in $cm^2 \cdot min^{-1} \cdot atm^{-1/4}$). Figure 8 shows typical oxygen permeation data as a function of pO_2 inside the cell, along with the least square-fittings to Eq. 3 (with r = 0.5) and Eq. 4 (with r = 0.25). The fitting is much better to Eq. 4 than to Eq. 3, indicating that bulk diffusion plays the controlling role in the oxygen permeation process in the $Sr_2Fe_2CoO_y$ membrane of ≈1 mm thickness at high temperatures (>650°C).

In Fig. 9, the steady-state oxygen permeation flux of a 2.9-mm-thick pellet specimen at 900°C, determined by solving Eq. 2, is plotted as a function of pO_2 inside the gas-tight cell. Figure 9 shows that j_{O_2} increases dramatically in the range between pO_2 = 0.21 and ~10^{-3} atm, and its slope becomes flatter when pO_2 inside the cell is reduced further. Results on oxygen permeability at various pO_2 gradients and temperatures show that, as expected, j_{O_2} increases with temperature and pO_2 gradients. At 900°C, oxygen permeability was ~2.5 $scc \cdot cm^{-2} \cdot min^{-1}$ for a 2.9-mm-thick specimen and increased as membrane thickness decreased.

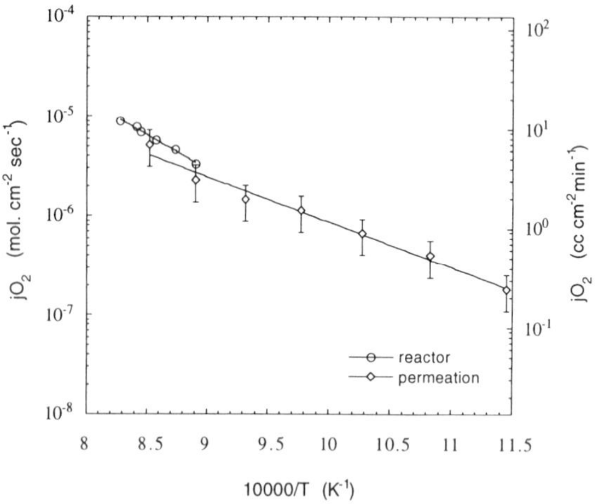

Fig. 10. Temperature dependence of oxygen permeation flux obtained from methane conversion reactor and gas-tight cell.

Sintered thin-wall tubes of $Sr_2Fe_2CoO_y$ were tested in a methane conversion reactor for more than 1000 h [6,7]. An oxygen permeation flux of ~10 scc·cm^{-2}·min^{-1} was observed in a 0.75-mm-thick tubular membrane [6]. Oxygen permeation flux, determined from reactor experiments and with a gas-tight electrochemical cell, have been plotted in Fig. 10 as a function of reciprocal temperature. For comparison, the oxygen permeation flux of the disk membrane was normalized to that of a 0.75-mm-thick membrane. Oxygen permeation data obtained from two independent experiments are in good agreement.

CONCLUSIONS

The phase components of $Sr_2Fe_{3-x}Co_xO_y$ are strongly dependent on cobalt content. The $Sr_2Fe_3O_y$ sample is single-phase material, while the high-cobalt-content samples ($x \geq 0.6$) contain multiple phases. The concentration of a perovskite phase in the high-cobalt-content samples increases with increasing cobalt content. In a reducing argon environment, the single-phase $Sr_2Fe_3O_y$ sample has the same crystal structure as the air-sintered sample, while a brownmillerite phase was observed in the cobalt-containing samples. Cobalt oxide has also been found in the argon-annealed cobalt-containing samples.

The total conductivity of $Sr_2Fe_{3-x}Co_xO_y$ increases with increasing temperature and cobalt content in the sample. At 900°C, the conductivity of $Sr_2Fe_2CoO_y$ in flowing air and argon environments is ~20 and ~6 S·cm^{-1}, respectively. The conductivities of $Sr_2Fe_{2.4}Co_{0.6}O_y$ and $Sr_2Fe_3O_y$ are ~10 and ~0.6 S·cm^{-1}, respectively, at 900°C in air. The ionic transference numbers do not have a strong dependence on temperature above 600°C in air. A higher total conductivity but lower ionic transference number was observed in the sample with lower cobalt content. The transient conductivity behavior of this sample, after the surrounding atmosphere is switched back and forth, indicates that the phase transition in the materials is reversible.

The oxygen permeation mechanism in the sintered $Sr_2Fe_2CoO_y$ material was found to be bulk controlled. At 900°C, an oxygen permeability of ~2.5 scc·cm^{-2}·min^{-1} was observed in a 2.9-mm-thick membrane. This permeability value agrees with values obtained in the methane conversion reactor experiments.

ACKNOWLEDGMENT

Work at Argonne is supported by the U.S. Department of Energy, Federal Energy Technology Center, under Contract W-31-109-Eng-38.

REFERENCES

1. T. Takahashi and H. Iwahara, *Energy Convers.*, **11**, 105 (1971).
2. B. C. H. Steele, *Mater. Sci. Eng. B-Solid State M.*, **13**, 79 (1992).
3. N. Q. Minh, *J. Am. Ceram. Soc.*, **76**, 563 (1993).
4. R. DiCosimo, J. D. Burrington, and R. K. Grasselli, *J. Catal.*, **102**, 377 (1992).
5. K. R. Kendall, C. Navas, J. K. Thomas, and H.-C. Loye, *Solid State Ionics*, **82**, 215 (1995).
6. U. Balachandran, S. L. Morissette, J. T. Dusek, R. L. Mieville, R. B. Poeppel, M. S. Kleefisch, S. Pei, T. P. Kobylinski, and C. A. Udovich, *Proc. Coal Liquefaction and Gas Conversion Contractor Review Conf.*, S. Rogers et al., eds., **Vol. 1**, pp. 138-160, U.S. Dept. of Energy, Pittsburgh Energy Technology Center (1993).
7. U. Balachandran, T. J. Dusek, S. M. Sweeney, R. B. Poeppel, R. L. Mieville, P. S. Maiya, M. S. Kleefisch, S. Pei, T. P. Kobylinski, C. A. Udovich, and A. C. Bose, *Am. Ceram. Soc. Bull.*, **74**, 71 (1995).
8. Y. Teraoka, H. M. Zhang, S. Furukawa, and N. Yamozoe, *Chem. Lett.*, **1985**, 1734 (1985).
9. Y. Teraoka, T. Nobunaga, and N. Yamazoe, *Chem. Lett.*, **1988**, 503 (1988).
10. B. Ma, J.-H. Park, C. U. Segre, and U. Balachandran, *Mater. Res. Soc. Symp. Proc.*, **393**, 49 (1995).
11. B. Ma, U. Balachandran, C.-C. Chao, J.-H. Park, and C. U. Segre, *Ceram. Trans. Series,* **73**, 169 (1997).
12. U. Balachandran, J. T. Dusek, P. S. Maiya, R. L. Mieville, B. Ma, M. S. Kleefisch, and C. A. Udovich, presented at 11th Intersociety Cryogenic Symp., Energy Week Conf. & Exhibition, Houston, Jan. 28-30, 1997.
13. S. Pei, M. S. Kleefisch, T. P. Kobylinski, J. Faber, C. A. Udovich, V. Zhang-McCoy, B. Dabrowski, U. Balachandran, R. L. Mieville, and R. B. Poeppel, *Catal. Lett.*, **30**, 201 (1995).
14. H. Iwahara, H. Uchida, K. Morimoto, and S. Hosgoi, *J. Appl. Electrochem.*, **19**, 448 (1989).
15. A. Yoshiasa, K. Ueno, F. Kanamaru, and H. Horiuchi, *Mater. Res. Bull.*, **21**, 175 (1986).
16. H. M. Rietveld, *J. Appl. Cryst.*, **2**, 65 (1969).
17. C. Greaves, A. J. Jacobson, B. C. Tofield, and B. E. F. Fender, *Acta Cryst.*, **B31**, 641 (1975).
18. H. J. M. Bouwmeester, H. Kruidhof, A. J. Burggraaf, and P. J. Gellings, *Solid State Ionics*, 53-56, 460 (1992).
19. S. Dou, C. R. Masson, and P. D. Pacey, *J. Electrochem. Soc.*, **132**, 1843 (1985).

Electrochemical Devices and Sensors

ZIRCONIA ELECTROLYTE BASED SOLID OXIDE FUEL CELLS

Srikanth Gopalan and Subhash C. Singhal
Westinghouse Electric Corporation
Science & Technology Center
1310 Beulah Road
Pittsburgh, PA 15235

ABSTRACT

The fabrication and performance of zirconia electrolyte based tubular solid oxide fuel cells (SOFC) have been successfully demonstrated. This paper outlines the electrochemical aspects of fuel cell operation, materials requirements for the various cell components and the fabrication processes presently adopted. Current efforts aimed at materials cost reduction, processing and cell power enhancement are summarized.

INTRODUCTION

Yttria-stabilized zirconia (YSZ) electrolytes have applications in many electrochemical technologies including solid oxide fuel cells (SOFCs), oxygen separation and automotive sensors. SOFC technology offers significant advantages over conventional fossil power generation technology. These include reduced emission of NO_x and SO_x, noiseless operation, higher efficiency, modularity and the ability to operate on various fuels. Due to the high temperature of operation (~1000°C), SOFCs produce high quality exhaust heat which can be used for a bottoming power cycle to further increase the overall system efficiency. SOFCs can also be operated under pressurized conditions and the pressurized SOFC - gas turbine (GT) combined cycle power plants can achieve operating efficiency close to 70% [1,2].

In the current technology, the SOFC is fabricated with all the active cell components deposited as thin layers on a tube which also functions as the cathode (air electrode). The fabrication processes and the materials used in tubular SOFCs have been reviewed previously [3] and are only briefly summarized in this paper. This paper reviews the electrochemical aspects of cell performance and the studies that are currently underway to further reduce cost and enhance performance of the tubular SOFCs.

DESIGN AND FABRICATION OF TUBULAR SOFCs

The air electrode tube in the state-of-the-art tubular SOFC, which is made of doped lanthanum manganite, is extruded in green form and sintered to between 30 and 35% porosity. This level of porosity in the air electrode tube is necessary from the standpoint of transporting oxygen in the gas phase to the air electrode - electrolyte interface where it is reduced to oxygen ions. The schematic design of the cell is shown in Figure 1. The active length, which is the length of the cell that electrochemically oxidizes fuel, in the current commercial prototype is 150 cm and the outer diameter (OD) of the cell is 2.2 cm.

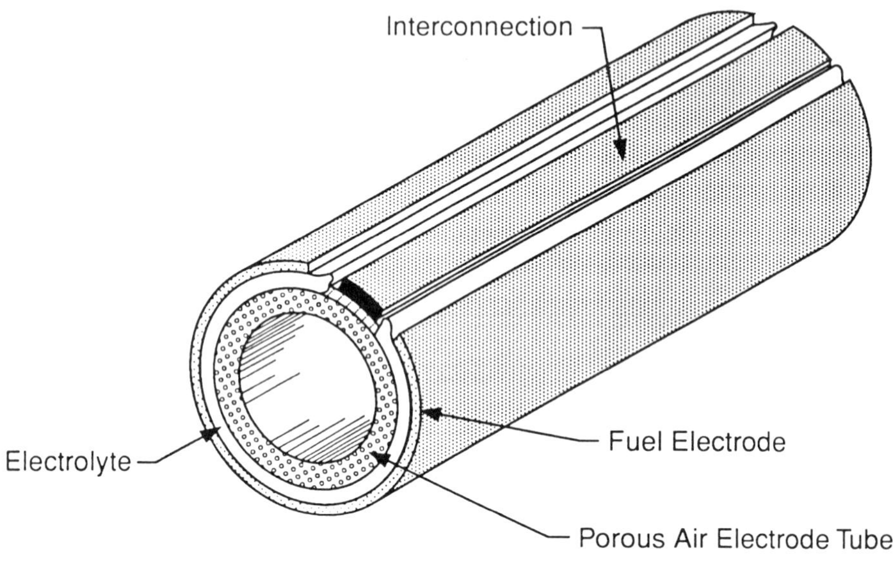

Fig. 1. Tubular solid oxide fuel cell design.

The fabrication of the cell involves following steps: (i) The interconnect, which is made from lanthanum chromite powder containing calcium aluminate, is first deposited on the air electrode tube by a plasma spray process [4], (ii) the electrolyte which is 10 mol% YSZ is then deposited by the electrochemical vapor deposition (EVD) process. The EVD process which utilizes the slight electronic conductivity of zirconia in conjunction with its ionic conductivity deposits high quality gas tight films, ~40 μm in thickness [5], (iii) the gas tight tube is then dipped into a slurry containing a mixture of Ni and YSZ and sintered to form the Ni-YSZ cermet fuel electrode. The replacement of the EVD process which was used earlier to deposit the fuel electrode by the simpler sintering technique has resulted in major cost reduction. Table I summarizes the materials and fabrication processes for the commercial prototype tubular SOFCs.

Table I: Materials and processes used in state-of the-art Westinghouse cells

Component	Material	Thickness	Fabrication Process
Air Electrode	Doped $LaMnO_3$	2.2 mm	Extrusion-Sintering
Electrolyte	YSZ	40 μm	Electrochemical vapor deposition (EVD)
Interconnection	Doped $LaCrO_3$	85 μm	Plasma Spraying
Fuel Electrode	Ni-YSZ cermet	100 μm	Dip coating followed by sintering

CELL PERFORMANCE

The tubular SOFC technology has been validated by a large number of electrical tests. The cells are usually tested in a fuel of composition 89% H_2 + 11% H_2O at 85% fuel utilization. The performance degradation in the present generation of tubular cells is about 0.1% per 1000 hours of operation. The voltage - current density (VJ) curve and power - current density curve of a typical 2.2 cm OD cell at 1000°C are displayed in Figure 2.

The cell, in the test assembly, comprises two nickel bus bars (one on the interconnect and one on the fuel electrode) which constitute two equipotential surfaces. Thus the terminal voltage measured across the bus bars is constant and not a function of cell axial position. The fuel is introduced at one end of the cell which is closed and flows towards the open end of the cell. As the fuel flows along the length of the cell, the electrochemical reaction converting fuel to oxidized species (H_2O in the case of H_2 fuel) takes place. Thus, the composition of the gas phase on the fuel side and the Nernst potential vary as a function of cell axial position. The terminal voltage is related to the local Nernst potential, ohmic loss and polarization through the following equation:

$$V_T = V_N(x) - J(x)R_i - \eta(x) \quad (1)$$

where,

V_T = terminal voltage; not a function of position x

$V_N(x)$ = Nernst potential at position x; depends on the fuel composition at position x,

$J(x)$ = current density at position x

R_i = resistance index of the cell ($\Omega.cm^2$); not a function of position x

$\eta(x)$ = sum of the polarization losses on the air electrode and fuel electrode sides of the cell.

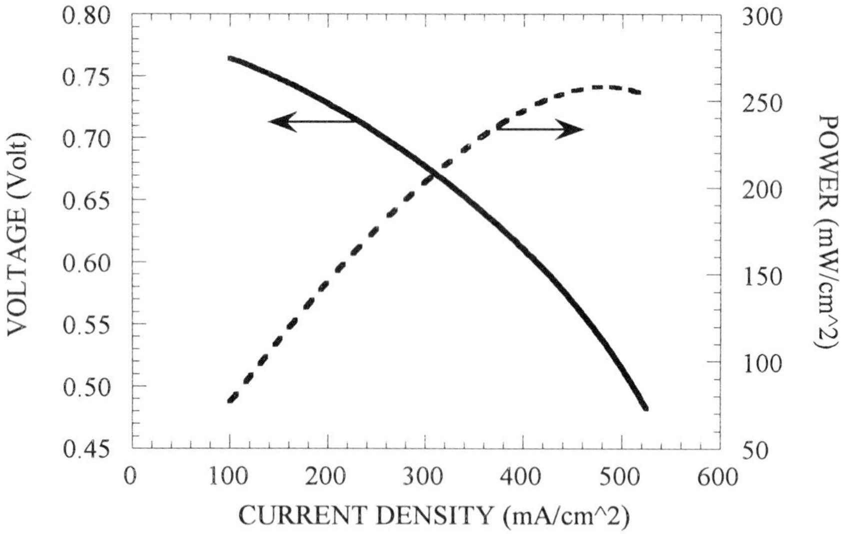

Fig. 2. Voltage - current density and power-current density curves of a typical 2.2 cm OD cell.

Figure 3 compares the VJ curves for a tubular cell at 1 atm and 15 atm operating pressure conditions. At 15 atm operation, the entire assembly is pressurized, i.e. there is no pressure differential between the air and fuel sides of the cell. The benefit of operating under pressurized conditions is clear from this comparison. The VJ curves in Figure 3 were fitted with a finite element model that uses the mass transport parameters on the air and fuel electrode sides and the internal resistance of the cell as fitting parameters. The open circles and diamonds in Figure 3 represent the experimental points and the two dashed lines are the results of the finite element model. Using the fitting parameters obtained from Figure 3, the thermodynamic Nernst potential and concentration polarization on

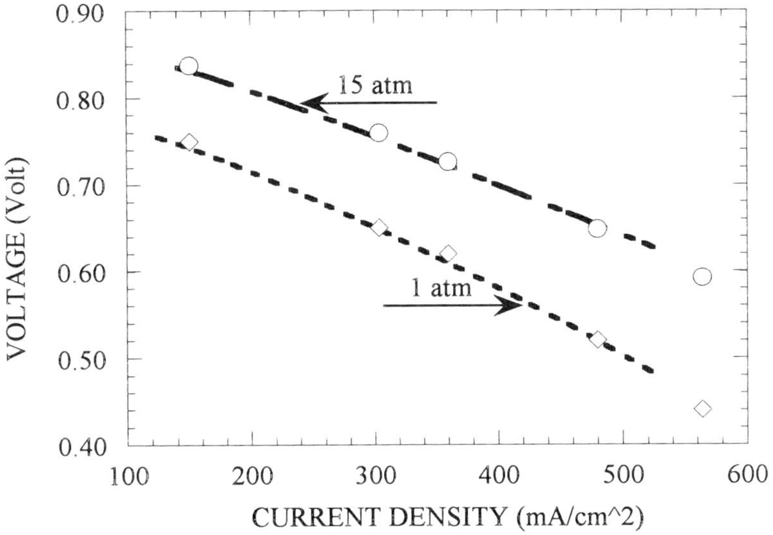

Fig. 3. Voltage - current density plot for a 150 cm active length cell at 1 atm and 15 atm operating pressure at 1000°C.

the air electrode side were calculated as a function of cell axial position for 1 atm and 15 atm operation.

Figure 4 shows the calculated Nernst potential, ohmic loss and air electrode and fuel electrode polarizations as functions of axial position for an average current density* of 300 mA/cm^2 and 1 atm operation. As prescribed by Equation 1, it can be seen from Figure 4 that the difference between the Nernst potential and the sum of the ohmic loss and polarizations equals the terminal voltage. Figure 5 compares the Nernst potentials and air electrode polarizations for 1 and 15 atm operation as functions of axial position for an average current density of 300 mA/cm^2, where the other voltage components are omitted for the sake of clarity. It is seen from Figure 5 that the origin of the performance enhancement obtained under pressurization consists of both an increase in the Nernst potential and a decrease in concentration polarization on the air electrode side. Since the air electrode is much thicker than the fuel electrode (2.2 mm versus 100 μm), the concentration polarization is much higher on the air electrode side, constituting

* Note that the current density and hence the ohmic loss $J(x)R_i$ varies as a function of axial position. The calculations in Figure 4 are for an axial position averaged current density of 300 mA/cm^2.

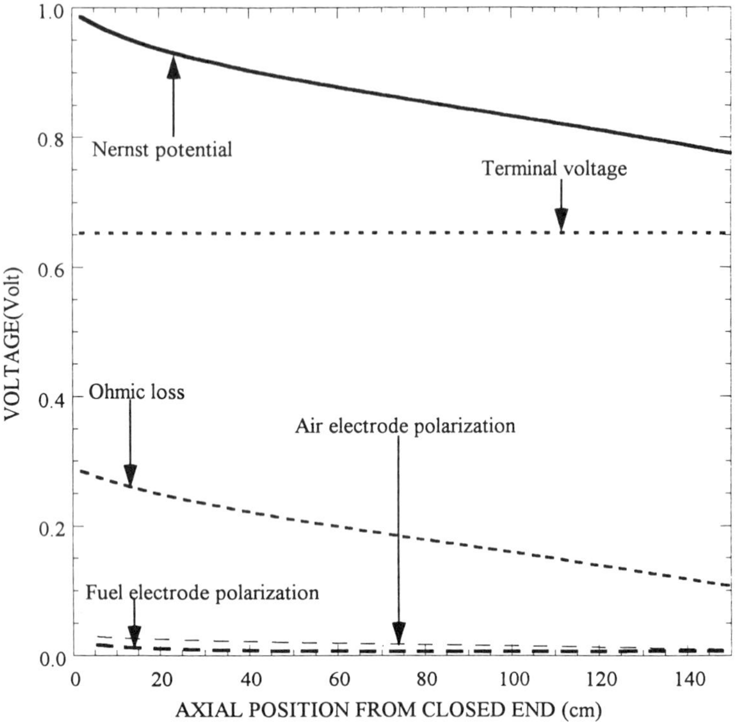

Fig. 4. Variation of Nernst potential V(Nernst), air electrode polarization loss V(AE), fuel electrode polarization loss and ohmic loss as a function of cell axial position from the closed end.

over 80% of the term $\eta(x)$. The calculation above takes into account only the concentration polarization. However, it is likely that the charge transfer reaction kinetics could also be enhanced under pressurized operation, thus decreasing the activation polarization.

EFFORTS TO REDUCE CELL COST

Various efforts are currently underway to reduce cell cost. These can be classified under three categories: (i) reducing cost of materials, (ii) search for cheaper alternate processes and (iii) design of cells which result in a higher volumetric stack power density (W/cm^3).

A major materials cost reduction currently under investigation is that of the air electrode. Since the air electrode serves as the mechanical support on which other

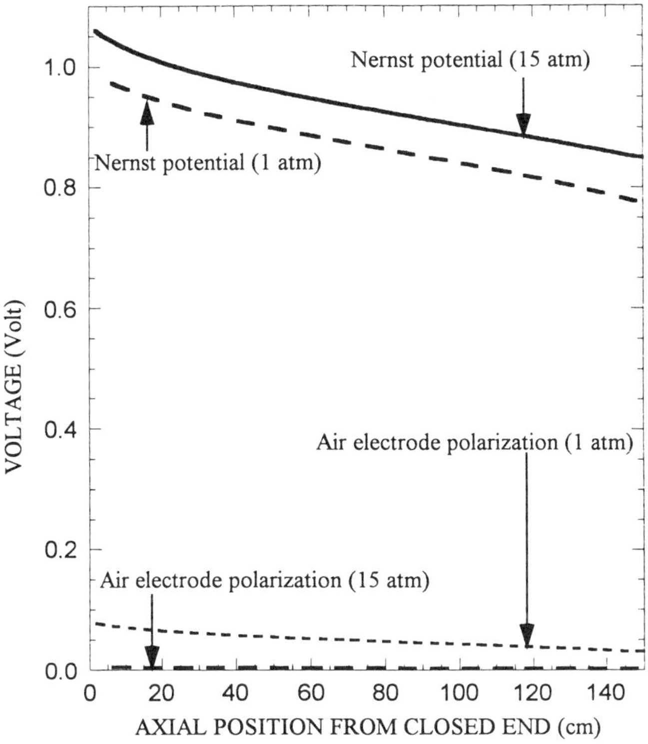

Fig. 5. Nernst potential and air electrode polarization as a function of cell axial position from the closed end of the cell.

cell components are fabricated, it also accounts for over 90% of the weight of the cell. Presently high purity component oxides such as La_2O_3 (99.9% purity) and MnO_2 are used to synthesize the air electrode. Use of lower purity La_2O_3 (98% to 99%) could result in substantial cost savings. Another strategy that is being explored to reduce air electrode material cost is to reduce the rare earth content of the air electrode material without compromising the electrical conductivity and the thermal expansion coefficient.

On the alternate processes front, efforts to substitute the EVD process for depositing the fuel electrode with the coating and sintering process described earlier have been successful. Figure 6 compares the performance of a 2.2 cm OD cell consisting of a coated and sintered fuel electrode with that of an EVD deposited fuel electrode cell at a current density of 300 mA/cm^2 at 1000°C. It is seen that the performance of the sintered fuel electrode cell is comparable to that

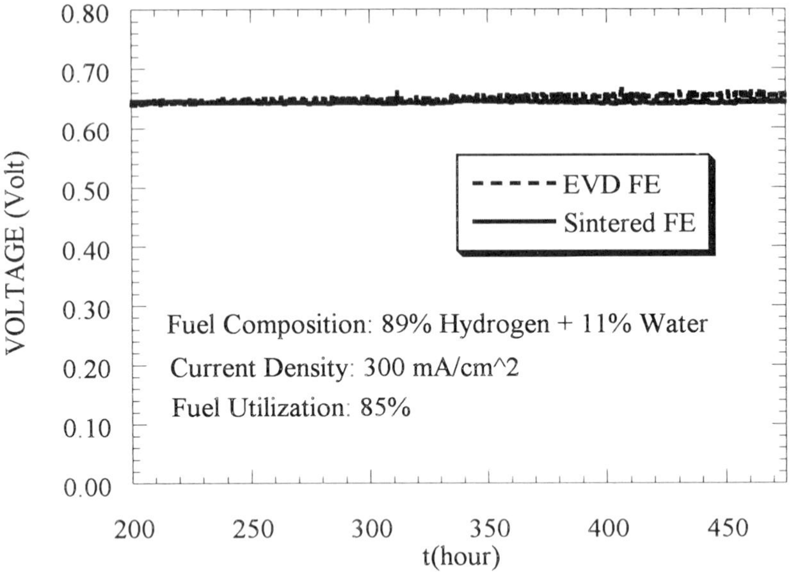

Fig. 6. Relative performance of a typical cell with a sintered fuel electrode and a cell with an EVD deposited fuel electrode at 1000°C.

of the EVD deposited fuel electrode. Several cells that incorporate the dip coated and sintered fuel electrodes have been tested to date. Coating and sintering will be the process for deposition of the fuel electrode in future production cells. Thus the number of cell components requiring the electrochemical vapor deposition (EVD) will be reduced to one, namely the electrolyte. Alternative processes such as electrophoretic deposition (EPD) and electrochemical deposition (ECD) followed by sintering are being explored for the electrolyte deposition step.

Efforts to develop cells that would result in a higher volumetric stack power density (W/cm^3) include alternate cell designs and improvement of the air electrode/electrolyte polarization by modification of that interface. Figure 7 compares the stack power density of an alternate geometry cell (shown as an inset in Figure 7) with the current 2.2 cm OD tubular geometry. Due to the flattened geometry of this cell, the packing efficiency is higher for these cells than the current 2.2 cm OD tubular cells resulting in increased stack power densities. Thus significant cost savings can be realized by application of these cells in future generators.

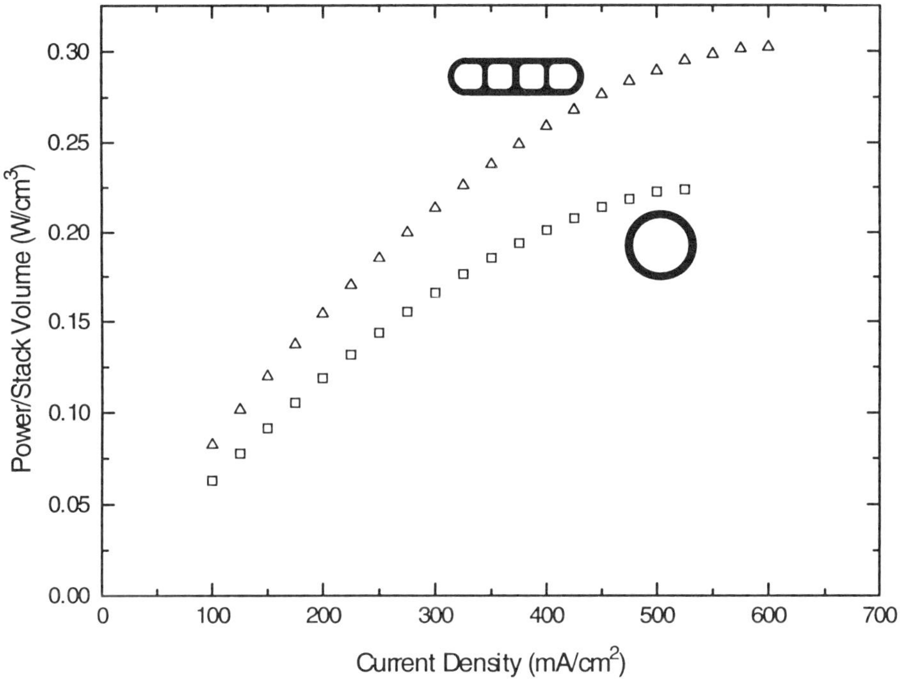

Fig. 7. Comparison of volumetric stack power densities for a tubular and an alternate geometry cell.

WESTINGHOUSE SOFC POWER GENERATION SYSTEMS

Many progressively larger SOFC power generation systems have been built and tested since 1984. Of these, three of the latest systems are highlighted here.

Two 25 kW systems, each consisting of 576 50-cm active length state-of-the-art cells were built. One of these systems operated at the Southern California Edison Company's (SCE) Highgrove Generating Station in Grand Terrace (near San Bernardino), California, under a program with the United States Department of Defense's Advance Research Projects Agency (ARPA). After successful completion of all objectives at SCE where it operated for 5,582 hours on various types of fuel, this unit has now been relocated to the University of California - Irvine. At this site, the unit continues to run on pipeline natural gas producing 21.8 kW power.

The other 25 kW system which was built for a consortium of Osaka Gas and Tokyo Gas, after completion of all objectives was decommissioned in February 1997 after 13,194 hours of operation.

Recently, a 100 kW SOFC power generation system was built and delivered to a consortium of Dutch and Danish utilities (EDB/ELSAM). This consists of 1,152 2.2 cm diameter, 150 cm active length cells in a series-parallel combination. It is currently providing about 110 kW AC to the Dutch power grid and is located at a district heating system owned by NUON, one of the members of the five Dutch gas and electricity distribution utilities that comprise the EDB. The net efficiency of this system taking into account recovery of part of the fuel energy in hot water is 75%. The maximum power output of this system is 160 kW. At maximum power the fuel efficiency of the system will approach 80%.

All of the units to date have operated at ambient pressure of 1 atm. Future units in the 250 kW to integer MW range will be operated under elevated pressures in combination with a gas turbine.

SUMMARY

Westinghouse tubular geometry fuel cells have exhibited significant improvements in electrical performance and structural reliability. Improvements in performance have been demonstrated by running the cells under pressurized conditions. From analysis of the data it is concluded that the improvement in performance of pressurized cells has its origins in an increase in Nernst potential and a decrease in air electrode concentration polarization. Significant cell cost reductions have been achieved by reducing the number of EVD steps in cell fabrication to one. When current materials and processing cost reduction programs prove successful, further cost reductions are expected.

ACKNOWLEDGMENTS

The authors acknowledge the contributions of many colleagues whose work is reviewed in this paper. The development of SOFC technology has been supported by the U.S. Department of Energy (DOE), the Gas Research Institute (GRI), Electric Power Research Institute (EPRI), and many utility and commercial sources. The pressurized SOFC testing is supported in part by Ontario Hydro Technologies and New Energy Development Organization (NEDO) of Japan and its participating electric power companies.

REFERENCES

[1] W.G. Parker and F.P. Bevc, "SureCELL™ Integrated Solid Oxide Fuel Cell/Combustion Turbine Power Plants for Distributed Power Applications," pp.

275-278 in *Proceedings of the 2nd International Fuel Cell Conference*, New Energy and Industrial Technology Development Organization, Kobe, Japan, 1996.

[2] S.E. Veyo and W.L. Lundberg, "Tubular SOFC and SOFC/Gas Turbine Combined Cycles-Status and Prospects," pp. 776-779 in *Fuel Cell Seminar Abstracts*, Courtesy Associates, Inc., Washington, DC, 1996.

[3] S.C. Singhal, "Recent Progress in Tubular Solid Oxide Fuel Cell Technology," pp. 631-651, *Proceedings of the Fifth International Conference on the Science and Technology of Zirconia*, Edited by S.P.S. Badwal, M.J. Bannister and R.H.J. Hannik, Technomic Publishing Company, Lancaster, PA 1993.

[4] L.H. Kuo, S.D. Vora and S.C. Singhal, "Plasma Spraying of Lanthanum Chromite Films for Solid Oxide Fuel Cell Interconnection Application," *Journal of the American Ceramic Society*, **80** [3] 589-93 (1997).

[5] U.B. Pal and S.C. Singhal, "Electrochemical Vapor Deposition of Yttria-Stabilized Zirconia Films," *Journal of the Electrochemical Society,* **137** [9] 2937-2941(1990).

PROPERTIES OF WO_3 COATINGS FOR LARGE AREA ELECTROCHROMIC DEVICES

J.P. Cronin, S.R. Kennedy, A. Agrawal, T.J. Gudgel, Y.J. Yao and J.C.L. Tonazzi
Donnelly Corporation, Advanced Technology Center, 4545 East Fort Lowell Road, Tucson, AZ 85712, USA
D.R. Uhlmann
University of Arizona, Department of Materials Science and Engineering, Tucson, AZ 85712, USA

ABSTRACT

A novel wet chemical method for forming thick WO_3 coatings is described. The method is applicable to large area applications. By chemical modifications to the precursor solution and control of the firing conditions the microstructure of the coatings could be changed. The lithium intercalation properties of the coatings are discussed in terms of microstructure, electrochemistry, and electrochromic behavior. A large area WO_3 coating (35 × 40 cm) was prepared and incorporated into an electrochromic device. The electrochromic performance of the device is described.

INTRODUCTION

The growing concern over the depletion of fossil fuels has led to increased interest in energy conservation. Much attention has been directed towards conservation of energy in buildings, both commercial and residential, while maintaining the comfort of occupants. One of the primary shortcomings of conventional windows is their high thermal loss compared with other building elements. Surface coatings for glass, mechanical shading devices, and modifications of building design and site orientation have contributed to reducing thermal loss and solar energy gain through windows. To further optimize the energy efficiency, electrochromic technology holds outstanding promise[1].

Electrochromic (EC) windows allow either manual or automated control over the amount of transmitted solar radiation. The degree of coloration is controlled by the potential imposed across the device. Furthermore, EC devices can be made with nonvolatile memory so that the flow of electricity can be interrupted and a desired state of transmission retained. Such windows can be adjusted to reduce glare and daylight heating needs or to provide night-time insulation and privacy. The dynamic control of transmittance is important because of the changing characteristics of the incident solar radiation. It has been estimated that the output of the sky can change in a random manner by 100% during the course of the day[2].

Unlike those of low-E coatings, the optical properties of EC coatings can be modulated rapidly in response to changes in weather conditions. In this way, daylight can be used for optimal performance during the winter months, while during the summer, it can be minimized to prevent solar gain. The advantage of EC windows for optimizing both lighting and cooling are summarized in Figure 1.

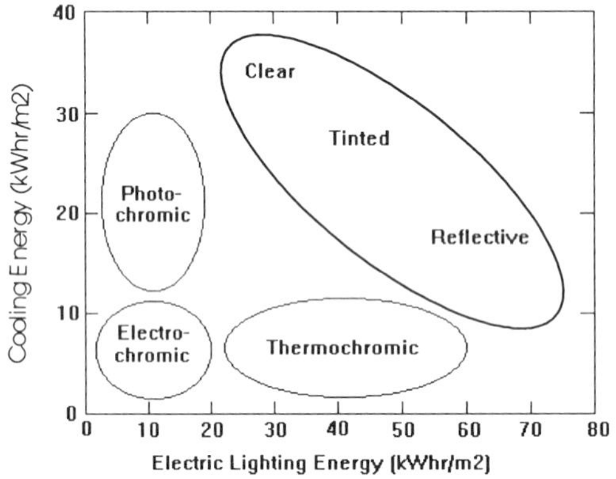

Figure 1
Diagram of projected cooling energy and electric lighting requirements taken from reference 3.

A schematic of an EC device is shown in Figure 2. The device is composed of two transparent conductors -- one with an electrochromic thin film coating and the other with a thin film counter electrode or ion storage layer. The two electrodes are separated by an electrolyte layer. In the case of a liquid electrolyte a cavity is formed through the use of a sealing material.

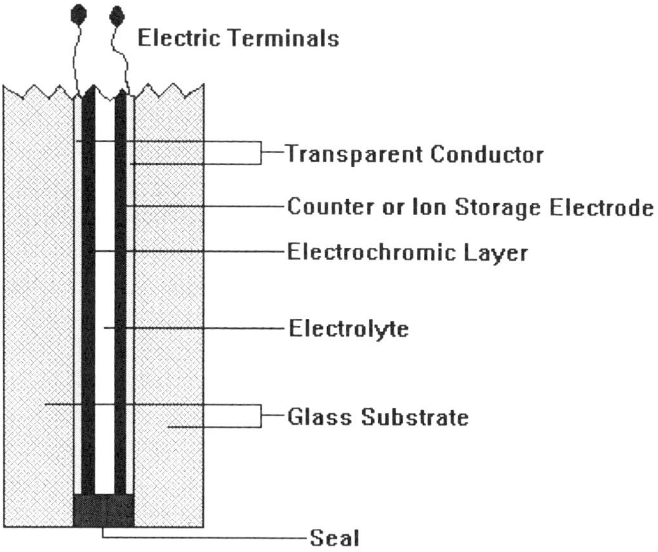

Figure 2
Schematic diagram of an electrochromic window device

For large area applications such as architectural glazing one needs to evaluate the individual components of the EC device particularly in relation to ease of forming and cost [4]. Table I lists focus areas that need to be developed to apply current EC technology to large area applications.

Table I. Focus areas for developing current EC technology for large area applications

Device component	Areas of improvement	Expected Results
Transparent conductors	• Higher conductivity • Reduce cost	Faster response times Better uniformity for color and bleach
Thin films	• Cost effective ways to coat large areas uniformly • Improve optical modulation efficiency of EC films	Low cost
Electrolyte	• Improve processability and / or conductivity of electrolyte	Enhanced performance
Sealants	• Hermetic sealants with good adhesion to the substrates and chemical compatibility with the device components.	Improved device life time for large area applications
Device assembly	• High throughput, low cost methods	Low cost

In this paper, we will focus on tungsten oxide coatings, which are used as electrochromic layers. Coatings of WO_3 have been deposited by many different methods[5,6,7,8,9] with wet chemical deposition being offered as a low cost procedure[10]. However, this is only true if the precursor material is readily available at low cost and has a reasonable working life-time. It is also important that the coating be deposited in sufficient thickness, by a single deposition step, to give the desired modulation range. The WO_3 films discussed were prepared using a novel peroxotungstic ester[11] (PTE) precursor dissolved in ethanol and deposited onto transparent conductive glass by dip or spin coating. This wet chemical method offers the advantage of stable precursor and precursor solution, single step deposition of thick coatings and the ability to modify the coating properties by simple changes to the chemistry of the coating solution and/or the heating method [12]. Four WO_3 coatings were prepared with different microstructures and their lithium insertion/extraction properties described. The proton intercalation properties have been described elsewhere[13].

EXPERIMENTAL METHODS

Solution Synthesis

Tungsten oxide precursor solutions were prepared by dissolving peroxotungstic ester in reagent alcohol (Fisher, histological grade). In the case of chemical modifications to the solutions with oxalic acid dihydrate as a drying control additive (DCCA), the acid was added in varying molar ratios to the amount of PTE in the solution.

The solution was deposited onto fluorine-doped tin oxide (TEC glass Libby Owens Ford) or ITO (Donnelly Applied Films) by either spin or dip coating. Film thickness was determined using profilometry (Tencor Instruments Alpha-Step 200) by etching a step in the coating with a 1M NaOH aqueous solution.

Ultraviolet-Visible-Near Infrared (UV-Vis-NIR) Spectroscopy

Transmission and reflectance spectra were measured between 200nm and 2,000nm using a Shimadzu UV 3100 spectrophotometer. For measurements of transmittance versus time the wavelength was set at 550 nm and a slit width of 20 nm.

Electrochemical Measurements

Cyclic voltammetry (current versus potential) was conducted using a Solartron SI 1286 Electrochemical Interface. Measurements were conducted in a three electrode cell with a platinum foil as the counter electrode, a non-aqueous $Ag/AgNO_3$ (0.01 M $AgNO_3$ in acetonitrile) reference electrode and 0.1 M $LiClO_4$ in propylene carbonate (PC) as the electrolyte. Cyclic voltammograms were run at a scan speed of 10 mV/s between potential limits determined for the experiment. Total immersed film area in all cases was constant at 7.5 cm^2. Integration of charge under a cyclic voltammogram curve was executed by the instrument software.

RESULTS AND DISCUSSION

Formation of WO_3 thin films

WO_3 films cast from a solution of 0.37 g PTE per ml of ethanol were deposited by dip coating onto ITO (15Ω/sq.). The deposited films were fired at a heating rate of 5 degrees/min to either 250°C or 350°C to obtain amorphous or crystalline films. The films were 300nm thick and after firing were extensively cracked rendering them unsuitable for most optical applications. Using a different peroxo-precursor, namely peroxotungstic acid, Yamanaka, et al[14] reported cracking at around 300nm for their system. The most likely cause of cracking are stresses generated during the decomposition of the peroxo-precursor at around 100°C and solvent removal. To reduce these stresses two approaches were taken. The first was the addition of oxalic acid dihydrate (OAD) to the precursor solution as a drying control chemical additive (DCCA) and the second reducing the rate of solvent removal through control of the firing atmosphere. Oxalic acid was chosen as the DCCA because of its compatibility with the precursor solution, its low carbon content and clean burn out properties. It was dissolved in the precursor solution up to 1:1 OAD to PTE molar ratio. This combined approach resulted in a dramatic improvement in the coating quality and final crack-free thickness as seen in Figure 3. Coatings of one micron from a single dip were obtained which were crack free after 250°C or 350°C firing.

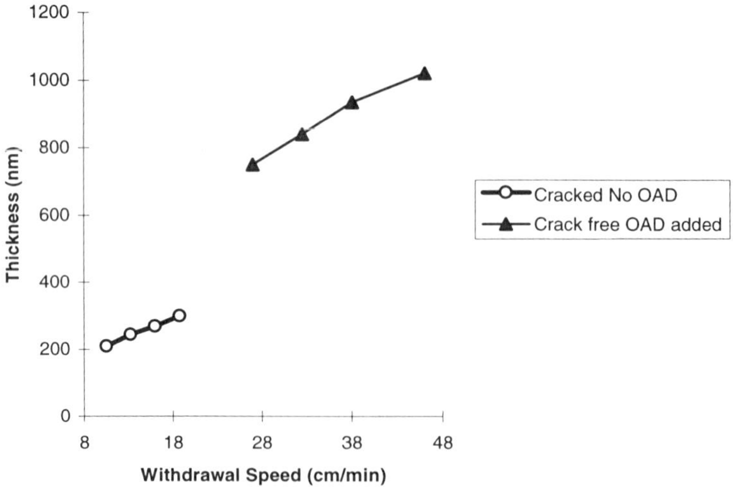

Figure 3
Plot of dip coating withdrawal rate versus thickness of fired WO_3 coatings

Effect of OAD Addition on the Microstructure of Thick WO_3 Films

The microstructure of the films was examined using X-ray diffraction (XRD), Rutherford back scattering (RBS) and transmission electron microscopy (TEM) the details of which have been reported elsewhere[13]. Wide angle X-ray diffraction patterns for the crystalline films (350°C) are shown in Figure 4. The 250°C films were amorphous exhibiting no diffraction peaks, except those from the crystalline ITO substrate and are not shown in the figure. Included in Figure 4 for reference are the diffraction patterns for ITO, triclinic tungsten oxide and WO_3 powder (randomly orientated) purchased from Cerac Inc., Milwaukee, Wisconsin. The triclinic tungsten oxide was the best match that could be found in the Joint Committee on Powder Diffraction Standards (JCPDS) library of spectra. The JCPDS spectra is shown by vertical lines with the height of the lines proportional to the intensity of the peaks.

The diffraction pattern (fig.4a) for the film cast from solution without OAD and heated to 350°C shows three dominant diffraction lines, at d = 3.84Å (hkl=001), 3.76 (020) and 3.65 (200). This pattern resembles that of the triclinic

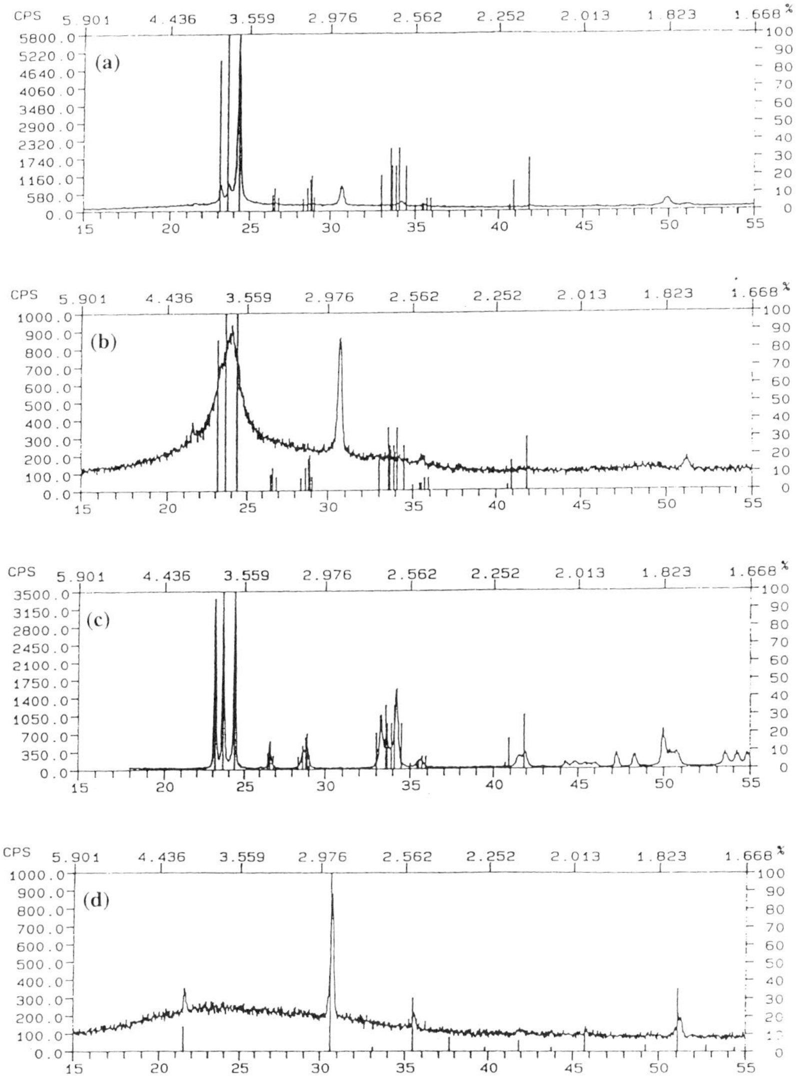

Figure 4

X-ray diffraction patterns of: (a) WO_3 350^{0}C on ITO coated glass; (b) WO_3 350^{0}C with OAD/PTE molar ratio 1:1 on ITO coated glass; (c) Cerac WO_3 powder; and (d) ITO coated glass. Vertical lines in (a), (b) and (c) represent the pattern from triclinic WO_3 (JCPDS 32-1395); and in (d) indium oxide (JCPDS 6-416).

form of WO_3 with preferred orientation at hkl = 200. The diffraction pattern (fig 4b) of the 350°C film from the OAD solution shows pronounced broadening of the diffraction pattern between 22 and 25°. This broadening was enough to engulf the lines at d = 3.76 and 3.84, causing the film to appear nearly X-ray amorphous. From these 350°C diffraction patterns, the average grain size was calculated with the Scherrer formula[15]. Using the d=3.65 line the grain size was calculated to be 19.7nm and 6.5nm for the coatings derived without and with oxalic acid, respectively. The degree to which the grain size could be modified was examined by preparing coatings from precursor solutions with different concentrations of OAD. The results are shown in Figure 5, where grain size in the fired coating is plotted as a function of the molar ratio of OAD to PTE in the precursor solution. At a molar ratio of 1:1 there is a dramatic drop in grain size.

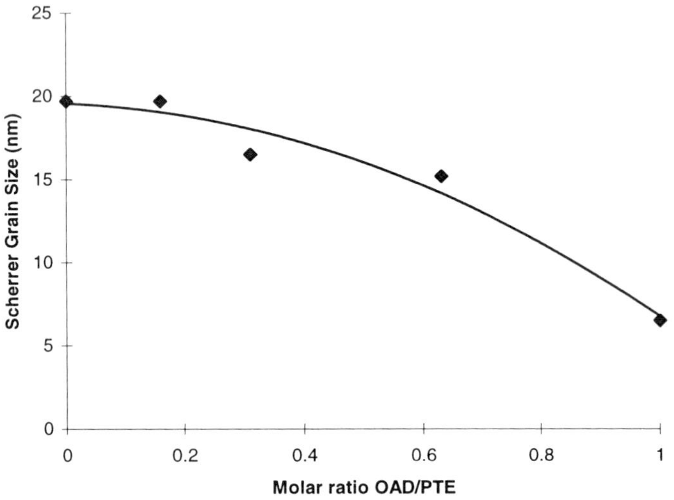

Figure 5
Calculated Scherrer grain size (X-ray diffraction) of 350°C WO_3 films as a function of the molar ratio of OAD to PTE in the precursor solution

From TEM analysis the average grain size was calculated to be 20 nm for the no-OAD film and around 7 nm for the PTE/ OAD (1:1) film, correlating well with the diffraction data. However, most interesting the TEM micrographs showed the presence of numerous small (~5 nm) crystallites in the 250°C OAD sample[16]. These crystallites were embedded in an amorphous matrix. The relatively small volume percent of these grains can explain the lack of diffraction activity. The grains did not appear to have formed epitaxially from the ITO surface; rather, they were scattered throughout the bulk of the film. In addition, micrographs of room-temperature dried OAD solution did not reveal any crystallites, suggesting that the grains were not formed in solution. A possible conclusion of this is that the OAD, upon subliming or burning out of the film, catalyzes the formation of the crystalline phase at lower temperatures.

The density of the films was determined by RBS analysis and was found to decrease with increasing concentration of OAD in the casting solution. This is shown in Figure 6 for the coatings heated to 250°C with the lowest density being obtained for a 1:1 OAD/PTE molar ratio.

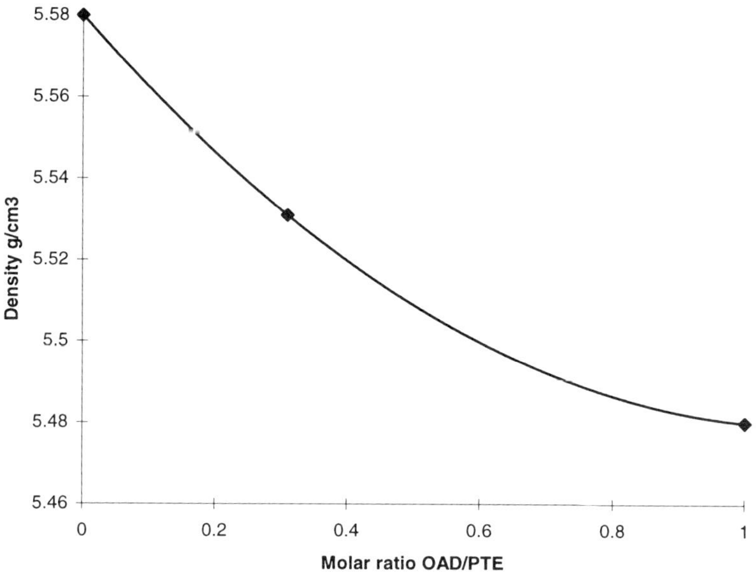

Figure 6
Density of WO_3 films heated to 250°C as a function of OAD in the precursor solution

Electrochemical Response

The optical and electrochromic properties of WO_3 thin films have been shown to vary with the preparation method[17,18]. The differences were related directly to the microstructures and morphologies of the films. Among the results reported were ranging electrochromic activity (high or low charge capacities), ranging reaction kinetics and altering of the transmission spectra of both the colored and bleached films. Specific in the literature are reports that crystalline WO_3 does not reversibly intercalate alkali metal ions[17].

The lithium intercalation properties of the four different WO_3 films listed in Table II were examined in a three electrode configuration. The coatings were labeled according to their firing temperature and whether OAD was present in the precursor solution.

Table II. Summary of precursor solution composition, firing temperature, microstructure, density and thickness for the different WO_3 films.

Sample	Firing Temp. °C	Molar Ratio OAD/PTE	Microstructure	Density g/cm^3	Thickness nm
WO_3 (A-250)	250	0.0	Amorphous	5.58	300
WO_3 (OAD-250)	250	1:1	Amorphous/Nano-crystalline ≅ 5nm	5.48	300
WO_3 (C-350)	350	0.0	Crystalline ≅ 20nm	6.02	280
WO_3 (OAD-350)	350	1:1	Crystalline ≅ 7nm	5.57	268

The films were subjected to 10 mV/s cyclic voltammetry between -0.6V for the anodic reaction and 1.2V cathodic (versus $Ag/AgNO_3$ reference). The CV plots for all the 250 and 350°C films are shown in Figures 7 and 8. The addition of OAD to the precursor solution greatly enhanced the electrochemical activity of the WO_3 film at 250°C and to a lesser extent at 350°C. The absolute value of the current density is an indication of the electrochemical activity of the electrode. These values are listed in Table III along with the density of each coating.

Figure 7

Linear sweep voltammogram of WO3 films fired to 250C. Sweep rate 10mV/sec.

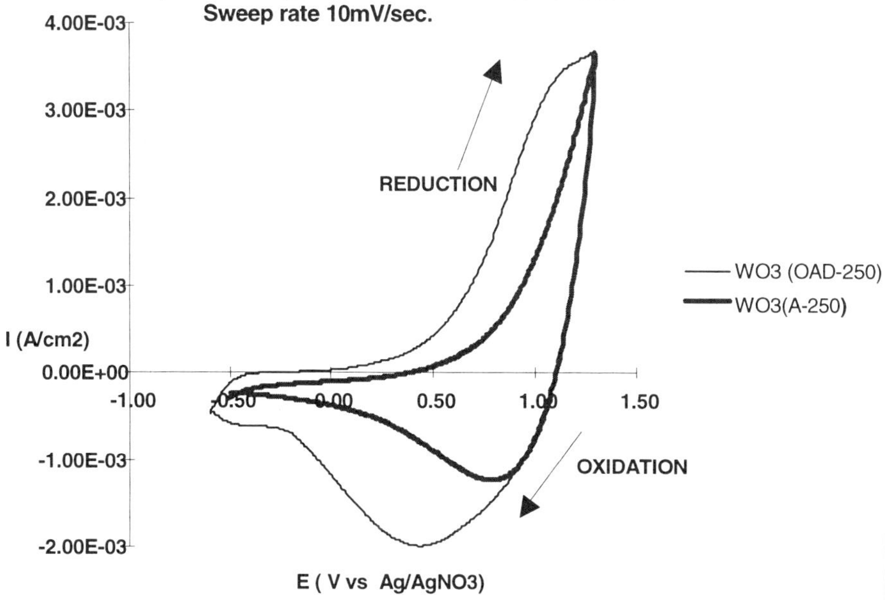

Figure 8

Linear sweep voltammogram of WO3 films fired to 350C. Sweep rate 10mV/sec.

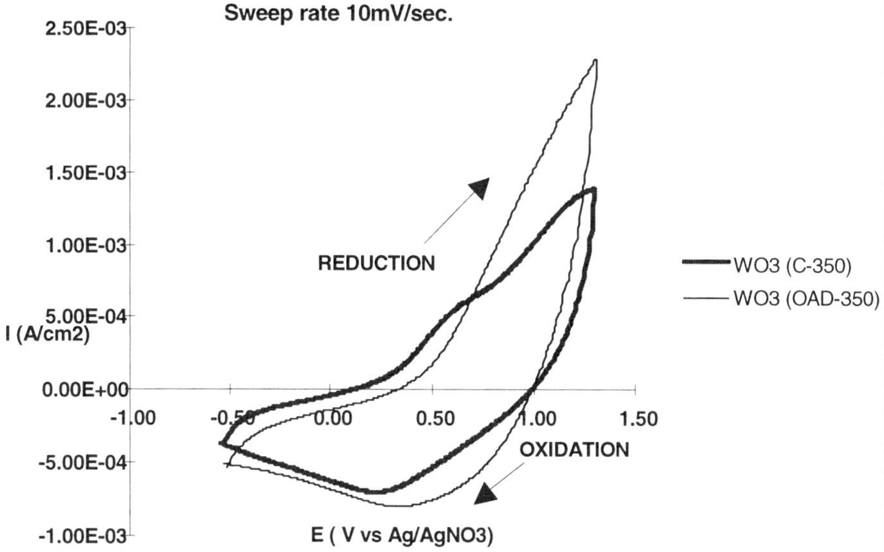

Table III. Maximum current generated during the reduction process at a scan rate of 10mV/sec between the potential limits of -0.6 to 1.3V vs. $Ag/AgNO_3$.

Sample	Thickness (nm)	Max Current Density Reduction i_p (A/cm^2)	Density g/cm^3
WO_3 (A-250)	300	4.85×10^{-4}	5.58
WO_3 (OAD-250)	300	4.91×10^{-4}	5.48
WO_3 (OAD-350)	268	2.87×10^{-4}	5.57
WO_3 (C-350)	280	1.75×10^{-4}	6.02

The rate of diffusion of Li^+ ions is sensitive to the microstructure of the coating. Nagai et al.[19] reported for a WO_3 film with a density of 5.84 g/cm^3 the diffusion coefficient of lithium at room temperature was 3×10^{-12} cm^2/s. A decrease in density to 3.38 g/cm^3 improved the diffusion coefficient by three orders of magnitude to 1.5×10^{-9} cm^2/s. Whereas the density variations in Table III are not as extreme as the above examples they still should result in faster lithium diffusion for the WO_3 (OAD-250) coating. This behavior is reflected in the maximum peak current values in Table III which decrease significantly upon densification and crystallization.

Electrochromic properties

The WO_3 films were colored by applying a step potential of -1.2 volts in 0.1M $LiClO_4$ with respect to the reference electrode. The voltage was applied until the current decreased to zero (steady state). Coloring efficiencies at 550nm of the coatings were calculated from the difference between the optical densities in both states divided by the charge consumed. The values are shown in Table IV.

Table IV. Bleached and colored optical density, cathodic charge and optical coloring efficiency for WO_3 coatings

Sample	Thick. (nm)	Bleached Optical Density at 550nm 0.6 V versus $Ag/AgNO_3$	Colored Optical Density at 550nm -1.3 V versus $Ag/AgNO_3$	Cathodic Charge (C/cm^2)	Optical coloring efficiency (cm^2/C)
WO_3 (A-250)	300	0.11	1.24	0.0448	25
WO_3 (OAD-250)	300	0.11	1.25	0.0431	26
WO_3 (OAD-350)	268	0.10	1.10	0.0511	20
WO_3 (C-350)	280	0.14	0.69	0.0530	10

It is reported [21] that the efficiency of amorphous WO_3 is greater than that of the crystalline material. This is evident from Table IV and Figure 9 where the optical efficiency of the 250°C films are similar and the efficiency decreases on heating to 350°C.

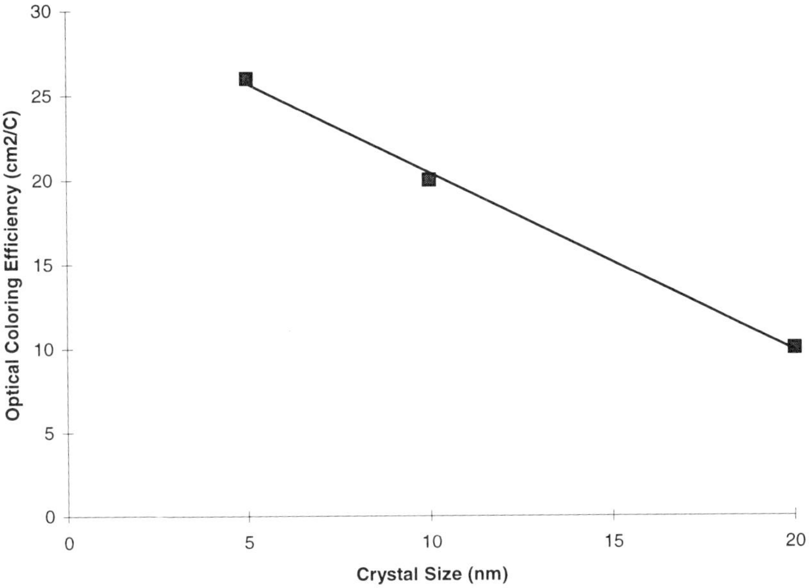

Figure 9
Optical coloring efficiency versus crystallite size for WO_3 coatings

The kinetics of coloring and bleaching were examined for three of the WO_3 films at 550nm under a color potential of -1.3 volts and a bleach potential of 0.6 volts with respect to $Ag/AgNO_3$. The data is listed in Table V with sample WO_3(OAD-250) having the fastest kinetics for coloring and bleaching.

Table V. The films on ITO were colored in 0.01M $LiClO_4$ propylene carbonate using $Ag/AgNO_3$ as the reference electrode.

Sample	Bleached State %T (550nm)	Colored State %T (550nm)	Color (Seconds)		Bleach (Seconds)	
			T_{50}	T_{90}	T_{50}	T_{90}
WO_3 (A-250)	77.1	5.8	47	154	97	316
WO_3 (OAD-250)	77.6	5.6	36	124	54	99
WO_3 (OAD-350)	78.7	8.0	59	171	100	294

T_{50} = time for the coating transmission to change 50% of the total difference between the fully bleached and fully colored states

T_{90} = time for the coating transmission to change 90% of the total difference between the fully bleached and fully colored states

The addition of OAD to the precursor solution sufficiently modifies the microstructure to enhance the bleach rate and appears to have little if any effect on the color rate. The advantage to this processing method is that it allows one to produce crystalline WO_3 films which are chemically more stable than amorphous material [21] without the penalty of slower kinetics.

Consequences of the microstructure are also evident in the optical properties of the films as well. While the transmission spectra of the films in both the colored and bleached states exhibit similar features as shown in Figure 10, the details of the mode of optical modulation are different. Amorphous tungsten oxide films are known to modulate their transmission primarily by absorption in the film. The crystalline materials, however, also exhibit a pronounced reflection in the near IR region as a contribution to the decreased transmission in this region[22]. The effects of the film crystallinity can be seen in the reflection spectra for the films shown in Figure 11 for the bleached and colored films respectively. The pronounced reflectivity in the near IR region is clearly present for the 350°C films which is not seen for the 250°C films. This result is consistent with the crystallinity data obtained via XRD and TEM.

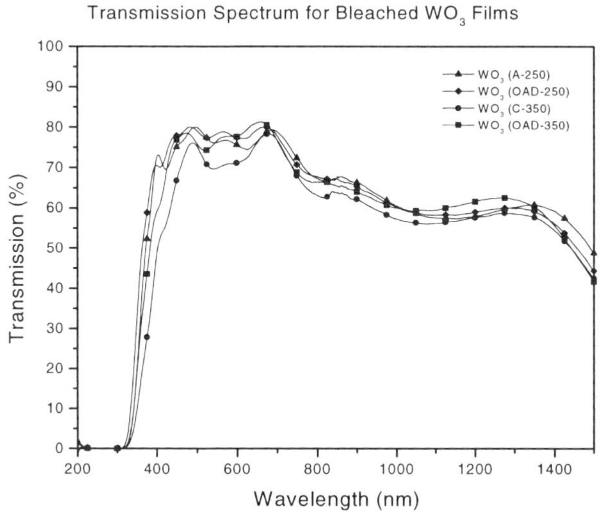

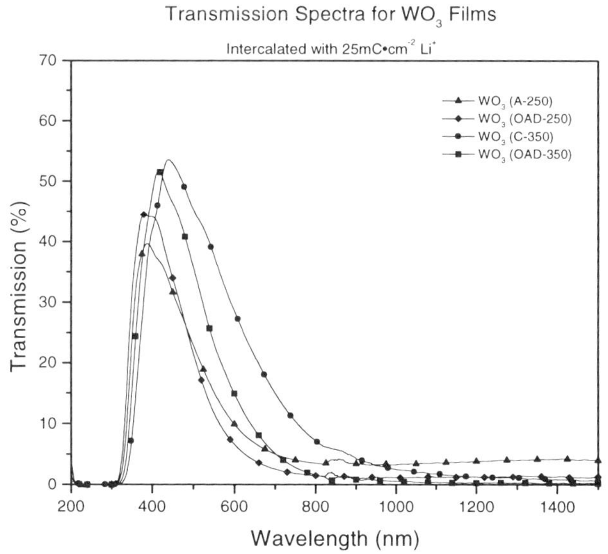

Figure 10
Transmission spectra for amorphous and crystalline WO_3 in the bleached and colored state. For the colored spectra the films were intercalated with $25mC/cm^2$ of Li^+

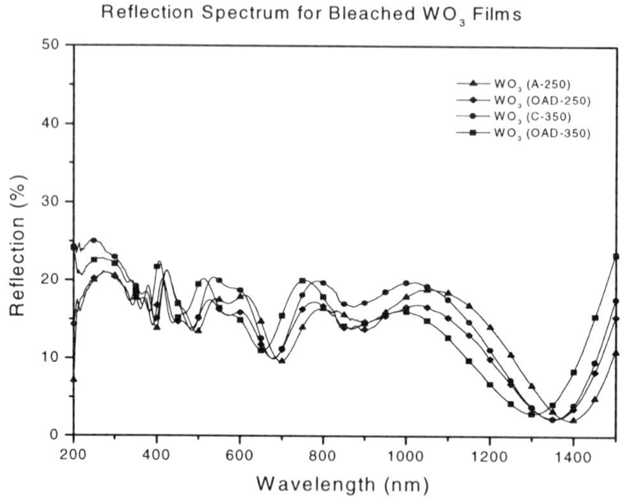

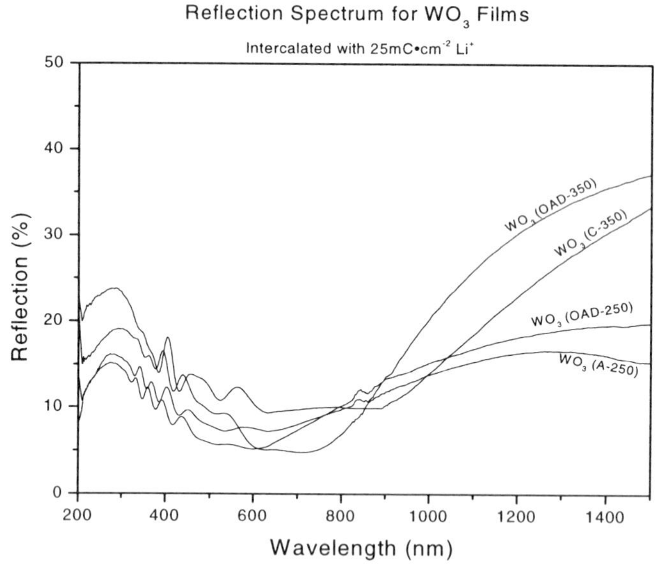

Figure 11

Reflectance spectra for amorphous and crystalline WO_3 in the bleached and colored state. For the colored spectra the films were intercalated with 25mC/cm^2 of Li^+

Large-Area Applications

A WO_3 film for a large-area device was deposited by dip coating the peroxotungstic ester solution onto a piece of SnO_2-coated glass. The amorphous coating was 35 × 40 cm in size. The uniformity of this film across the substrate was determined to be about +/- 60Å, and the average thickness was 520nm. This coating was then assembled into one large device. This device was examined as to its electrochromic kinetics with the test beam probing the center of the device. The kinetic plot may be seen in Figure 12. The window had a high-end transmission of 73% and a low-end photopic transmission (after 200 seconds of coloration at 1.3 V) of 7%. The device was cycled several-thousand times with little effect on the coloration kinetics and transmission values.

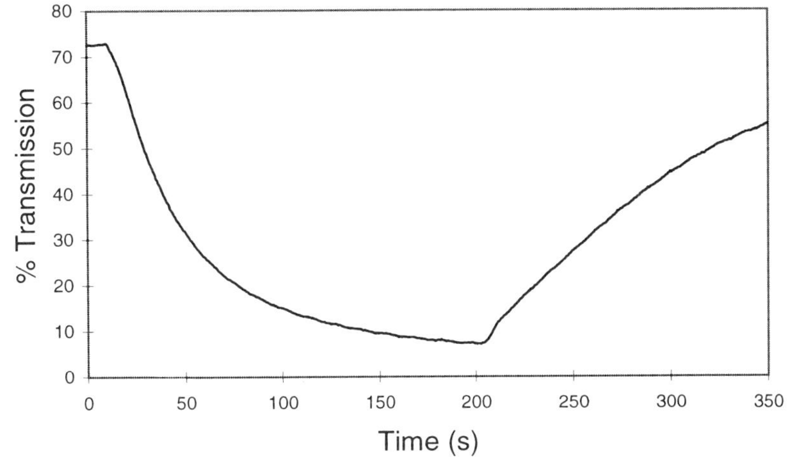

Figure 12
Color Kinetics (Photopic) of Large-Area Window (Colored at 1.3 V and Bleached at -0.3V)

SUMMARY

WO_3 thin films for large area applications were prepared by a peroxotungstic ester wet chemical method. By a combination of additions of oxalic acid dihydrate to the precursor solution and modifications to the heat treatment the microstructure of the WO_3 films could be controlled. A film [WO_3 (OAD-250)] prepared at 250°C which was partially crystalline showed good electrochromic properties under Li^+ insertion/extraction. Its advantages for large area applications include its partial crystalline nature which imparts chemical stability, good optical modulation/efficiency, and fast electrochromic kinetics.

ACKNOWLEDGMENTS

We thank Donnelly Corporation for their support throughout this research. We also thank the Materials Science and Engineering Department at the University of Arizona, Tucson, Arizona, for their assistance and use of various analytical instruments.

REFERENCES

1) J. L. Warner, M. S. Reilly, S. E. Selkowitz, D. K. Arasteh and G. D. Ander, "Utility and Economic Benefits of Electrochromic Smart Windows", LBL-32368. (1992).

2) S. E. Selkowitz, and S. LaSourd, "Amazing Glazing", Progressive Architecture 75, 108. (1994).

3) S. E. Selkowitz and C. M. Lampert, in: Large-Area Chromogenics: Materials and Devices for Transmittance Control, Eds. C. M. Lampert and C. G. Granqvist, (SPIE, Optical Engineering Press, Bellingham, Washington 1990) 22.

4) J. P. Cronin, A. Agrawal, Proc. of the Norbert Kreidl Symposium Present State and Future Prospects of Glass Science and Technology, Glastech. Ber. Glass Sci. Technol. 70C (1997) 237.

5) F. G. K. Baucke, C. Bohnke and G. Robert, Solid State Ionics 6 (1982) 121.

6) D. K. Benson, C. E. Tracy, J. S. E. M. Svensson and B. E. Liebert, Optical Materials Technology for Energy Efficiency and Solar Energy Conversion VI, SPIE, 823 (1987) 72.

7) R. Hurdich, Electron. Lett. 11 (1975) 142.

8) C. G. Granqvist, Handbook of Inorganic Electrochromic Materials, Elsevier, Amsterdam, (1995).

9) P. M. S. Monk, R. J. Mortimer and D. R. Rosseinsky, Electrochromism: Fundamentals and Applications, VCH, Weinheim, (1995).

10) A. Agrawal, J. P. Cronin and R. Zhang, Sol. Energy Mater. Sol. Cells 31 (1993) 9.

11) J. P. Cronin, D. J. Tarico, A. Agrawal and R.L. Zhang, US Patent # 5,457,218, Oct. 10, (1995)

12) J. P. Cronin, A. Agrawal, D.J. Tarico, and Juan Carlos L. Tonazzi, US Patent # 5,525,264, Jun.. 11, (1996)

13) J. P. Cronin, D. J. Tarico, J. C. L. Tonazzi, A. Agrawal and S. R. Kennedy, Sol. Energy Mater. Sol. Cells 29 (1993) 371.

14) K. Yamanaka, H. Oakamoto, H. Kidou and T. Kudo, Japanese Journal of Applied Physics, 25, 9 (1986) 1420.

15) P. Scherrer, Gott. Nachr., 2, (1918) 98.

16) M. Denesuk, J. P. Cronin, S .R. Kennedy, K.L. Law, G.F. Neilson and D. R. Uhlmann, Optical Materials Technology for Energy Efficiency and Solar Energy Conversion XIII, Volker Wittwer, Claes G. Grangvist, Carl M. Lampert, Editors, Proc. SPIE 2255, (1994) 52.

17) S. Badilescu, P. V. Ashrit, N. Minh-Ha, G. Bader, F. E. Girouard and V. V. Truong, Thin Solid Films, 250 (1994) 47.

18) P. Judeinstein and J. Livage, Sol-Gel Optics I, Proc. SPIE 1328 (1990) 345.

19) J. Nagai, T. Kamimori and M. Mizuhashi, Proc. SPIE 502 (1984) 59.

20) J. S. E. M. Svensson and C. G. Granqvist, Thin Solid Films 126, (1985) 31.

21) T. C. Arnoldussen, J. Electrochem. Soc. 128 (1981) 117.

22) S. F. Cogan, T. D. Plante, M. A. Parker and R. D. Rauh, Sol. Energy Mater. Sol.Cells.14, (1986) 185.

ON-LINE ELECTROCHEMICAL SENSORS IN MOLTEN METAL PROCESSING TECHNOLOGY: A REVIEW

Girish. M. Kale[*] [$] and Rajnish Kurchania

School of Process, Environmental and Materials Engineering, Department of Mining and Mineral Engineering, University of Leeds, Leeds LS2 9JT, UK

ABSTRACT

In order to facilitate the intelligent processing of metals, there is an urgent need to continuously, accurately and instantaneously analyse dissolved non-metallic or metallic species in molten metal at elevated temperature using on-line electrochemical sensors. Much of the development was devoted to oxygen sensors for the metallurgical and automotive industries. However, the analysis of many more elements is required. In this article, the application of solid state sensors for the analysis of sodium, lithium, copper, magnesium, aluminium, phosphorous, sulphur, silicon, calcium, strontium, chromium, antimony and arsenic using solid electrolytes will be reviewed. Solid electrolytes, such as stabilised-zirconia and beta-alumina are best suited materials for the development of electrochemical sensors, mainly due to their excellent ionic conductivity and refractory nature. These electrolyte materials not only offer the possibility of an instantaneous detection, but may also meet the industrial requirements of reliability, ruggedness, and compatibility with microelectronics.

INTRODUCTION

Most of the ferrous and non-ferrous metals throughout the world are produced by pyrometallurgical processing of complex ores containing significant proportion of gangue minerals. Most of the ferrous and non-ferrous metal ores are either complex oxides or sulphides. The upgraded oxide ores are directly reduced to metal at elevated temperatures, whereas upgraded sulphide ores are roasted to oxides and subsequently reduced to metal. It is generally difficult to selectively reduce the desired metal oxide without reducing the gangue mineral oxides for the thermodynamic and kinetic restrictions. Most of the refractory oxides (Al_2O_3, Cr_2O_3, CaO, MgO etc.) almost entirely form part of slag. However, the non-refractory (P_2O_5, As_2O_3, Sb_2O_3, Bi_2O_3, SnO_2, FeO etc.) mineral oxide partially get

[*] Member of the American Ceramic Society

[$] Author for Correspondence

reduced and result in the formation of metal solution. Therefore, almost all the metals at the end of the pyrometallurgical processing still contains significant amount of dissolved impurity elements either non-metallic or metallic in nature. These impurities are generally removed by well known metal refining techniques, during the metal refining process. Metal refining process uses expensive refining agents and therefore the extent to which the refining needs to be carried out must be accurately controlled as this can influence significantly the economics of the metal production and its market potential. Currently, there is an increasing emphasis on recycling of metals and metallurgical industrial waste, in order to minimise land, environment and ground water pollution. One of the ways of minimising pollution is to smelt metallurgical waste to further recover metals. This will certainly give rise to a more complex molten metal solution compared to that produced during the primary metal production process. Refining such complex metal solution efficiently would be extremely challenging to metallurgical industries. Therefore, it is necessary to develop new more efficient technology for continuous monitoring and controlling the refining process.

Currently the technique used in metallurgical industries to monitor metal refining process is by discretely sampling the metal, casting the sample in the form of disk followed by either destructive or non-destructive chemical analysis. A short list of disadvantages of this procedure is as follows:
(i) slow (ii) discontinuous (iii) ex-situ (iv) unreliable (v) measures total impurity content instead of the dissolved impurity content, and (vi) needs expensive infra-structure.

A method of on-line analysis of molten metal can provide an inexpensive solution to the above mentioned deficiencies of the currently adopted industrial practice. Solid state electrochemical sensors have a tremendous potential of selectively and continuously sensing or detecting the dissolved non-metallic and metallic impurities in molten metals at their respective metal refining temperatures, ranging from as low as 400°C to as high as 1600°C depending on metal being refined.

The main purpose of this article is to review the current understanding of the applications of the solid state electrochemistry to design and develop the electrochemical sensors for on-line monitoring of dissolved impurities in molten metals at elevated temperatures during the metal refining process.

THEORY

There are three important components in any type of solid state sensor whether it be potentiometric or amperometric. The heart of the system is the solid electrolyte, which should be extremely carefully selected in order to sense the desired species in molten metals. The choice of electrolytes is wider for low temperature applications whereas it is extremely limited for high temperature applications such as sensing in molten metals. This is because of the corrosive nature of molten metals and rapid kinetics of undesired chemical reactions between the solid electrolyte material and either reference or working electrode (molten metal).

The operating principle of potentiometric sensors is relatively simpler compared to that of an amperometric sensor. In a potentiometric sensor, an open circuit voltage is a measure of the chemical potential gradient of conducting species across the solid electrolyte. The potentiometric sensors are extremely sensitive in the concentration region of the solute below 5000 ppm. In an amperometric sensor, a standard potential is applied across the solid electrolyte, which results in to an ionic current through the solid electrolyte. This ionic current, relative to pure metal as a standard state, is proportional to the relative concentration of conducting species at the working electrode. These type of sensors are more sensitive in the higher concentration region typically between 10-80 mol % of solute.

In most of the metal refining operations, the concentration of impurities is below 5000 ppm, therefore in the subsequent discussion only potentiometric sensors will be included.

Sensors Without Auxiliary Electrode

A general schematic representation of a potentiometric sensor, in which the cationic species to be sensed in molten metal is same as the conducting species in a solid electrolyte, is shown below:

$$M, \mu^{I,}{}_{i}^{n+} \; // \; \text{Solid Electrolyte } (i^{n+}) \; // \; \mu^{II,}{}_{i}^{n+}, M \qquad (I)$$

(Measuring Electrode) (Reference Electrode)

Similarly, a general representation of a potentiometric sensor capable of sensing anionic species in molten metal incorporating an anionic conducting solid electrolyte is as follows:

$$M, \mu^{I}{}_{j}{}^{m-} \text{ // Solid Electrolyte } (j^{m-}) \text{ // } \mu^{II}{}_{j}{}^{m-}, M \qquad \text{(II)}$$

(Measuring Electrode) (Reference Electrode)

In the electrochemical cell (I) and (II), μ_i^{n+} and μ_j^{m-} are the chemical potential of the species i and j respectively. The superscripts I and II denote the chemical potential of conducting species at the measuring and reference electrode, respectively. The solid electrolyte in the electrochemical cell (I) has a transport number greater than 0.99 for i^{n+} cations whereas that in cell (II) has a transport number of j^{m-} anions greater than 0.99. The electrical leads in the above cells are denoted by M, which are identical on other side of the electrolyte in order to avoid any thermoemf contributions to the measured open circuit voltage. The schematic diagram of sensor assembly is shown in Figure1.

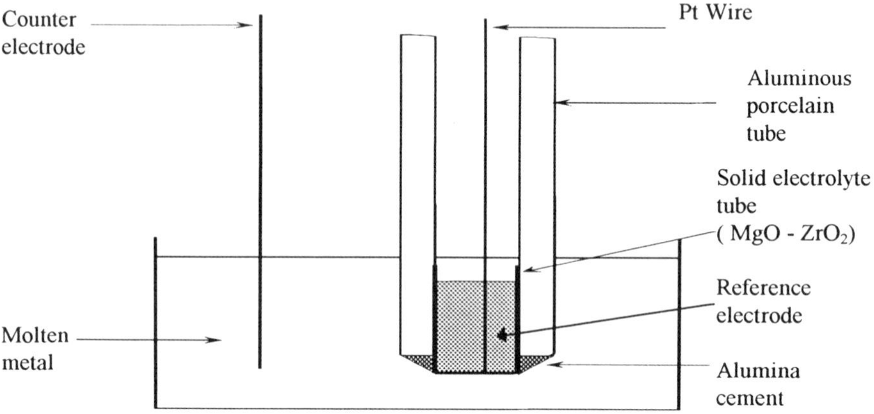

Figure 1 Schematic diagram of oxygen sensor without auxiliary electrode.

The open circuit voltage of the electrochemical sensors described above is related to the relative chemical potential of the conducting species at the reference and working electrode by the Nernst equation:

$$E = -RT/ZF \int_{\mu^{I}_{k}}^{\mu^{II}_{k}} t_{(ion)} \, d\ln\mu_k \qquad (1)$$

where E is the measured emf in volts, R is the universal gas constant in J/mole.K , T is the temperature in K, Z is the number of electrons participating in the electrode reaction, F is the Faraday constant in Coulomb/mole, $t_{(ion)}$ is the transport number of conducting species, and μ_k is the chemical potential defined as $RT\ln a_k$, where a_k is the thermodynamic activity of species "k".

Sensors With Auxiliary Electrode

There are no solid electrolyte materials known to be stable at elevated temperatures, which conducts ions such as phosphorous, sulphur, silicon, magnesium, antimony, arsenic etc. Therefore one needs to develop an alternate methodology compared to that described earlier. Application of an appropriate auxiliary phase in conjunction with a suitable solid electrolyte has been successfully demonstrated by several investigators to sense gaseous species such as carbon-dioxide and sulphur-dioxide[1-6]. There are no solid electrolytes which conduct these molecules. An analogues principle can be applied for designing solid state electrochemical sensor for sensing those metallic and non-metallic dissolved species in molten metals which are not mobile in any known materials. The schematic diagram of sensor assembly is shown in Figure 2.

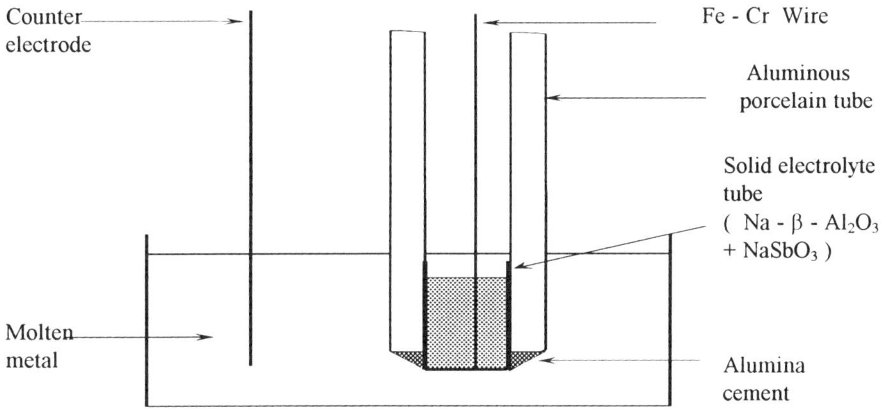

Figure 2 Schematic diagram of antimony sensor with $NaSbO_3$ as auxiliary electrode

The combination of solid electrolyte and the auxiliary electrode is extremely crucial for designing such sensors. As a rule, the solid electrolyte and auxiliary phase are chosen such that the conducting ion is common to both the phases and the auxiliary phase contains the ion to be sensed. As an example, an antimony sensor[7] could be schematically represented as:

$$M, Na_{ref} // Na\text{-}\beta\text{-}Al_2O_3 + NaSbO_3 // \underline{Sb}_{melt}, M \qquad (III)$$

The auxiliary phase in cell (III) is hetrogeneously doped in sodium-β alumina, which is an excellent sodium ion conductor. The auxiliary phase can also be chosen such that it forms a homogeneous solid solution with the solid electrolyte material. For example, in the case of arsenic sensor, solid electrolyte used is a novel potassium ion conductor with arsenic dissolved in it to form a homogeneous phase. The arsenic sensor[8] can be schematically represented as:

$$M, K_{ref} // K_3Al_2As_3O_{12} // \underline{As}_{melt}, M \qquad (IV)$$

At the left hand electrode of cell (III) and (IV), a constant reference chemical potential of Na and K respectively is established at a fixed temperature. Similarly, chemical potential of Na and K at the right hand electrode of cell (III) and (IV) is established as a result of :

$$NaSbO_3(s) \rightarrow \underline{Na}_{melt} + \underline{Sb}_{melt} + 3\underline{O}_{melt} \qquad (2)$$

and

$$2K_3Al_2As_3O_{12}(s) \rightarrow 2Al_2O_3(s) + 6\underline{As}_{melt} + 9O_2(g) + 6\underline{K}_{melt} \qquad (3)$$

respectively.

From equations (2) and (3), it can be seen that for an oxygen saturated melt, the chemical potential of Na or K at the right hand electrode of cell (III) and (IV) will change with the change in the concentration (or activity) of antimony or arsenic in molten metal respectively. Therefore, although the electrochemical sensor represented by cell (III) and (IV) responds to change in the chemical potential gradient of Na and K across the electrolyte, the emf of such sensor can be easily correlated to the change in concentration of antimony and arsenic respectively at the measuring electrode. Based on the similar concepts, an oxygen sensor[9] has also been designed having the following configuration:

$M, Na_{ref} // Na\text{-}\beta\text{-}Al_2O_3 // \underline{O}_{melt}, M$ (V)

On writing the appropriate electrode reactions for cell (III) and (IV), the open circuit voltage can be shown to vary linearly as a function of logarithm of concentration of the sensing species akin to that for cell (I) and (II).

SPECIFIC APPLICATIONS

On the basis of the theory outlined above, there have been a number of attempts to develop sensors for measuring dissolved impurities in liquid metals at elevated temperatures for on-line monitoring of the metal refining process. In the following sections, development of sensors for oxygen, lithium, sodium, copper, aluminium, and phosphorous, sulphur, silicon, magnesium, chromium, calcium, strontium, antimony and arsenic will be briefly reviewed.

Sensors Without Auxiliary Electrode

The solid state electrochemical sensors of this type can be schematically represented as cell (I) and (II).

Oxygen sensors:

The most successful and well understood sensor of this type is oxygen sensor based on well known oxygen ion conductors such as calcia, magnesia or yttria stabilised zirconia. It is very important to measure dissolved oxygen in molten metal which normally enters into the process stream during the oxygen refining of metals. Solubility of oxygen in molten metals usually decreases with decreasing temperature at the same time the thermodynamic stability of metal-oxides increases with decreasing temperatures. This causes blisters formation during the casting of metal or formation of brittle oxide inclusions in the cast metal leading to either poor quality of casting or damaging the mechanical properties. In case of steel, the dissolved oxygen reacts with carbon in steel to form carbon mono-oxide gas thereby giving rise to blow holes in casting and also results in oxidising the alloying elements in steel during the solidification. Therefore, it is extremely important to deoxidise the molten metal before casting it in various shape and sizes.

There have been large number of investigations in last 15-20 years on measuring oxygen in liquid metals using stabilised zirconia as a solid electrolyte [10-24]. Most of the sensors have been developed using chromium and chromium sesqui-oxide ($Cr\text{-}Cr_2O_3$) as a reference electrode mainly because the oxygen potential for chromium-chromium oxide is very close to the oxygen potential in molten steel. Since the difference in the oxygen potential across the solid electrolyte is small, the sensor generates a proportionately small emf, thereby increasing the sensitivity of the sensor. Other reference electrode system commonly explored is molybdenum-molybdenum oxide (Mo-MoO_2). Recently, Ranganathan and Janke[11] proposed the use of a ternary reference electrode for designing oxygen sensors. However, the possibility of using the ternary reference electrode system successfully for designing the oxygen sensors appears to be very remote, because of the practical difficulty in maintaining stable three phase equilibrium at the reference electrode at high temperature and low oxygen potential. In all the oxygen sensors using different reference electrodes, it is required to make sure that oxygen potentials at two electrodes are well within the ionic conduction domain[25] of stabilised zirconia solid electrolyte at the temperature of measurement. If the oxygen potential falls outside the conduction domain, significant electronic conduction is set-up in the electrolyte, which gives rise to semi-permeability flux of molecular oxygen from the electrode having higher oxygen potential to that having lower oxygen potential. This effect reduces the measured emf proportionally depending on the reduction of the transport number of oxygen ions in the solid electrolyte. Hence, the emf must be corrected to take care of the oxygen semi-permeability effect, as outlined by Schmalzrid [26,27].

In order to measure dissolved oxygen in non-ferrous metals, the reference electrode systems described earlier for oxygen sensors in steelmaking, may not be suitable because of the presence of relatively large oxygen potential gradient across the solid electrolyte. Reference electrode systems, having higher oxygen potential compared to nickel-nickel oxide (Ni-NiO), would be more suitable. However, this needs to be more thoroughly investigated. Further, dip-type oxygen probes used in steel industry are not warmly welcomed by the non-ferrous metallurgical industry because they can not be used for continuously monitoring the refining process over a long period of time typically in excess of two hours. Therefore, it is necessary to focus on two main aspects: (i) the reference electrode system for oxygen sensors for non-ferrous metallurgical industries, and (ii) development of long life sensors typically lasting between two to eight hours.

Sodium Sensors:

The solid electrolyte material which is more well studied for its physical, chemical and electrical properties next to stabilised zirconia is the well known sodium beta alumina (Na-β-Al_2O_3). This materials exhibits very high sodium ion conduction above 573 K and also has a significant homogeneity range (88.9 to 90.5 mol% Al_2O_3)[28]. Monitoring sodium is extremely important in non-ferrous metals such as zinc, lead, copper and aluminium. Sodium is used to remove elements such as arsenic and antimony from zinc and lead; oxygen, antimony and arsenic from copper; and phosphorous from iron. Sodium is also used as an alloying element to improve the properties of lead and to aluminium-silicon alloys as grain modifier to improve its mechanical properties. However, sodium is detrimental to aluminium based aero-space alloys, as it has a tendency to precipitate hard intermetallics of alloying elements along the grain boundaries giving rise to poor mechanical properties. As a result of this, there have been some investigations conducted by Fray and co-workers[29-31] and Pelton and co-workers[32-34], to develop solid state sensors for measuring sodium in molten non-ferrous metals. Since the industrial application of sodium sensors is at significantly low-temperature typically between 573 to 1073 K, the most challenging task in designing sodium sensors is to develop a sodium reference electrode system having rapid equilibrium kinetics. It has been found recently that the reference electrode system consisting of $Na_2Fe_2O_4$+Fe_2O_3 (below 1020 K) and $Na_3Fe_5O_9$+Fe_2O_3 (1025-1323 K)[30,35], has been found to equilibrate rapidly enough to be considered as potential reference electrode system in conjunction with sodium ion conducting solid electrolyte.

In case of the use of potassium ion conducting solid electrolyte such as K-β-Al_2O_3 or $K_3Al_2As_3O_{12}$ a biphasic mixture of $K_2O.9Fe_2O_3$ (K-β-Fe_2O_3) and Fe_2O_3 is found to be suitable reference electrode for temperatures ranging between 550 to 1050 K.

Lithium Sensors:

Lithium is the lightest metal in the periodic table having the density of 0.53 gm cm^{-3} and forms alloys with aluminium and magnesium. As a result of this, lithium is used very widely to produce light weight alloys for aero-space industries, automotive industries and also to produce high strength, light weight structural alloys. Lithium is also a very expensive and reactive metal (£55,000 per ton), produced generally by molten salt electrolysis. For example the addition of few percent of lithium to aluminium reduces the net weight by about 15% without any appreciable change in its mechanical properties. To prevent the excessive loss of lithium during alloying and casting, a special care has to be taken in the form of providing an inert atmosphere. In the event of not having an inert atmosphere, excess of lithium is added to form over specification alloys and the excess of lithium is allowed to oxidise during the subsequent process to give rise to alloys with required lithium contents. Both these techniques are expensive and the cost can be cut down significantly by on-line monitoring of lithium contents using electrochemical sensors. A number of lithium ion conductors have been investigated because of their applications in rechargeable micro-batteries for use in electronic devices. However, most of them are unstable in contact with molten Al-Li-Mg alloys at temperatures above 950 K. Lithium ion exchanged Na-β-Al_2O_3 has been used to design the sensors for monitoring lithium in molten aluminium and its alloys[32-34,36,37]. Ion exchange of lithium and sodium is a reversible process at higher temperatures. This gives rise to the problem of cross sensitivity, between lithium in aluminium alloys and sodium that is inadvertently present in aluminium. Sodium enters into aluminium because aluminium is produced by the molten salt electrolysis of alumina dissolved in cryolite (Na_3AlF_6). In order to overcome the problem of cross-sensitivity, lithium beta spodumene ($Li_{3.6}Si_{0.6}P_{0.4}O_4$) was found to be most suitable as a lithium ion conducting solid electrolyte[38]. Since the oxygen partial pressure in Al-Li-Mg alloys is likely to be extremely low ($\approx 10^{-58}$ atm), the long term stability of lithium beta spodumene and any other oxide lithium ion conductors needs to be thoroughly investigated. Further, the application of lithium sensors is around 950 K, which suggests that there is also a greater need to develop novel reference electrode systems having rapid equilibrium kinetics.

Copper Sensors:

Two main applications of copper sensor in molten system is in copper-tin (Cu-Sn) solder and copper matte ($CuFeS_2$). Attempts have been made to design copper sensors using Cu^+ ion exchange Na-β-Al_2O_3 by Little and Fray[39] and subsequently using copper superionic conductors ($CuZr_2P_3O_{12}$) by Yao and Fray[40]. Both the electrolytes have copper in monovalent state, which easily oxidises to divalent state depending on the oxygen potential in molten metal at high temperature. This gives rise to mixed electronic and ionic conduction in the above mentioned solid electrolyte material and hence affects adversely the performance of the sensor. Therefore, it is not possible to develop a potentiometric sensor for copper, however an amperometric sensor could prove to be better option. To the best of our knowledge, there is only one study reported in the literature on development of amperometric sensors for copper in matte to measure copper concentration between 35 to 75 wt%, at 1473 K[41]. Hence, a more detailed investigation needs to be conducted on developing amperometric sensors for copper.

Magnesium Sensors:

Magnesium is added to aluminium-lithium alloys or pure aluminium to increase its strength. The concentration of magnesium therefore needs to be controlled precisely in order to obtain desired properties of aluminium alloys. There is no known solid electrolyte to date which conducts magnesium ions and therefore it is impossible to develop solid state sensor for monitoring magnesium in molten alloys. In order to circumvent this problem, attempts have been made to develop sensors using fused salt containing magnesium chloride ($MgCl_2$) as the electrolyte[42]. Since the molten salt electrolyte is difficult to handle and is highly corrosive, Zhang and Fray[42] designed the sensor by impregnating molten salt electrolyte into porous magnesium oxide (MgO) thimbles. The response of this type of magnesium sensors was found to be reasonably satisfactory. However, the long term stability of this type of magnesium sensor still needs to be carefully investigated. Further, these alkali and alkaline earth metal chloride salts are highly hygroscopic and therefore limits the shelf life of the sensors which are already fabricated. The availability of a solid magnesium ion conducting solid electrolyte would greatly improve the prospects of developing a reliable solid state magnesium sensor.

Aluminium Sensors:

Aluminium is used as an alloying element in zinc along with other minor elements to produce different grades of galvanising alloys. The aluminium content in galvanising alloys varies from 0 wt% to as high as 55 wt%. Aluminium in galvanising alloy serves as a modifier and controller of variety of properties of galvanising layer. In a continuous hot-dip galvanising process of steel, the aluminium content of galvanising bath keeps changing due to the formation of insoluble intermetallic compounds of Fe and Al. Since the properties of galvanising layer depends strongly on the aluminium contents of galvanising bath, it is extremely important to continuously monitor the aluminium contents of the bath. There is no known solid aluminium ion conductor and hence the possibility of developing a solid state aluminium sensor is remote. Attempts have been made in recent past to develop aluminium sensors, similar to the magnesium sensors, based on molten salt electrolyte[43,44]. However, these sensors also suffer from similar problem as described for the magnesium sensors. Sensors based on solid electrolytes are likely to be more durable and easy to handle by less-skilled plant operators and hence future research should be focused more on developing new solid aluminium ion conductors and applying these materials to design the aluminium sensors.

Sensors With Auxiliary Electrode

Phosphorous and Sulphur Sensors:

There are no known ionic conductors of phosphorous and sulphur hence attempts have been made in past to develop sensor for phosphorous and sulphur in molten metals, in particular steel and copper, using a solid electrolyte coupled with suitable auxiliary electrode. Cationic conductors such as Na-β-Al_2O_3 and Ca-β-Al_2O_3 administered with a thin layer of Na_3PO_4 and $2CaOP_2O_5$ and $3CaOP_2O_5$, respectively as the auxiliary electrode have been used to sense phosphorous in molten copper, aluminium, tin, cast iron and steel[23]. The oxygen potential in a liquid metal containing dissolved phosphorous is extremely low because of highly negative standard free energy of formation for the reaction:

$$2P(s) + 5/2\ O_2(g) \rightarrow P_2O_5(s) \qquad (4)$$

If the activity of phosphorous is further reduced in molten metal due to the formation of liquid solution the equilibrium oxygen partial pressure for the above reaction decreases. Also, phosphorous dissolved in liquid metal is extremely reactive at high temperature because of greater stability of phosphate compounds at low oxygen potentials. Therefore it is more easy to design phosphorous sensors which are of disposable-type, rather than long life sensors. Because of the highly aggressive nature of melts containing dissolved phosphorous, no reliable measurements are available in the literature for phosphorous dissolved in molten metals using electrochemical sensors. In case of sulphur containing melts, the environment is not as aggressive as phosphorous. There have been some measurements[45] of sulphur in molten metals using Ca-β-Al_2O_3[45] as a solid electrolyte and calcium sulphide (CaS) as the auxiliary electrode. However, calcium sulphide (CaS) reacts with moisture in the atmosphere readily according to the reaction:

$$CaS(s) + 2H_2O(l) \rightarrow Ca(OH)_2(s) + H_2S(g) \qquad (5)$$

and hence poses difficulty in its use as an auxiliary electrode. Calcium sulphide (CaS) is a stable component at low oxygen potentials and high temperatures and hence can be formed in-situ on the solid electrolyte surface according to the reaction:

$$\underline{CaO}_{\ \beta\text{-}Al_2O_3} + \underline{S}_{\ metal} \rightarrow CaS(s) + \underline{O}_{\ metal} \qquad (6)$$

when a solid state sensor incorporates Ca-β-Al_2O_3 as a solid electrolyte. Similar reaction takes place when Na-β-Al_2O_3 is used as solid electrolyte forming sodium sulphide as the auxiliary electrode. Results obtained by Swetnam et al[46] for the measurement of sulphur in molten iron using the above approach are shown in Figure 3. The measured emf of the sensor was found to agree well with the available thermodynamic data at 1673 K in molten iron. Based on the standard thermodynamic data available in the literature[47] on CaO and CaS, the following reaction:

$$CaS(s) + 1/2\ O_2(g) \rightarrow CaO(s) + 1/2\ S_2(g) \qquad (7)$$

will be triggered when the oxygen content of molten metal is greater than 100 ppm at 1800 K.

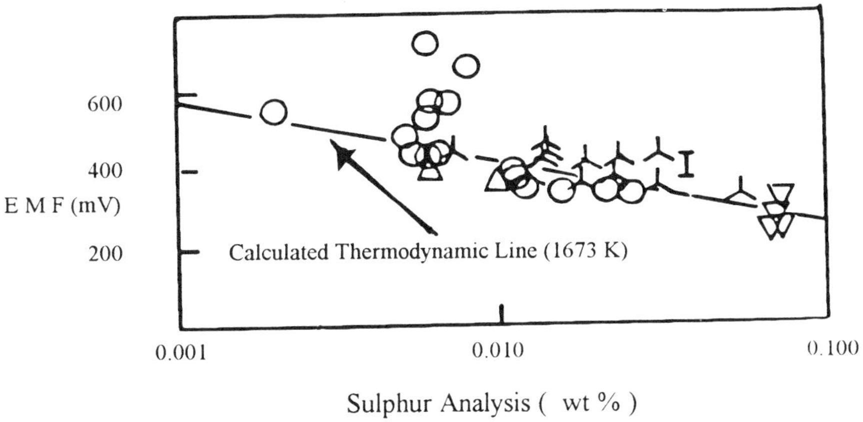

Figure 3 Plot of sulphur sensor emf versus analysed sulphur. Each symbol denotes a separate industrial trial [46].

Hence, it is likely that the sulphur sensor based on above approach will sense oxygen in molten iron or steel instead of sulphur above 100 ppm oxygen at 1800 K. A similar reversal in sensing mechanism is likely to occur while sensing phosphorous in liquid metals. Therefore, there is a strong need to design sensors for phosphorous and sulphur with an auxiliary electrode, which is thermodynamically stable in highly oxidising conditions at high temperature in molten metals containing sulphur and phosphorous.

Silicon Sensors:

Attempts have been made to develop a solid state sensor for silicon in molten metals using Ca-β-Al_2O_3 in conjunction with calcium silicate ($CaSiO_3$) as the auxiliary phase[23], stabilised zirconia in conjunction with zirconium silicate ($ZrSiO_4$) as the auxiliary phase[48-51], magnesia stabilised zirconia (MgO-ZrO_2) in conjunction with magnesium silicate (Mg_2SiO_4) as the auxiliary phase[52], magnesia stabilised zirconia (MgO-ZrO_2) incorporating SiO_2+CaF_2 mixture as the auxiliary electrode[53], and CaO-Al_2O_3-SiO_2-molten slag as a liquid electrolyte[54]. Among the above mentioned variations of the silicon sensor, the results obtained using $ZrSiO_4$ as an auxiliary electrode with stabilised zirconia as a solid electrolyte appears to be less reliable because $ZrSiO_4$ does not co-exist with stabilised zirconia above 1723 K according to the phase diagram of ZrO_2-MgO-SiO_2[55-58]. According to the phase diagram of ZrO_2-MgO-SiO_2, Mg_2SiO_4 appears to co-exist with magnesia stabilised zirconia and hence probably is the more appropriate auxiliary phase for developing silicon sensors using magnesia stabilised zirconia as the solid electrolyte. It should be noted that the silicon sensors developed using stabilised zirconia as the solid electrolyte can not be used for sensing silicon above the oxygen potentials dictated by the following reaction:

$$Si(s) + O_2(g) \rightarrow SiO_2(s) \tag{8}$$

at the measuring temperature. This is because if the oxygen potentials are greater than that dictated by the above reaction at the temperature of measurement then stabilised zirconia, which is primarily an oxygen ion conductor will respond to the oxygen content of the metal in preference to silicon via the auxiliary electrode.

Magnesium Sensors:

As mentioned earlier, there are no Mg^{2+} ion conductors therefore the only method of designing magnesium sensors is to either use molten salt containing Mg^{2+} ions as the electrolyte or to use a known solid electrolyte in conjunction with an appropriate auxiliary phase containing magnesium. The use of molten salt electrolyte is outlined earlier. Since molten salts are ionic in nature, the transport of ionic species is significantly rapid compared to that in the solid electrolytes. However, it is difficult to handle molten salts and therefore attempts have been made to design magnesium sensors for molten aluminium using Na-β-Al_2O_3 as the solid electrolyte by Laros et al[59] and using magnesium fluoride (MgF_2) as a solid electrolyte by Fergus and co-workers[60,61]. Na-β"-Al_2O_3 contains approximately 1 to 2 wt% MgO in order to stabilise the structure at higher temperatures. Because of the presence of magnesium oxide (MgO) dissolved in $Na_2O.5.5Al_2O_3$(Na-β"-Al_2O_3), MgO acts as an auxiliary phase enabling MgO stabilised Na-β"-Al_2O_3 to be used as solid electrolyte for sensing magnesium in molten aluminium. The electrode reaction taking place at the electrolyte and molten metal interface is as follows:

$$\underline{MgO}_{\text{Na-}\beta''\text{-}Al_2O_3} \rightarrow \underline{Mg}_{\text{molten Al}} + \underline{O}_{\text{molten Al}} \tag{9}$$

It should be noted that Na-β"-Al_2O_3 is an excellent sodium ion conductor, even better than Na-β-Al_2O_3. Furthermore, sodium is inherently present in molten aluminium and hence there is a strong possibility that the magnesium sensor based on Na-β"-Al_2O_3 stabilised with magnesium oxide will exhibit cross-sensitivity to change in sodium concentration in aluminium containing magnesium alloys. A slightly more improved version of magnesium sensor is based on magnesium fluoride as the solid electrolyte[60,61]. Magnesium fluoride is a well known fluorine (F^-) ion conductor. A schematic representation of the magnesium sensor based on magnesium fluoride solid electrolyte is as follows:

$$Mg(l) \,//\, MgF_2(s) \,//\, \underline{Mg}_{\text{molten Al}} \qquad \text{(VI)}$$

At the reference electrode, an equilibrium fluorine potential is established according to the reaction:

$$MgF_2(s) \rightarrow Mg(l) + F_2(g) \qquad (10)$$

Exactly same reaction occurs at the interface between molten metal containing magnesium at reduced activity and MgF_2 solid electrolyte.

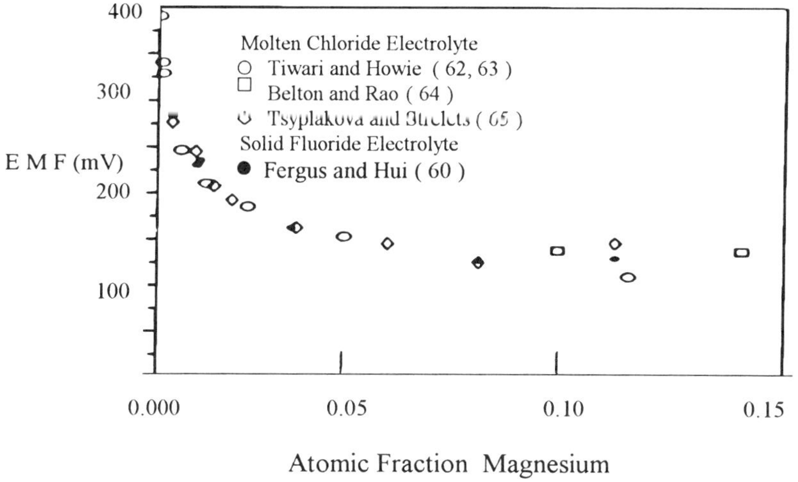

Figure 4 The output of electrochemical magnesium sensors showing the voltage for the cell Mg(l) //electrolyte // $\underline{Mg}_{Al(l)}$ using molten chloride and solid fluoride electrolytes[66].

The difference in the chemical potential of F^- is proportional to the difference in the chemical potential of magnesium across the solid electrolyte which gives rise to the open-circuit voltage of the sensor. A comparison of the results of magnesium sensor using molten salt electrolyte and magnesium fluoride solid electrolyte is reproduced in Figure 4. It can be clearly seen that the emf of the sensor is extremely sensitive to change in concentration of magnesium in the alloy below 200 ppm.

Calcium and Strontium Sensors:

Magnesia stabilised Na-β"-Al_2O_3 has been used as solid electrolyte to design sensors for sensing strontium and calcium for in molten aluminium by Laros et al[59]. No auxiliary electrode of any kind was used in conjunction with MgO-stabilised Na-β"-Al_2O_3 solid electrolyte. The results obtained by Laros et al [59] have shown that these sensor exhibited a Nerstian behaviour having a slope of -2.303(RT/2F) for the variation of the measured emf of the sensor with the logarithm of mole fraction of calcium and strontium in the molten aluminium. However, it is not very clear how a sensor based on MgO-stabilised Na-β"-Al_2O_3, can sense calcium, strontium in aluminium.

Low Oxygen Sensors:

Most of the oxygen sensors are based on either CaO, MgO or Y_2O_3 stabilised zirconia as a solid electrolyte. However, in a melt such as phosphorous deoxidised copper or aluminium deoxidised steel the oxygen content of molten metal is extremely low and often falls outside the ionic conduction domain of stabilised zirconia solid electrolyte[25]. Therefore, there is a need to develop sensors, which are capable of measuring such low oxygen potentials without triggering the electronic conductivity in the solid electrolyte. Na-β-Al_2O_3, which is a sodium ion conductor, has sodium oxide (Na_2O) at a reduced activity and hence can be exploited as a solid electrolyte to design sensors for measuring low oxygen potentials in molten metals. The emf of the following electrochemical cells:

$$\text{Pt, Fe + FeO // Na-}\beta\text{-}Al_2O_3\text{ // Ni + NiO, Pt} \qquad \text{(VII)}$$

$$\text{Pt, Ni + NiO // Na-}\beta\text{-}Al_2O_3\text{ // Cu + }Cu_2O\text{, Pt} \qquad \text{(VIII)}$$

has been successfully measured by Choudhury[67]. Elrefaie and Smeltzer[68] also successfully measured the chemical potential of oxygen for Ni+$Al_2O_3(\alpha)$+ $NiAl_{2.54}O_{4.81}$ three phase equilibrium, relative to Ni+NiO, using Na-β-Al_2O_3 and $CaO(ZrO_2)$ as the solid electrolyte separately. Their results from the two independent experiments agreed well with each other. Later on, Kumar and Fray[14] also measured successfully the standard free energy of formation of aluminium oxide from elements:

$$2Al(l) + 3/2O_2(g) \rightarrow Al_2O_3(s) \quad (11)$$

using Na-β-Al_2O_3 and Ca-β-Al_2O_3 as a solid electrolyte. Their measured oxygen potentials were in excellent agreement with the standard thermodynamic data reported in the literature[47]. The oxygen potentials for the reaction (11) are significantly lower than the limiting values for the ionic conduction domain of stabilised zirconia. Hence, it is possible to design sensors using β-Al_2O_3 solid electrolyte for measuring extremely low oxygen potentials prevailing in molten metals such as phosphorous deoxidised copper and aluminium deoxidised steel. However, care must be taken to ascertain that no additional phases are formed between the constituents of the melt and Na_2O or CaO present in β-Al_2O_3.

Chromium Sensors:

Sensors for measuring chromium in molten iron and steel have been investigated by Iwase and co-workers and Janke[69-71]. Chromium sensors studied by above investigators can be reported schematically as follows:

$$M + M_xO_y \,//\, (MgO, CaO, Y_2O_3)\, ZrO_2 \,//\, Cr_2O_3 + \underline{Cr}_{\,melt} \quad (IX)$$

where Cr_2O_3 is used as an auxiliary electrode and the solid electrolyte is an oxygen ion conductor. This type of sensor has been used successfully for measuring chromium content from 0.5 wt% to as high as 40 wt% in molten metal. The oxygen potentials for the $Cr+Cr_2O_3$ equilibrium lies outside the ionic conduction domain of stabilised zirconia. Therefore, the main problem of electronic conduction in stabilised zirconia used as solid electrolyte at the oxygen potentials outside its ionic conduction domain still remains unsolved. Therefore, such sensors may need frequent calibration before using them in unknown melts containing chromium. A better option would be to use a β-Al_2O_3 type solid electrolyte in conjunction with a suitable ternary auxiliary phase to design a chromium sensor for sensing chromium in molten steel. For example, a sensor of the type:

$$\alpha\text{-}Al_2O_3 + Ca\text{-}\beta\text{-}Al_2O_3 \,//\, Ca\text{-}\beta\text{-}Al_2O_3 \,//\, CaCr_2O_4 + \underline{Cr}\text{ melt} \quad (X)$$

could be a possibility.

Antimony and Arsenic Sensors:

Antimony and arsenic need to be measured in molten non-ferrous metals such as zinc, lead, tin and copper. These metals are generally removed from the non-ferrous metal during the refining stage. In order to control the refining process accurately and avoid the transfer of base metal into dross due to over refining, it is important to determine the concentration of antimony and arsenic in molten metal continuously by on-line analysis. Solid state sensors for antimony and arsenic can provide an ideal solution to the current industrial practice. Attempts have been made to develop sensors for antimony and arsenic dissolved in non-ferrous metals using Na-β-Al_2O_3 as a solid electrode and sodium antimony oxide ($NaSbO_3$) as the auxiliary phase[7,72]. The emf of the antimony sensor has been empirically correlated to the concentration of antimony in molten zinc and copper. A more rigorous thermodynamic analysis of the emf of the sensor could not be done because of the non-availability of thermodynamic data of $NaSbO_3$ and also due to the presence of trivalent and pentavalent antimony ions in $NaSbO_3$. A typical plot showing the variation of the emf of the antimony sensor as a function of the logarithm of the mole fraction of antimony in molten tin at 733 K is shown in Figure 5.

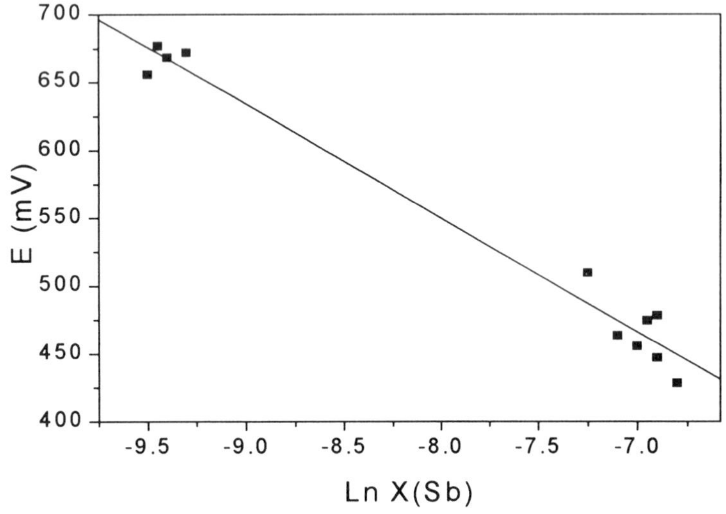

Figure 5 Sn-Sb sensor testing results[72].

In the above measurements, it is assumed that molten metal is saturated with oxygen and hence the oxygen content of the melt does not influence the measured emf of the sensor. An antimony sensor based on stabilised zirconia electrolyte investigated by Fergus et. al[73] can be schematically represented as follows:

$$Sb\text{-}Zn(l) + ZrSb_2(s) \,//\, YSZ \,//\, Zn(l) \,//\, YSZ \,//\, Sb(s) + ZrSb_2(s) \qquad (XI)$$

where yttria stabilised zirconia solid electrolyte conducting oxygen ion is denoted by YSZ. The phase diagram of zinc-antimony-zirconium used to identify a suitable Zr-Sb intermetallic compound as an auxiliary electrode was hypothetically constructed by Fergus and Hui[73]. There are at least six intermetallic compounds along the Zn-Zr binary and there are at least two intermetallic compounds, $ZrSb_2$ and Zr_2Sb on the Sb-Zr binary. It is likely that Zr-Sb intermetallic compounds will be stable at higher oxygen potentials whereas Zr-Zn intermetallic compounds will be more stable at extremely low oxygen potentials. Therefore, in the antimony sensor used by Fergus et. al[73] $ZrSb_2$ auxiliary phase will not be stable in contact with the dilute alloys of zinc containing antimony.

Although the measurements of Fergus et al[73] are encouraging and shows good agreement with the thermodynamic data, the instability of Sb-Zn(l) + $ZrSb_2$(s) electrode raises doubts about the long-term stability of their antimony sensor. Further, they have also pointed out that the lowest possible antimony concentration in molten zinc, which could be detected by using their sensor, is 7 at% antimony in zinc. This is significantly higher than the antimony concentration required to be sensed either in metal refining process or in hot-dip galvanising process. Hence, the antimony sensor developed by Fergus based on cell (XI) can at best be used as a disposable-type sensor. Further, the antimony sensor represented by cell (XI) will have to undergo significant modification before a commercial probe can be fabricated for an on-line analysis of molten zinc containing antimony.

A schematic representation of arsenic sensor was shown[7,8] earlier (cell IV). Solid electrolyte used is a novel potassium ion conductor, $K_3Al_2As_3O_{12}$, incorporating arsenic in its lattice. Therefore, this particular material not only acts as solid electrolyte conducting potassium ions, but also at the same time acts as an auxiliary phase. Attempts have been also made recently to use calcia stabilised zirconia ($CaO\text{-}ZrO_2$) as the solid electrolyte and calcium arsenate ($Ca_3As_2O_8$) as an auxiliary phase to measure arsenic in copper[74].

The results obtained in molten zinc at 793 K using an arsenic sensor of the type represented by cell (IV) are shown in Figure 6. The sensor incorporates a novel reference electrode for potassium consisting of $K\text{-}\beta\text{-}Fe_2O_3$+$Fe_2O_3$ biphasic mixture.

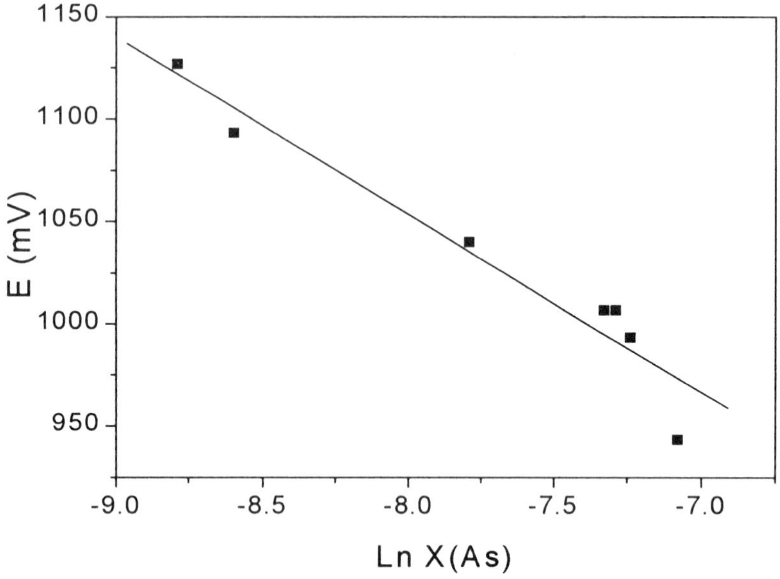

Figure 6 Zn-As sensor testing results[8].

CONCLUSIONS

In order to produce high quality metal from its ore or by recycling scrap metal and metallurgical industry waste, it is very important not only to have a good understanding of the refining process but also to have a high degree of control over the refining process. The ever increasing demand of high grade metal and depleting resources, the metal refining process in future is likely to become more and more complex. The cost of refining process and thereby the cost of metal production can be reduced significantly by :

(i) minimising the use of expensive refining agents
(ii) accurately monitoring of impurity levels and
(iii) minimising the generation of drosses

which can be easily achieved by developing in-expensive, reliable, rugged, long life sensors for on-line process control. At the present moment, it appears that a lot is known about the development of sensors for on-line process control in metallurgical industries. However, the technology is still in its infant stage and there lies a big challenge to take this sensors technology from its infant stage to a mature stage, at which time, it could be implemented in the industries. Metals industries still have and will continue to have a great future because of the special status of metals in mankind. Sensors technology, therefore, will remain an exciting area of research and development for scientists from variety of disciplines including Metallurgy, Materials Science, Chemistry, Physics, Electronics Engineering and Software Engineering to name a few.

REFERENCES

[1]S. Choi, W. Chung and D. Lee, "SO_2 Sensing Characteristics of NASICON Electrolytes", *Sensors and Actuators B*, **35-36** 263-266 (1996).

[2]K.T. Jacob, M. Iwase and Y. Waseda, "A Galvanic Sensor for SO_3/SO_2 Based on the $CaF_2/CaSO_4$ Couple", *Solid State Ionics*, **23** 245-252 (1987).

[3]S. Ikeda, S. Kato, K. Nomura, K. Ito, H. Einaga, S. Saito, Y. Fujita, "CO_2 Sensor Using Solid Electrolytes with Zirconium Phosphate Framework", *Solid State Ionics*, **70/71** 569-571 (1994).

[4]M. Holzinger, J. Maier and W. Sitte, "Fast CO_2-Selective Potentiometric Sensor with Open Reference Electrode", *Solid State Ionics*, **86-88** 1055-1062 (1996).

[5]Y. Sakoda, Y. Sakai and T. Manabe, "Detection of CO_2 Using a Solid-State Electrochemical Sensor Based on Na ionic Conductors", *Sensors and Actuators B*, **15-16** 166-170 (1993)

[6]G.M. Kale, A.J. Davidson, and D.J. Fray, "Investigation into an Improved Design of CO_2 Sensor", *Solid State Ionics*, **86-88** 1107-1110 (1996).

[7]G.M. Kale, A.J. Davidson, and D.J. Fray, "Solid-State Sensor for Measuring Antimony in Non-Ferrous Metals", *Solid State Ionics*, **86-88** 1101-1105 (1996).

[8]G.M.Kale and D.J. Fray, "Novel Solid-State Sensor for Measuring Arsenic in Molten Metals", *126th TMS Annual Meeting and Exhibition*, Orlando, Florida (USA) February (1997).

[9]R.V. Kumar and D.J. Fray, "Improvements in or Relating to Electrolytic Sensors", British Patent Application 2350 Y, 1988.

[10]M. Iwase, T. Kanazana and T. Mori, "Kinetics of Oxygen Transfer as a Solid-Oxide Electrolyte-Mo, MoO_2 Electrode Interface", *Solid State Ionics*, **2** 331-336 (1981).

[11]S. Ranganathan and D. Janke, "Thermodynamics of Some New Oxide Reference Materials for Electrochemical Oxygen Sensors", *Scripta Metallurgica et Materialia*, **29** 305-310 (1993).

[12]H. Keqin, X. Yuhua, W. Weijiang, L. Qingguo, L. Wen, and H. Clauss "Investigations on Commercially Available MgO-PSZ Electrolytes with Y_2O_3-FSZ

Coating for Low-Oxygen Determination in Liquid Steel", *Solid State Ionics*, **70/71** 563-568 (1994).

[13]S. Tu and D. Janke, "EMF Sensor Control in Vacuum Decarburization and Deoxidation of Steel Melts", *ISIJ International*, **35** 1362-1367 (1995).

[14]R.V. Kumar and D.J. Fray, "Application of Novel Sensors in the Measurement of Very Low-Oxygen Potentials", *Solid State Ionics*, **70/71** 588-594 (1994).

[15]W.L. Worrell and Q. Liu, "Development of an Extended-Life Oxygen Sensor for Iron and Steel Melts", *Solid State Ionics*, **40/41** 761-763 (1990).

[16]C.B. Alcock, B. Li, J.W. Fergus and L. Wang, "New Electrochemical Sensors for Oxygen Determination", *Solid State Ionics*, **53/56** 39-43 (1992).

[17]D. Janke, "A New Immersion Sensor for the Rapid Electrochemical Determination of Dissolved Oxygen in Metallic Melts", *Solid State Ionics*, **3/4** 599-604 (1981).

[18]W. Weijiang, T. Shaohua, H. Shiqiang and L. Qungguo, "Development of Long-Term Oxygen Sensor in Molten Copper Using MgO-PSZ Electrolyte", *Solid State Ionics*, **70/71** 606-609 (1994).

[19]D. Janke, "Oxygen Probes Based on Calcia-Doped Hafnia or Calcium Zirconate for Use in Metallic Melts", *Metallurgical Transactions B*, **13B** 227-235 (1982).

[20]M. Iwase, M. Tanida, A. Mclean and T. Mori, "Electronically Driven Transport of Oxygen from Liquid Iron to $CO+CO_2$ Gas Mixtures through Stabilised Zirconia", *Metallurgical Transactions B*, **12B** 517-524 (1982).

[21]D. Janke, "Solid Electrolyte Probe for Continuous Oxygen Measuring in Liquid Steel", *Stahl u Eisen*, **103** 57-58 (1983).

[22] D. Janke, K. Hugen, P. Hammerschmid, D. Dittert, H. Lindenberg and L. Weber, "Continuous Measuring Oxygen Probes for Steel Melts", *Stahl u Eisen*, **107** 77-78 (1987).

[23]R. V. Kumar and D. J. Fray, "Development of High-Temperature Solid-State Electrochemical Sensors", *The Institute of Mining and Metallurgy London* 315-340 (1989).

[24]D.J. Fray, "The Use of Solid Electrolytes as Sensors for Applications in Molten Metals", *Solid State Ionics*, **86-88** 1145-1054 (1996).

[25]J. W. Patterson, "Conduction Domain for Solid Electrolytes", *Journal of the Electrochemical Society*, **118** 1033-1039 (1971).

[26]H. Schmalzried, "ZrO2 as an Electrolyte for Electrochemical Studies at High Temperatures", *Z. Elektrochem.*, **66** 572-576 (1962).

[27]H. Schmalzried, "Ion and Electron Conduction in Binary Oxides and Their Investigation With Electromotive-Force Measurements", *Z. Physik. Chem.*, **38** 87-102 (1963).

[28]G.M. Kale, "Na_2O-Al_2O_3 System-Activity of Na_2O in (α+β) and (β+β") -Alumina", *Metallurgical Transactions B*, **23B** 833-839 (1992).

[29]R.J. Brisley and D.J. Fray, "Determination of the Sodium Activity in Aluminium and Aluminium Silicon Alloys Using Sodium Beta Alumina", *Metallurgical Transactions B*, **14B** 435-440 (1983).

[30]L. Zhang, D.J. Fray, J.C. Dekeyser and F. De Schutter, "Reference Electrode of Simple Galvanic Cells for Developing Sodium Sensors for Use in Molten Aluminium", *Metallurgical Transactions B*, **27B** 794-800 (1996).

[31]J. C. Dekeyser, F. De Schutter, C. Van der Poorten, L. Zhang and D. J. Fray, "An Electrochemical Sodium Sensor for Aluminium Melts", *Sensors and Actuators B*, **24-25** 273-275 (1995).

[32]A.A. Dubrell and A.D. Pelton, "Solid Electrolyte Probes to Monitor the Lithium and Sodium Content of Liquid Aluminium", *CIM Bulletin*, **80(902)** 92 (1987)

[33]A.A. Dubrell and A.D. Pelton, "Electrochemical Probes to Monitor the Sodium and Lithium Content of Liquid Aluminium", *Journal of Metals*, **39 (10)** A48 (1987).

[34]A.A. Dubrell and A.D. Pelton, "Probes for the Continuous Monitoring of Sodium and Lithium in Molten Aluminium", *Journal of Metals*, **36 (12)** 63 (1984).

[35]G.M. Kale and S. Srikanth, "Electrochemical Determination of the Gibbs Energy of Formation of $Na_3Fe_5O_9$ and $Na_2Fe_2O_4$ Employing Na-β-Al_2O_3 Solid Electrolyte", *J. Am Ceram. Soc.* (1998) [In Press].

[36]P.C. Yao and D.J. Fray, "Determination of the Lithium Content of Molten Aluminium Using a Solid Electrolyte", *Metallurgical Transactions B*, **16B** 41-46 (1985).

[37]D.J. Fray, "Possible Uses of Sensors in the Aluminium Industry", *Materials Science and Technology*, **3** 61-65 (1987).

[38]D.J. Fray, "The Determination of Li", U.K. Patent Application No. 7922927, 1979.

[39]J.A. Little and D.J. Fray, "Preparation and Properties of Copper Beta Alumina", *Electrochimica Acta.*, **25** 957-964 (1981).

[40] P.C. Yao and D.J. Fray, "The Preparation and Properties of the Solid-State Ionic Conductor, $CuZr_2(PO_4)_3$, *Solid State Ionics*, **8** 35-42 (1983).

[41]A.J. Davidson and D.J. Fray , Unpublished Research, (1996).

[42]L Zhang and D.J. Fray, "Electrochemical Sensor for Measuring Magnesium Content in Molten Aluminium", *J. Appl. Electrochemistry*, **26** 269-275 (1996).

[43]S. Yamaguchi, N. Fukatsu, H. Kumira, K. Kawamura, Y. Iguchi, and T. O-Hashi, "Development of Al Sensor in Zn Bath for Continuous Galvanising Processes", *Galvatech '95' Conference Proceedings*, 647-655 (1995).

[44]S. Yamaguchi, H. Makino, A. Sakatoku and Y. Iguchi, "Phase Stability of Dross Phases in Equilibrium with Liquid Zn Measured by the Al Sensor", *Galvatech '95' Conference Proceedings*, 787-794 (1995).

[45]D. Gozzi and P. Graniti, "Sulfur Determination in Carbon-Saturated Iron by Solid-State Electrochemical Sensor", *Metallurgical Transactions B*, **25B** 561-568 46 (1994).

[46]M.A. Swetnam, L. Holt, S.R. Witek and J.D. Usher, "On-line Measurement of Sulfur in Molten Iron with a Ceramic Electrochemical Cell", *Steelmaking Conference Proceedings*, **78** 199-203 (1995).

[47]L.B. Pankratz, "Thermodynamic Properties of Elements and Oxides", *US Bureau of Mines*, **672** (1982).

[48]M. Iwase, "Rapid Determination of Silicon Activities in Hot Metal by Means of Solid State Electrochemical Sensors Equipped with an Auxiliary Electrode", *Scandinavian Journal of Metallurgy*, **17** 50-56 (1988).

[49]D. Janke, "Recent Development of Solid Ionic Sensors to Control Iron and Steel Bath Composition", *Solid State Ionics*, **40/41** 764-769 (1990).

[50]D. Janke, "Electrochemical Silicon Sensor Equipped with Auxiliary Electrode, $ZrO_2 + ZrSiO_4$", *Solid State Ionics*, **40/41** 773-775 (1990).

[51]K. Huang, W. Wu and Q. Liu, "A New Electrochemical Sensor for Rapid Determination of Silicon in Carbon Saturated Iron", *Solid State Ionics*, **53/56** 24-29 (1992).

[52]K. Gomyo, I. Sakaguchi, Y. Sinya, V.I. Lakshmanan, A. McLean and M. Iwase, "Solid State Sensor for Silicon in Molten Metals by Zirconia-Based Electrolyte", *Solid State Ionics*, **70/71** 551-554 (1994).

[53]C. Furuta, T. Nagatsuka, K. Iwasaki, N. Saito and M. Sasabe, "Development of an Electrochemical Sensor for Determination of Si Contents in Molten Metal", *Solid State Ionics*, **40/41** 776-778 (1990).

[54]K. Ichihara, D. Janke and H.J. Engell, "A New Silicon Sensor for Hot Metal Measurements", *Transactions ISIJ*, **27** B-294 (1987).

[55]E. M. Levin and H.F. McMurdie, "Phase Diagrams for Ceramists 1975 Supplement", Editor M.K. Reser, pp. 245, Fig. 4578, Am. Ceram. Soc., Columbus, OH (1975).

[56]D. Viechnicki and V.S. Stubican, "Mechanism of Decomposition of the Cubic Solid Solutions in the System ZrO_2-MgO", *J. Am. Ceram. Soc.*, **48**, 292-297 (1965).

[57]W.C. Butterman and W.R. Foster, "Zircon Stability and the ZrO_2-SiO_2 Phase Diagram", *Amer. Mineral*, **52**, 880-885 (1967).

[58]N.L. Bowen and O. Anderson, "The Binary System: MgO-SiO_2", *Am. J. Sci.*, **37**, 487-500 (1914).

[59]S. Larose, A. Dubreuil and A.D. Pelton, "Solid Electrolyte Probes for Magnesium, Calcium and Strontium in Molten Aluminium, *Solid State Ionics*, **47** 287-295 (1991).

[60]J.W. Fergus and S. Hui, "Solid Electrolyte Sensors for Measuring Magnesium in Molten aluminium, *Metallurgical Transactions B*, **26B** 1289-1291 (1995).

[61]J.W. Fergus, S. Hui, W.F. Gale and Z.A.M. Abdo, "Origins of Anomalous Output of Magnesium Sensor", *Materials Science and Technology*, **13** 533-536 (1997).

[62]B.L. Tiwari and B.J. Howie, "Determination of Magnesium in Molten Aluminium Alloy Using an Electrochemical Sensor", pp 895-902, Light Metals 1989, ed P.G. Campbell (Warrendale, PA: TMS, 1988).

[63]B.L. Tiwari, "Thermodynamic Properties of Liquid Al-Mg Alloys Measured by the EMF Method", *Metallurgical. Transaction A*, **18A** 1645-1651 (1987).

[64]G.R. Belton and Y.K. Rao, "A Galvanic Cell Study of Activities in Mg-Al Liquid Alloys", *Trans. Metall. Soc. AIME*, **245** 2189-2193 (1969).

[65]M.M. Tsyplakova and Kh. L. Strelets, "Study of the Thermodynamic Properties of the Magnesium-Aluminium System by the EMF Method", *J. Applied Chem. USSR*, **42(11)** 2354-2359 (1969).

[66]J.W. Fergus, "Electrochemical Magnesium Sensors for Aluminium Processing", *JOM Journal of the Minerals Metals and Materials Society*, **47(11)** 36-40 (1995).

[67]N.S. Choudhury, "Electromotive Force Measurements on Cells Involving Beta-Alumina Solid Electrolyte", *J. Electrochem. Soc.*, **120** 1663-1667 (1973).

[68]F.A. Elrefaie and W.W. Smeltzer, "Thermodynamics of Nickel-Aluminum-Oxygen System Between 900 and 1400 K", *J. Electrochem. Soc.*, **128** 2237-2242 (1981).

[69]K.S. Goto, M. Sasabe, Y. Iguchi, M. Iwase and T. Onoo-Ye, "Solid State Sensors Incorporating Auxiliary Electrode for Rapid Determinations of Chromium in Molten Iron", *Solid State Ionics*, **40/41** 770-772 (1990).

[70]T.K. Inouye, H. Fujiwara and M. Iwase, "Activities of Chromium in Molten Copper at Dilute Concentrations by Solid-State Electrochemical Cell", *Metallurgical Transactions B*, **22B** 475-480 (1991).

[71]M. Heinz and D. Janke, "Electrochemical Sensing of Oxygen, Chromium and Other Elements in Nickel-Based Metals", *Memoires et Etudes Scientifiques Revue de Metallurgie*, **88** 239-249 (1991).

[72]G.M. Kale, "Measurement of Antimony in Molten Tin Employing Solid State Electrochemical Sensor", *J. Electrochem. Soc.*, (1998) In Press.

[73]J.W. Fergus and S. Hui, "Solid Electrolyte Based Galvanic Cell for Measuring the Antimony Concentration in Molten Zinc", *J. Electrochem. Soc.*, **143** 2498-2502 (1996).

[74]P. Fredriksson and S. Seetharaman, Private Communication, (1998).

SOLID ELECTROLYTE-BASED NO_X SENSORS

P. Thamboon, S. Yao, P. Gouma, and S. A. Akbar
Center for Industrial Sensors and Measurements (CISM)
The Ohio State University
Columbus, OH 43210

ABSTRACT

NO_X sensors using Na^+ Super Ionic Conductor (NASICON)-based solid electrolyte were investigated. The selectivity was enhanced by appropriate choice of auxiliary phases such as $NaNO_3$, $Ba(NO_3)_2$, $Tb(NO_3)_3$ and their binary combinations. Two reference electrode systems were constructed: one sealed from and the other open to the probing environment. Results were obtained on the EMF change as a function of NO_X at various temperatures. The interference tests to O_2 and CO_2 were examined. The deviation from the Nernst behavior and its relationship to microstructural defects are discussed.

INTRODUCTION

Nitrogen oxides (NO_X), commonly referred to NO and NO_2 gases, are of concern as air pollutants. NO_X gases contribute to the acid rain and the excessive ozone formation. These gases are mainly emitted from combustion and incineration processes. Therefore, efforts to find reliable, rugged, compact, on-site monitoring and inexpensive NO_X sensors are significantly increasing [1-6].

NO_X sensors reported so far are not selective. Solid electrolyte-type sensors show promise for overcoming the selectivity problem. Since the working principle of this type is based on electrochemical equilibrium and reactions, the activities of chemical species are related directly and quantitatively to the electrical signal using well-known laws. NASICON ($\underline{Na}^+$ Super Ionic Conductor)-based solid electrolyte has been widely used because of its high ionic conductivity, good sinterability and little volume change associated with the phase transition in the wide temperature range from room temperature to 1200 K [7-10].

In the present study, the sensors were made of the NASICON with two electrode compartments, sensing and reference electrodes. The selectivity to NO_X was enhanced by appropriate choice of auxiliary phases such as $NaNO_3$, $Ba(NO_3)_2$, $Tb(NO_3)_3$ and their binary combinations as the sensing electrode. Two reference electrode systems were constructed: one sealed from and the other open to the probing environment. Synthetic air was used as a reference gas.

For the sealed system NASICON was protected from NO_X at the reference side. The potentiometric cell arrangement is represented by:

NO_X, O_2, Nitrate, Pt / NASICON / Pt, O_2 (1)

In the open system, both the electrodes expose to the same test gas. Yttria Stabilized Zirconia (YSZ) responding only to oxygen was attached to NASICON at the reference side. The cell arrangement is represented by:

NO_X, O_2, Nitrate, Pt / NASICON / YSZ / Pt, O_2, NO_X (2)

The electromotive force (EMF) between the sensing and the reference electrodes was measured as a function of NO and NO_2 concentrations at various temperatures. Interference gases, e.g., O_2 and CO_2 was examined in order to test the selectivity. Microstructural characterization techniques were used to gain an understanding of the relationship between the sensor morphology and the performance.

EXPERIMENTAL PROCEDURES

Sealed reference electrode system

NASICON, with typical composition $Na_3Zr_2Si_2PO_{12}$, was prepared by ball milling technique following the procedure reported by Miller *et al.* [11]. The sintered discs of 2 mm thick and 11 mm in diameter were obtained. NASICON was fixed on a quartz glass tube with an inorganic adhesive. The reference electrode using platinum black was pasted to the surface of the disc inside the tube, on which a platinum mesh with a Pt lead wire was connected and was pressed mechanically to establish the electrical contact. After the application of the layer of auxiliary phase on the sensing electrode, the sensor was covered by a platinum mesh and a Pt lead wire was attached. The schematic of this device is shown in Figure 1.

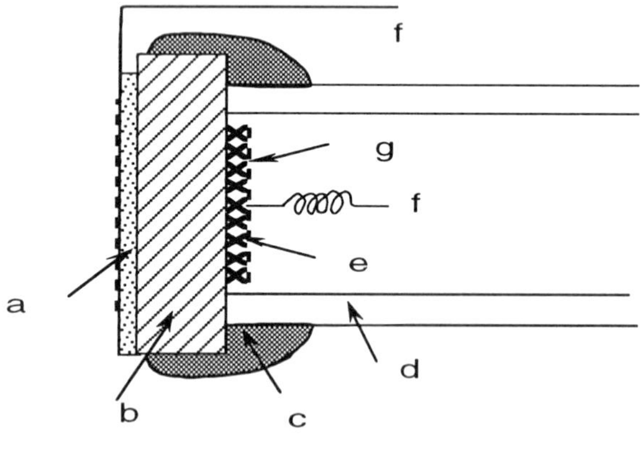

Figure 1. Structure of sealed reference electrode sensors
(a) Nitrate, (b) NASICON, (c) Inorganic adhesive, (d) Quartz glass tube, (e) Pt-black, (f) Pt-wire, (g) Pt-mesh

Open reference electrode system

YSZ powders with 10 mole% Y_2O_3 were obtained from Nextec Co., Worthington, OH. The powders were ball milled for 24 hours, calcined at 1000 °C for 5 hours, and sintered at 1550 °C for 10 hours. The discs of 2 mm thick and 9 mm in diameter were obtained. NASICON and YSZ discs were attached to each other by diffusion bonding at 1250 °C. The schematic of this device is shown in Figure 2.

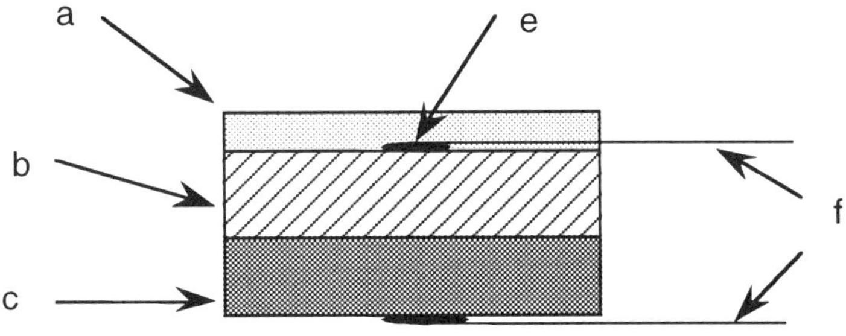

Figure 2. Structure of open reference electrode sensors
(a) Nitrate, (b), NASICON, (c) YSZ, (e) Pt paste, (f) Pt wire

EMF measuring procedures

Each sensor device was placed inside an electrical furnace to heat it up to a target temperature. The gas flow was controlled by means of flow meters. As a source of NO_X, NO (2000 ppm) and NO_2 (200 ppm) in N_2 balance were used. These gases were mixed with synthetic air and/or oxygen to obtain the sample gases at maximum concentrations of 1600 ppm and 160 ppm for NO and NO_2 respectively in a 100 cm^3/minute total flow rate. The electromotive force (EMF) between the sensing and the reference electrodes was then measured by a digital electrometer (HP 34410A Hawlett-Packard). Probing signals were collected using data acquisition software (HP34812A Benchlink).

Philips SEM XL30 FEG equipped with EDX detector, capable of performing energy dispersive x-ray spectrometry (EDX), was used to perform microstructural characterization and microanalysis.

RESULTS AND DISCUSSION

For both reference system, the possible reactions are as follows:

I. Sensing electrode

$$NO_2 + \frac{1}{2}O_2 + e^- \leftrightarrow NO_3^- \text{ (Nitrate)} \qquad \text{for } NO_2 \text{ sensing} \qquad (3)$$

$$NO + O_2 + e^- \leftrightarrow NO_3^- \text{ (Nitrate)} \qquad \text{for NO sensing} \tag{4}$$

II. Reference electrode

$$\frac{1}{2}O_2 + 2e^- \leftrightarrow O^{2-} \tag{5}$$

In NASICON

$$O^{2-} + 2Na^+ \leftrightarrow Na_2O \quad \text{(in NASICON)} \tag{6}$$

These reactions lead an expression for EMF, under a fixed oxygen partial pressure based on Nernst law as:

$$E = E_\theta + 2.303\frac{RT}{F}\log(P_{NO \text{ or } NO_2}) \tag{7}$$

where E_θ is a constant value determined by several factors such as the activity of Na_2O, the oxygen partial pressure and the temperature.

From equation (7), the slope of the EMF signal as a function of NO or NO_2 concentration is expected to be the same for all the representing one electron transfer in electrode reaction according to equations (3) to (5). Figure 3 shows the EMF as a function of NO_2 concentration at 450 °C of the sealed reference electrode system with $Ba(NO_3)_2$ as the auxiliary phase.

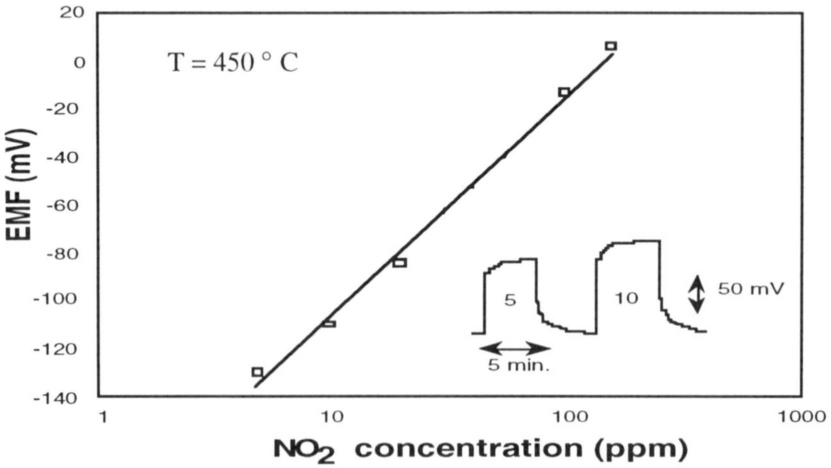

Figure 3. EMF signal as a function of NO_2 concentration at 450°C for the $Ba(NO_3)_2$ / NASICON/ sealed reference electrode system.

The calculated slope indicates 1.6 electron transfer in the reaction. The results from the other detected temperatures (400 to 550 °C) do not follow the assumption of one electron transfer either. The deviation from the Nernstian behavior is due to unexpected reactions on the sensing electrode, e.g., the competitive reactions of Pt / Nitrate and Pt / NASICON interface. The latter interface might lead to the formation of $NaNO_3$ according to Na^+ in NASICON and NO_2 reaction and thus, the decrease in Na_2O activity. These interface reactions depend strongly on the temperature as shown in Figure 4.

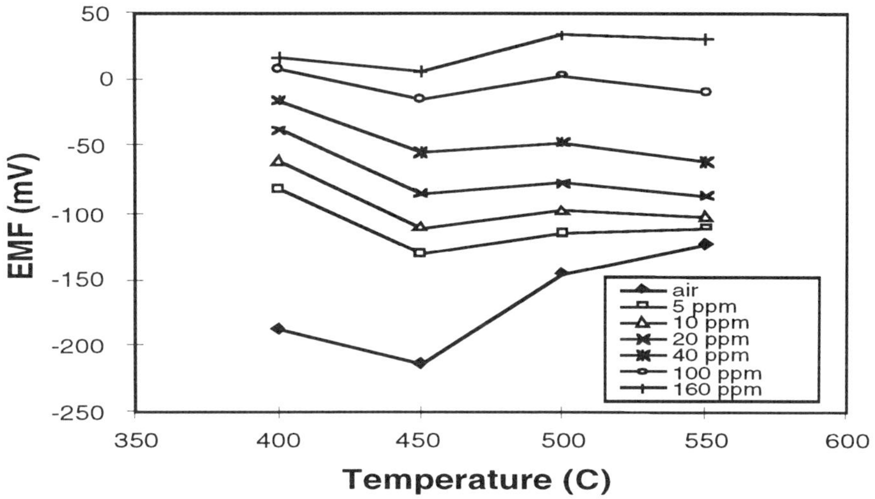

Figure 4. EMF signal as a function of temperature for different NO_2 concentrations for the $Ba(NO_3)_2$ / NASICON / sealed reference electrode system.

The EMF signal as a function of temperature for different NO_2 concentrations, are divided into two regions: the EMF decreases when temperature increases at lower temperatures while the EMF increases at higher temperatures. The Pt/NASICON interface is likely to dominate at relatively low temperatures. Yamazoe *et al.* shows a similar situation in the study of CO_2 sensing [12]. Furthermore, the drop of EMF signals at higher temperatures is observed due to the leakage problem of the sealed system. The leakage of NO_2 into the reference electrode forms the similar reaction as at the sensing electrode with unknown NO_2 concentration. As a result, the EMF becomes lower than the actual value.

For the open reference electrode system, the results shown here are from the $Tb(NO_3)_2$-$NaNO_3$ binary system. This binary phase shows a large change in the EMF upon NO and NO_2 concentration as shown in Figures 5 and 6 , respectively. The response time of the sensors for both the sealed and the open system is about 1 minute. The long recovery found in the open system is mainly due to the slow desorption and diffusion process. At high temperatures, these processes become faster resulting in a faster recovery.

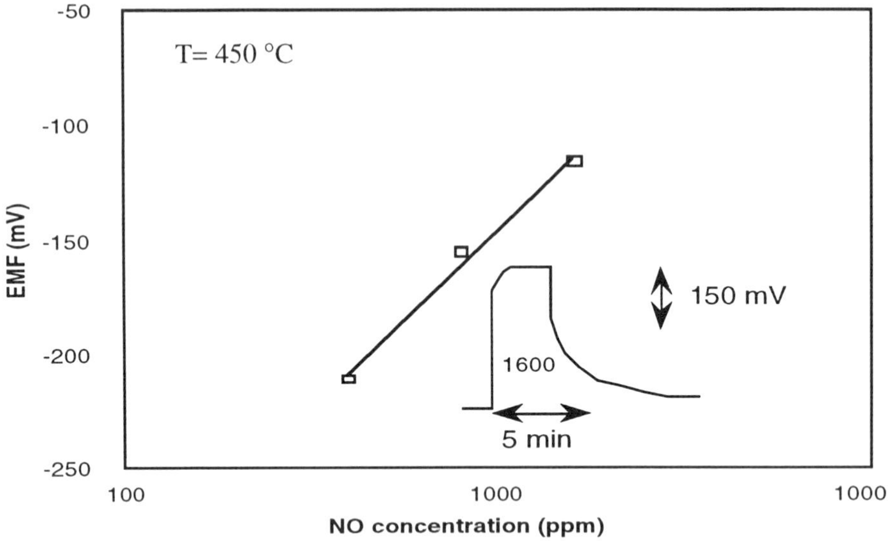

Figure 5. EMF signal as a function of NO concentration at 450°C for the $Tb(NO_3)_2$-$NaNO_3$ / NASICON : YSZ / open reference electrode system.

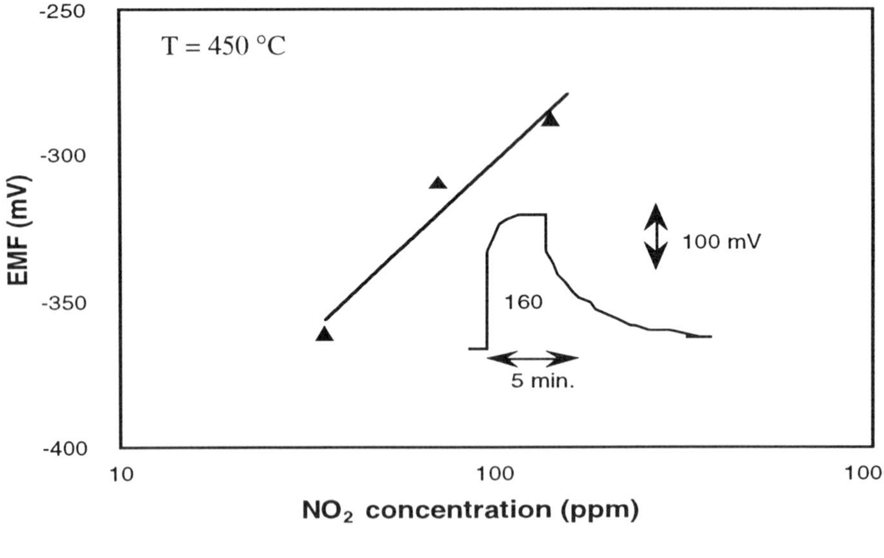

Figure 6. EMF signal as a function of NO_2 concentration at 450°C for the $Tb(NO_3)_2$-$NaNO_3$ / NASICON / YSZ / open reference electrode system.

The EMF signal as a function of temperature for the system shows similar behavior as the sealed system. The sensitivity of the sensors to O_2 and CO_2 are shown in Figure 7 and 8 respectively. The results are for the $NaNO_3$ auxiliary phase system at 290 °C. From the figures, the sensors show much higher sensitivity to NO_2, hence the selectivity is improved.

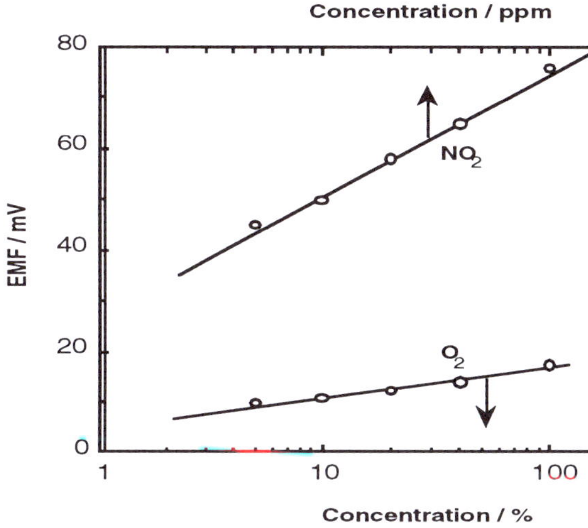

Figure 7. EMF signal as a function of NO_2 and O_2 concentrations at 290 °C for the $NaNO_3$ / NASICON/YSZ/ open reference electrode system.

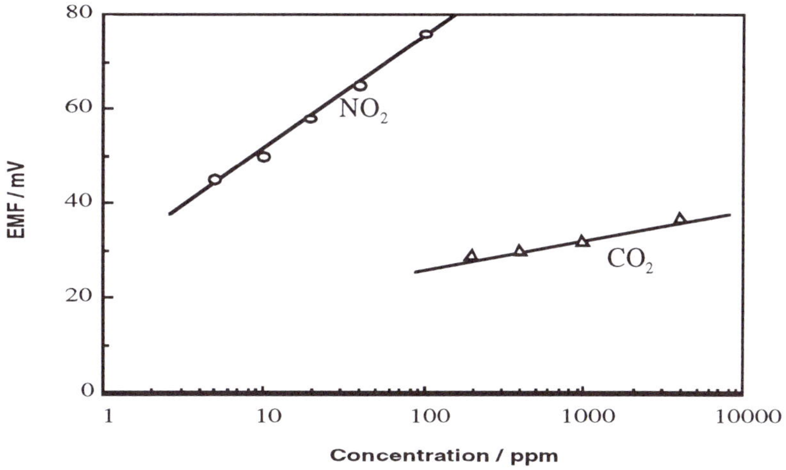

Figure 8. EMF signal as a function of NO_2 and CO_2 concentrations at 290 °C for the $NaNO_3$ / NASICON/YSZ/ open reference electrode system.

The results for others auxiliary phase system performances are summarized in Table I. Since the EMF signal of the sensors shows the temperature dependence, the optimum temperatures at which the largest change in the EMF per the decade change in NO_X concentration are given. The 90% response time is calculated as the time at which the sensor experiences the EMF change at 90% of equilibrium value upon the change in NO_X concentration. The response time of the sensors are less than one minute. The sensors showed reproducibility over the measurement period of 1 to 3 days. However, there is a possibility of the formation of $NaNO_3$ at the Pt/NASICON interface upon exposure to NO_X. As a result, the long-term drift of the EMF signal may occur due to the reduction of Na activity in NASICON.

Table I. Summarized sensor performances

Sensor system	Test gas	Optimum Temperature (°C)	$\Delta E/\Delta P$ (mV/log decade)	90% Response time (s)
1. Sealed system (/NASICON/)				
$Ba(NO_3)_2$	NO_2	450	96	25
2. Open system (/NASICON/YSZ/)				
I. $NaNO_3$	NO_2	270	50	50
II. Ba $(NO_3)_2$	NO	450	117	25
III. Ba $(NO_3)_2$-$NaNO_3$	NO_2 NO	500 500	125 43	25 10
IV.$Tb(NO_3)_3$	NO_2 NO	500 500	47 73	25 25
V. $Tb(NO_3)_3$-$NaNO_3$	NO_2 NO	450 450	123 158	50 50

BSE images viewed by SEM is based on atomic number contrast between various phases leading to the identification of different component in the system.

Figure 9 shows that there is bonding problem of NASICON and YSZ, manifested by the cracked interface. The interdiffusion layer between them is approximately 100 micron within YSZ. Figure 10 shows interdiffusion region in more details. The dark region is prior porosity filled with Na. This finding suggests Na diffusion from NASICON into YSZ.

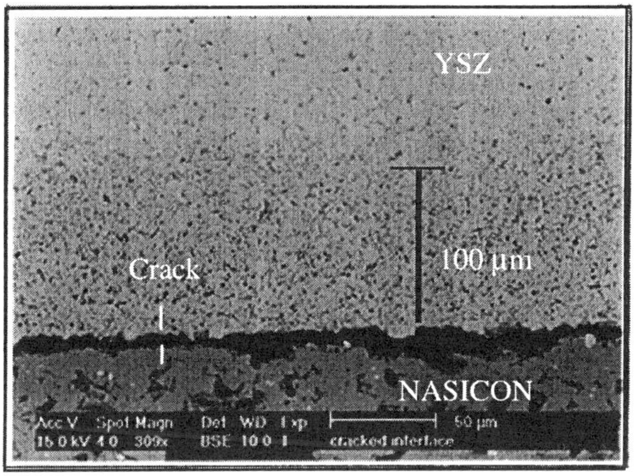

Figure 9. BSE image of the cracked interface of NASICON and YSZ. The interdiffusion region is approximately 100 micron within YSZ.

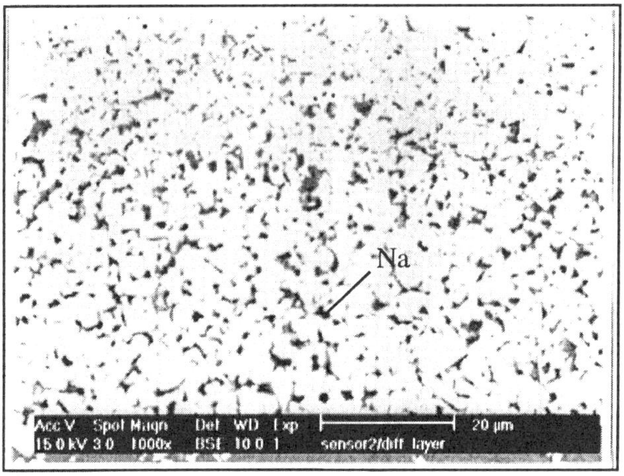

Figure 10. BSE image of the interdiffusion region at higher magnification. EDX analysis reveals Na diffusion from NASICON into YSZ.

SUMMARY

The sensors show good sensitivity toward NO_X. Cross-sensitivities to O_2 and CO_2 are negligible. The deviations from the Nernstian behavior are due to unexpected interface reactions and microstructural defects. The sealed reference electrode system is not favored for practical use due to the leakage problem and the loose Pt electrode contacts. The interface contacts of the open reference electrode system need to be improved.

ACKNOWLEDGMENTS

This study is sponsored by the NSF Center for Industrial Sensors and Measurements (CISM; grant EEC 9523358) and the Royal Thai Government.

REFERENCES

1. S. A. Akbar, C. C. Wang and L. Wang, "Ceramic Materials May Revolutionize Automotive Emissions Control", *Ceramic Industry*, June, 32-38 (1996).
2. P. T. Moseley, "Solid state Gas Sensors", Meas. Sci. Technol., **8**, 223-237 (1997).
3. T. Seiyama, "Chemical sensors-Current State and Future Outlook", *Chemical Sensor Technology* , **1**, 1-13 (1988).
4. Committee on New Sensor Technology, *Expanding the Vision of Sensor Materials*, National Academic Press, Inc., 1995.
5. W. Göpel, J. Hesse, J.N. Zemel (eds.), *Sensors: A Comprehensive survey,* **3**, Weinhelm, New York, 1995.
6. R. W. Cattrall, *Chemical Sensors*, Oxford University Press Inc., New York, 1997.
7. S. Yao, "Development of Novel Auxiliary materials for high performance solid electrolyte CO_2 and NO_X sensors", Ph.D. Dissertation, Kyushu University, Kyushu, Japan, 1994.
8. N. Rao, C. M. van den Bleek and J. Schoonman. "A novel temperature-gradient Na^+ -β'' alumina Solid Electrolyte Based SOx gas sensor without gaseous reference electrode", *Solid State Ionics*, **53-56**, 30-38 (1992).
9. A. L. Laskar and S. Chandra (eds.), *Super ionic solids and solid electrolytes*, Academic press, Inc., California, 1989.
10. T. Takahashi (ed.), *High conductivity Solid Ionic Conductors*, World Scientific Publishing Co. Pte. Ltd. Singapore, 1989.
11. G.R. Miller *et al.* , "Processing and properties of sodium B''- Alumina and NASICON ceramic electrolytes", *Fast Ion Transport in Solids ,* North-Holland Inc., New York, 83-97 (1979).
12. N. Yamazoe *et al.*, Sensors and Actuators B, **34**, 361-366 (1996).

HIGH TEMPERATURE SO_x SENSOR FOR FLUE GAS MONITORING USING K-β-Al_2O_3 AND THERMODYNAMIC STABILITY OF K-β-Fe_2O_3

J. E. Hayes and G. M. Kale*[,1]

School of Process, Environmental and Materials Engineering
Department of Mining and Mineral Engineering,
The University of Leeds, Leeds LS2 9JT, UK

ABSTRACT

A high temperature solid state potentiometric sensor for measuring SO_x gas in hot flue gases has been designed and tested. A K^+ ion conductor, K-β-Al_2O_3, was used as the solid electrolyte with K-β-Fe_2O_3 + Fe_2O_3 biphasic mixture as the solid reference electrode and K_2SO_4 as the auxiliary electrode. A NiO + $NiSO_4$ solid mixture was used to establish an equilibrium partial pressure of SO_x gas in a closed environment, which enabled the e.m.f. to be monitored at different temperatures. The data obtained from closed system experiment enabled the calculation of activity of K_2O in K-β-Fe_2O_3 and Gibbs energy of formation of K-β-Fe_2O_3 from the component oxides, K_2O and Fe_2O_3, for the first time. The e.m.f of the sensor was measured with varying partial pressure of SO_2 through an open system at 1007.5 K. The response time of the sensor to attain 90% of the steady-state e.m.f. was less than 60 s and was reproducible. The measured e.m.f. of the sensor was found to vary linearly as a function of logarithm of partial pressure of SO_3 and the slope of the relationship was in excellent agreement with that obtained by the theory.

INTRODUCTION

The recognised need for a reliable SO_x measuring device has led to many studies. Current SO_x detection methods use UV or IR absorption, gas chromatography or are chemical/electrochemical[1-3]. Problems relating to cost, bulkiness, lifetime and regular calibration requirements have prompted research into finding new techniques. Utilisation of solid electrolyte-based sensors have emerged as the most promising because of their high selectivity and high sensitivity in the low concentration region typically below 5000 ppm. Investigations in past

---

*Author for Correspondence
[1] Member of the American Ceramic Society

have focused on use of alkali-metal sulphates such as K_2SO_4[4] and later superionic conductors NASICON[5, 6, 7] and beta-aluminas[8, 9, 10] as solid electrolytes for developing SO_x sensor. To the best of our knowledge, there has not been any investigations into the use of K-β-Al_2O_3, which is also a good K^+ ion conductor, as a solid electrolyte having a layered structure, akin to Na-β-Al_2O_3, which is ideal for this purpose.

In this study, the solid state galvanic cell; Pt, K-β-Fe_2O_3 + Fe_2O_3 + O_2 ‖ K-β-Al_2O_3 ‖ SO_2 + O_2 + K_2SO_4, Pt, has been used to monitor e.m.f. in relation to temperature and partial pressure of SO_x. The first set of measurements were carried out to determine the thermodynamic stability of K-β-Fe_2O_3, which is not reported in the literature to date. These experiments were conducted in a closed system in which different partial pressure of SO_x was established by an equilibrium mixture of NiO + $NiSO_4$ over a range of temperatures. Development, fabrication, testing, and efficacy of the SO_x sensor are reported.

THEORY

Potassium ions are generated at both the working electrode and the reference electrode of the cell. Depending on the chemical potential gradient of K^+ ions across the solid electrolyte, there will be a net flow of ions from one side to the other, through the electrolyte. Ionic conductivity of the electrolyte is therefore of critical importance, and is provided for by the layered structure of K-β-Al_2O_3.

Oxidation of potassium ion occurs at the anode and therefore the anodic reaction is:

$$K_2O(\beta\text{-}Fe) \rightarrow 2K^+ + 2e^- + 1/2O_2(g) \quad (1)$$

Reduction of potassium ion occurs at the cathode therefore the cathodic reaction is:

$$2K^+ + 2e^- + SO_3(g) \rightarrow K_2SO_4(s) \quad (2)$$

Combining the two half-cell reaction gives the overall equilibrium reaction:

$$K_2O(\beta\text{-}Fe) + SO_3(g) \rightarrow K_2SO_4(s) \quad (3)$$

The e.m.f. of the cell is dependent upon the Gibbs energy change of Reaction (3), and hence only on partial pressure of SO_3. This can be explained by the manipulation of the Nernst equation:

$$\Delta G_{(3)} = \Delta G^0_{(3)} + RT\ ln[a_{K_2SO_4} / (a_{K_2O(\beta-Fe)} \cdot P_{SO_3})] \quad (4)$$

$$-nFE = \Delta G_{(3)} = \Delta G^0_{(3)} - RT\ ln\, a_{K_2O(\beta-Fe)} - RT\ ln\, P_{SO_3} \quad (5)$$

At a fixed temperature, the first and second term on the right-hand-side of Equation (5) will be constant. Therefore,

$$-nFE = [A] - RT\ ln\, P_{SO_3} \quad (6)$$

$$E = [B] + 2.3026(RT/nF)\ log\, P_{SO_3} \quad (7)$$

where ΔG is free energy change for net cell reaction, ΔG^0 is standard free energy change, R is the universal gas constant, 8.314 J/Kmol, T is temperature in Kelvin, $a_{K_2SO_4}$ is activity of K_2SO_4, $a_{K_2O(\beta-Fe)}$ is activity of K_2O in K-β-ferrite, *P* is the relative partial pressure, n is the number of electrons participating in the electrode reaction, F is the Faraday's constant, 96,500 C/mol, E is the measured e.m.f. of the sensor in millivolts, and A and B are constants. It can be clearly seen from Equation (7) that, at a fixed temperature, the measured e.m.f. of the sensor is expected to vary linearly with the logarithm of the partial pressure of SO_3.

EXPERIMENTAL

Materials

The reference electrode material used in the sensor is K-β-Fe_2O_3 ($K_2O{\bullet}9Fe_2O_3$). It is further mixed with Fe_2O_3 to form a two phase mixture so that the K_2O activity remains constant at a fixed temperature and partial pressure of oxygen. K-β-Fe_2O_3 was prepared by mixing $K_2(CO_3)$ and $Fe_2O_3(\gamma)$ in a 1:9 mole ratio and then calcining at 1273 K for 12 hours. The $K_2O{\bullet}9Fe_2O_3$ product was processed in the same way a second time to ensure complete homogeneity. The powder X-ray diffraction of the material indicated the complete formation of the single phase K-β-Fe_2O_3. This powder was mixed with pure Fe_2O_3 in a 2:1 ratio by weight to produce the biphasic reference material.

The material used for the auxiliary electrode was a high purity K_2SO_4. To apply this to the electrolyte tube, it was ground very finely and ethanol was added to make a paste. After administering potassium sulphate paste to the tube, it was left for the ethanol to evaporate.

K-β-Al_2O_3 powder was directly synthesized from the precursor powders $K_2(CO_3)$ and γ-Al_2O_3. They were mixed in a 1:8.5 mole ratio and ball milled with acetone to achieve uniform mixing. The slurry was dried and the powder was calcined at 1073 K for 2 hours and then at 1573 K for 6 hours. The complete formation of a single phase K-β-Al_2O_3 was confirmed by the powder X-ray diffraction. The K-β-Al_2O_3 powder was ground, transferred to tube moulds and pressed at 275 MPa in a cold isostatic press. Finally, the tubes were sintered for one hour at 1873 K to maximise strength and impermeability and to reduce the porosity.

For the closed system experiment, it was necessary to prepare NiO + $NiSO_4$ pellets. The anhydrous components were mixed in a 1:4 mole ratio, respectively and pelletised using a cold isostatic press. Compacting the powder decreased the volume and also reduced the extent of initial $NiSO_4$ decomposition.

Sensor Assembly

A schematic diagram of the SO_x sensor is shown in Figure 1. The construction of the sensor began with the attachment of the Pt wires to the electrolyte tube using Pt paste (Engelhardt type A4731) and heating to 573 K for 5 minutes, then 773 K for a further 5 minutes to completely burn the organic matter.

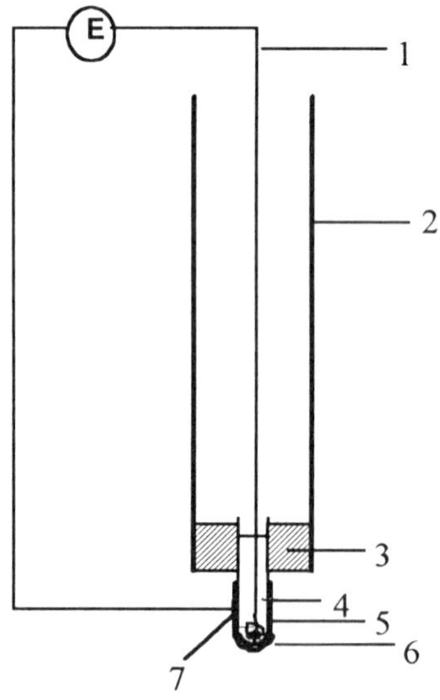

Figure 1.Schematic diagram of SO_x sensor: (1) Pt wire; (2) ceramic tube; (3) cement; (4) reference electrode material; (5) electrolyte tube; (6) auxiliary phase; (7) Pt paste.

The reference material was packed tightly inside the electrolyte tube and then the tube was cemented into position in an alumina tube using zirconia base ceramic adhesive. After the cement had completely set, the K_2SO_4 was applied to the outside of the electrolyte tube. The alumina tube and the outer Pt wire were then tightly fitted through a rubber bung so that the sensor was ready for use.

Measurements

A quartz apparatus consisting of a closed system was used for the first experiment which is shown in Figure 2.

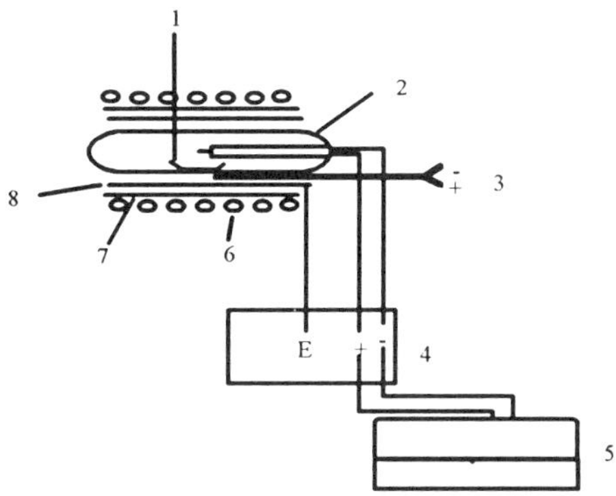

Figure 2. Schematic diagram of experimental set-up, closed system approach: (1) boat holding NiO + $NiSO_4$; (2) quartz apparatus for housing the sensor; (3) K-type thermocouple; (4) Keithley 614 electrometer; (5) Siemens X-t recorder; (6) furnace windings; (7) alumina furnace tube; (8) Faraday cage.

The NiO + $NiSO_4$ pellets were placed directly under the sensor tip so that the equilibrium concentration of SO_3 which emanated from the pellets according to the reaction:

$$NiSO_4(s) \rightarrow NiO(s) + SO_3(g) \quad (8)$$

would be instantaneously detected. The testing involved measuring the e.m.f. of the sensor as a function of temperature between 773 and 1023 K. The temperature of the sensor was measured using a room temperature compensated K-type thermocouple placed adjacent to the sensor tip. The Pt wires of the sensor were attached to the Keithley 614 high impedance (>10^{13} Ω) digital electrometer leads and the electrometer linked parallel to an x-t chart recorder for continuous e.m.f. recording. During this experiment, the steady state e.m.f. was recorded at each temperature of measurement. The activity of K_2O in K-β-ferrite + Fe_2O_3 biphasic mixture was calculated using the equation:

$$RT\, ln\, a_{K_2O(\beta-Fe)} = \Delta G^0_{(3)} + RT\, ln[1/P_{SO_3}] + 2EF \qquad (9)$$

where E is the measured e.m.f. of the sensor, $\Delta G^0_{(3)}$ is the standard Gibbs energy change for Reaction (3) and P_{SO_3} is the partial pressure of SO_3 obtained from the standard Gibbs energy change[11] for Reaction (8).

The second experiment involved the use of air + 5000 ppm SO_2 and air gas mixtures flowing through an open system to establish different partial pressures of SO_2 at fixed temperatures. The partial pressure was changed by adjusting the flow rate of each gas. The schematic diagram of the experimental rig is shown in Figure 3.

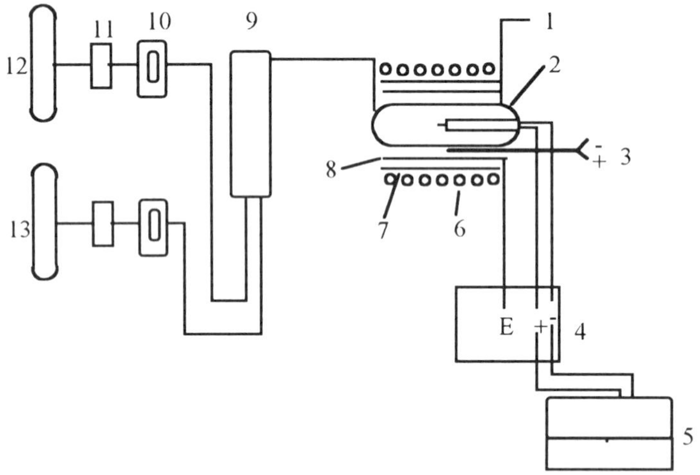

Figure 3. Schematic diagram of experimental set-up, open system approach:(1)gas outlet to water bubbler;(2)quartz apparatus for housing the sensor;(3)K-type thermocouple;(4)high impedance digital electrometer, Keithley 614;(5)Siemens X-t recorder;(6)furnace windings;(7)alumina furnace tube;(8)Faraday cage;(9)mixing column;(10)flow meter;(11)drying column;(12)5000 ppm SO_2 gas in air;(13)air.

Initial testing revealed that the sensor below 1000 K took approximately half an hour to reach a stable value each time after changing the partial pressure of SO_2, although nearly 75% of the stable e.m.f. value was always recorded within 60 s. However, for temperature above 1000 K the stable e.m.f was attained by the sensor instantaneously. Since the application of the sensor are in the atmosphere containing high oxygen partial pressure, air was deliberately used as a carrier gas for SO_2. The different flow rates used were as follows:

200 cc/min (5000 ppm SO_2 + air) only
100 cc/min (5000 ppm SO_2 + air) and 100 cc/min air
150 cc/min (5000 ppm SO_2 + air) and 50 cc/min air
50 cc/min (5000 ppm SO_2 + air) and 150 cc/min air

This cycle was repeated three times at 1007.5 K. The e.m.f. of the sensor at 1007.5 K was extremely stable over a long period of time, reproducible in each cycle and also reached equilibrium almost instantaneously after changing the gas composition. The e.m.f. values used to plot against $log\, P_{SO_3}$ were the average of six readings taken at an interval of half-an-hour immediately prior to changing the gas composition.

RESULTS AND DISCUSSION

The variation of the e.m.f. of the SO_x sensor to changing temperature in the first experiment is displayed in Figure 4.

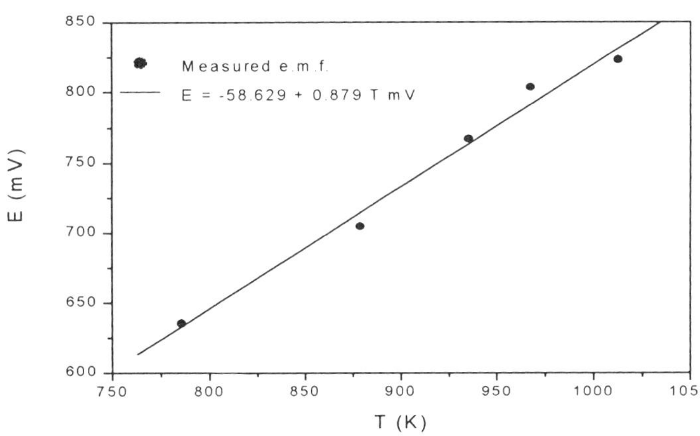

Figure 4. Variation of e.m.f. of SO_x sensor as a function of temperature in a closed system with NiO + $NiSO_4$ as an auxiliary electrode.

The results of this experiment were used to calculate activity of K_2O in Fe_2O_3 + K-β-Fe_2O_3 as a function of temperature using Equation (9). The partial pressure of SO_3 in Equation (9) is calculated at each temperature of measurement from the knowledge of the standard Gibbs energy change for the Reaction (8) reported in the literature[11].

The activity of K_2O in Fe_2O_3 + K-β-Fe_2O_3 calculated from the results obtained in this study is shown in Figure 5 as a function of temperature, alongside the data available for α-Al_2O_3 + K-β-Al_2O_3[12]. There is a good linear relationship between the temperature and the calculated activity of K_2O in K-β-Fe_2O_3.

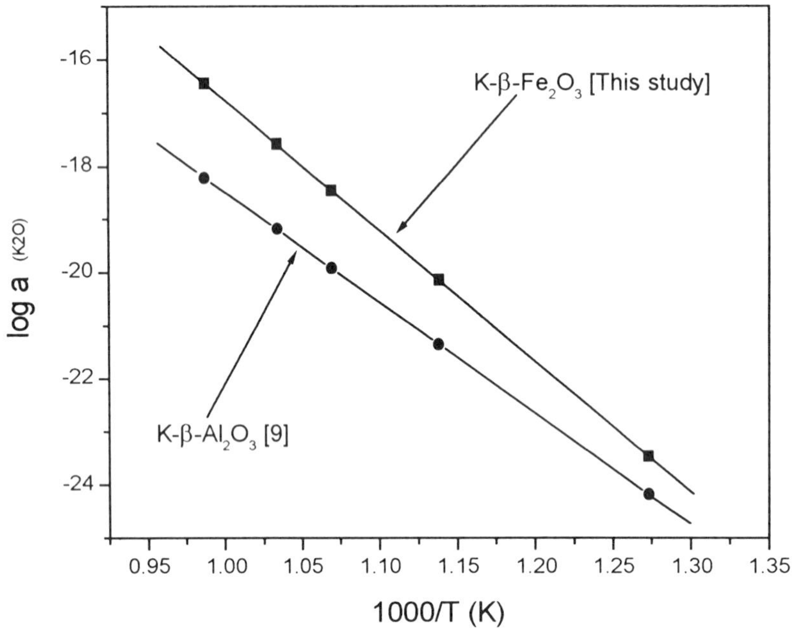

Figure 5. Comparison of the activity of K_2O in K-β-Al_2O_3 and K-β-Fe_2O_3.

The activity of K_2O in K-β-ferrite is higher by an order of magnitude compared to that in K-β-Al_2O_3 in the present temperature range of measurement. On examining the phase diagram of K_2O-Al_2O_3 system[13], it was noted that K-β-Al_2O_3 is stable up to about 2273 K whereas the phase diagram of K_2O-Fe_2O_3[13] indicates that K-β-Fe_2O_3 is stable only up to about 1723 K. Clearly, K-β-ferrite is less stable in relation to K-β-Al_2O_3 which suggests the activity of K_2O in K-β-ferrite must be higher than that in the corresponding K-β-alumina. This is in line with the results obtained in the present investigation. Further, we believe that the lower stability of K-β-ferrite relative to the K-β-alumina is because of the fact that the Fe^{3+} ions in oxides is thermodynamically unstable above 1623 K in air relative to Fe^{2+}[11] and hence β-alumina type structure breaks down. This type of reduction reaction does not occur in the case of Al^{3+} ions and hence the K-β-Al_2O_3 is stable up to 2273 K. The results obtained in the present investigation gives for the following reaction,

$$K_2O(s) + 9Fe_2O_3(s) \rightarrow K_2O{\bullet}9Fe_2O_3(\beta) \qquad (10)$$

$$\Delta G^0_{(10)} = RT\, ln\, a_{K_2O(\beta-Fe)} = -470{,}027.5 + 149.152T\ (\pm 1000)\ J/mol \qquad (11)$$

Rearranging Equation (11) gives:

$$log\, a_{K_2O(\beta-Fe)} = 7.791 - 24{,}552.5/T(K)\ (\pm 0.05) \qquad (12)$$

as an expression for the logarithm of the activity of K_2O in Fe_2O_3 + $K_2O{\bullet}9Fe_2O_3(\beta)$ as a function of temperature. Combining the Gibbs energy of formation of $K_2O{\bullet}9Fe_2O_3(\beta)$ given by Equation (11) with that for the formation of $K_2O{\bullet}9Al_2O_3(\beta)$ reported earlier by Kale and Jacob[12] gives for the exchange reaction:

$$K_2O{\bullet}9Fe_2O_3(\beta) + 9Al_2O_3(s) \rightarrow K_2O{\bullet}9Al_2O_3(\beta) + 9Fe_2O_3(s) \qquad (13)$$

$$\Delta G^0_{(13)} = 71{,}107.5 - 104.142\ T\ (\pm 1000)\ J/mol \qquad (14)$$

From the open system experiment, using the gas mixture containing air and 5000 ppm SO_2 in air, the response time of the sensor was found to be less than 60 s on average at 1007.5 K. The response time of the sensor was much longer at temperatures below 1000 K, probably because of the sluggish kinetics of the equilibrium reaction:

$$SO_2(g) + 1/2O_2(g) \rightarrow SO_3(g) \qquad (15)$$

in the absence of a catalyst. The response of the sensor to change in concentration of SO_2 in the gas mixture at 1007.5 K is shown in Figure 6.

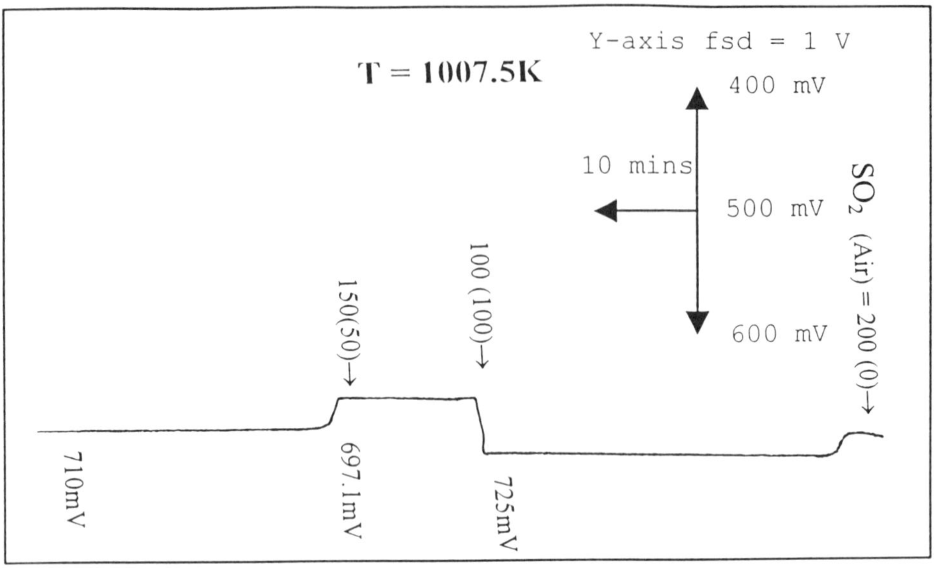

Figure 6. Response of SO_x sensor to change in air+5000 ppm SO_2 to air ratio at T = 1007.5 K.

The plot of the measured e.m.f. of the sensor at 1007.5 K against $log\, P_{SO_3}$ shown in Figure 7 reveals that the slope of the line is in excellent agreement with the theoretical slope, considering a two electron process. The variation of the measured e.m.f. of the sensor at 1007.5 K as a function of $log\, P_{SO_3}$ is given by the expression:

$$E = 996.15 + 94.849\, log\, P_{SO_3} \quad (16)$$

According to Equation (7), the slope of the line in Figure 7, which is equal to 94.849, can be used to calculate the value of n, the number of electrons involved in the electrode reactions, in order to confirm that whether the e.m.f. of the sensor obeys the Nernst relationship[14]. The results obtained in the present investigation give the value of n equal to 2.1 which is in excellent agreement with the theoretical value of 2, implying that the response of the sensor is Nernstian.

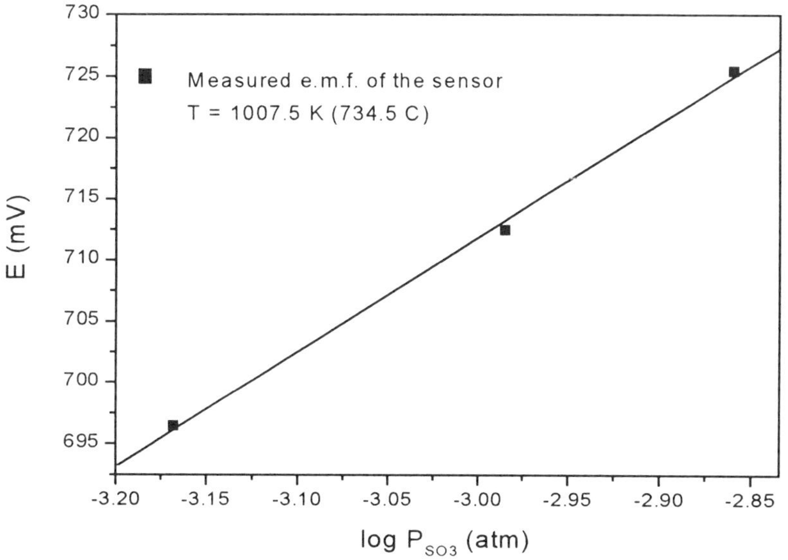

Figure 7. Variation of the e.m.f. of SO_3 sensor as a function of the logarithm of P_{SO_3} at 1007.5 K.

CONCLUSION

As a method to measure SO_x gas concentration in the flue gases at high temperature the sensor using K-β-Al_2O_3 solid electrolyte has produced valid results. The response of the sensor was found to be less than 60 s at 1007.5 K and the e.m.f. of the sensor confirmed nearly a Nernstian behaviour. The present investigation also enabled the determination of the thermodynamic stability of K-β-Fe_2O_3 ($K_2O{\bullet}9Fe_2O_3$) from the component oxides for the first time.

REFERENCES

[1]J. Janata and R. J. Huber in *Solid State Chemical Sensors*, Academic Press, Orlando, 1985.

[2]J. Fraden, "Chemical Sensors", pp. 532 - 546 in AIP *Handbook of Modern Sensors*, American Institute of Physics, New York, 1993.

[3]J. R. Bates and M. Campbell, "Gas Sensors and Analysers", pp. 127 - 178 in *Sensor Systems for Environmental Monitoring*, **Vol. 1**, M. Campbell (ed.), Blackie Academic and Professional, London, 1997.

[4]M. Gauthier and A. Chamberland, "Solid-State Detectors for the Potentiometric Determination of Gaseous Oxides", *J. Electrochem. Soc.*, **124** 1579-1583 (1977).

[5]H. Y. P. Hong, "Crystal Structures and Crystal Chemistry in the System $Na_{1+x}Zr_2Si_xP_{3-x}O_{12}$", *Mater. Res. Bull.*, **11** 173-182 (1976).

[6]S. Choi, W. Chung and D. Lee, "SO_2 Sensing Characteristics of NASICON Electrolytes", *Sensors and Actuators B*, **35-36** 263-266 (1996).

[7]D. J. Slater, R. V. Kumar and D. J. Fray, "A Bielectrolyte Solid-State Sensor which Detects SO_3 Independently of O_2. Implications for a SO_3 Sensor", *Solid State Ionics*, **86-88** 1063-1067 (1996).

[8]Y. Jianhua, Y. Pinghua and M. Guangyou, "A Fully Solid-State SO_x (x = 2, 3) Gas Sensor Utilising Ag-β"-Alumina as Solid Electrolyte", *Sensors and Actuators B*, **31** 209-212 (1996).

[9]N. Rao, O.T. Sorensen, J. Schoonman and C.M. van den Bleek, "Electrochemical SO_x and NO_x Sensors with Ag-β"-Alumina as Solid Electrolyte and Ag Solid as Reference Electrode", pp. 367-380 in *Beta Aluminas and Beta Batteries*, G. Staikov (ed.), Trans Tech Publications Ltd., Zurich, 1991.

[10]N. Rao, C. M. Van Den Bleek and J. Schoonman, "A Novel Temperature-Gradient Na^+-β"-Al_2O_3 Solid Electrolyte Based SO_x Gas Sensor without Gaseous Reference Electrode", *Solid State Ionics*, **53-56** 30-38 (1992).

[11]L. B. Pankratz, J. M. Stuve and N. A. Gokcen, p. 280, 284, 323, 324, in *Thermodynamic Data for Mineral Technology*, Bulletin No. 677, US Bureau of Mines, Washington DC, 1984.

[12]G. M. Kale and K. T. Jacob, "Thermodynamic Stability of K-β-Alumina", *Metall. Trans. B*, **20B** 687-691 (1989).

[13]L. P. Cook and H. F. McMurdie, p. 90, 213, 214, in *Phase Diagrams for Ceramists*, **Vol. 6**, The American Ceramic Society, Inc., Ohio, 1987.

[14]G. M. Kale, A. J. Davidson and D. J. Fray, "Investigation into an Improved Design of CO_2 Sensor", *Solid State Ionics*, **86-88** 1107-1110 (1996).

Processing

CERAMIC MEMBRANES FOR SEPARATION OF SODIUM

Glenn W. Hollenberg, Dean E. Kurath, and Kriston P. Brooks
Pacific Northwest National Laboratory
Box 999, Richland, WA 99352

Shekar Balagopal and Thom Landro
Ceramatec, Inc.
2425 S 900 W
Salt Lake City, UT 84119

ABSTRACT

Altlhough organic sodium membranes have been available for almost a centrury, ceramic sodium membranes, such as beta" alumina, are only a few decades old. NaSICON (**Na** **S**uper **I**on **Con**ductors) ceramic membranes, a recent development, offer unique capabilities for separating sodium from other soluble ions in solution. Researchers have successfully used NaSICON membranes for caustic recovery in electrochemical cells for partial salt splitting of radioactive tank waste with high levels of sodium. A more challenging application is the complete salt splitting of sodium sulfate from the paper and pulp industries, which potentially offers an environmentally and economically attractive method for addressing an industrial waste issue.

INTRODUCTION

NaSICON (**Na** **S**uper **I**on **Con**ductors) membranes are composed of ceramic crystallites with chemical compositions that result in unusually high transport rates for sodium ions near ambient temperature. In recent years, ongoing development efforts have been directed at using these unique electrochemical transport properties for several separation processes. The technology supporting these efforts is historically derived from both (1) the NaSICON materials development and (2) the cell engineering aspects of both beta-alumina-based-batteries and conventional organic (i.e., Nafion™) membranes.

The recent work at the Pacific Northwest National Laboratory (PNNL) and Ceramatec* has focused on producing membranes that could separate sodium from salt solution within an aqueous electrochemical cell to recycle caustics in nuclear waste processing and recycle both acids and caustics from industrial salt waste streams. By introducing a sodium ion conductor as a membrane between the cathode and anode, sodium ions can be transported from the anolyte through the membrane and reach the catholyte solution. By such transport, sodium can be separated from a variety of species (radioactive cations, anions, etc.).

Sodium separation membranes for use in electrochemical cells should possess several desirable properties. Low electrical resistance coupled with high sodium-transport efficiency ensure that most of the electrical energy is going to separate chemicals rather than ohmic (Joule) heating. The ability to achieve high current densities maximizes throughput. Cation and anion selectivity, i.e., transport numbers, is useful to ensure that pure products are produced that resist fouling. This paper will review the historical basis for using NaSICON and the recent cell engineering activities.

Table I lists a variety of sodium ion conductors. The discussion that follows describes the characteristics and uses of organic cation exchange membranes and β" alumina. Following that is a description of NaSICON membranes, including the background of their development, performance measurements, and potential applications. Then the potential applications of sodium ion conductors are discussed.

---

* Ceramatec Inc. is located in Salt Lake City, Utah.

Table I. Categories of Sodium Conducting Membranes

Identification	Composition	Conduction	Temperature °C	Stability	Issues
Nafion, etc.	Sulfonated Polymer	Hydrated ions	RT to 80	acid/base	fouling, selectivity, irradiation damage
β" Alumina	$(Na,Li)AlO_2$	cations (2D)	>250	non-aqueous	nonaqueous
Rare Earth NaSICON	$Na_5(RE)Si_2O_{12}$	cations (1D)	>RT	nonacidic	brittleness, corrosion
NZP	$Na_3Zr_2Si_2PO_{12}$	cations (3D)	>RT	nonacidic	brittleness, corrosion
Composite NaSICON	Multilayer	cations	RT to 80	acidic	TBD

where RE is Y, La, Dy, Nd, Gd, Sm, etc. and RT is room temperature

EARLY DEVELOPMENTS IN SODIUM ION CONDUCTORS

Organic Cation Exchange Membranes

Characteristics: As shown in Figure 1, these organic membranes are essentially porous polymeric films with sulfonated surface sites that control transport.[1-3] Although such membranes operate well with the pure sodium chloride solutions of the chlor-alkali industry, degradation of performance by fouling is a consideration in many waste streams. In addition, these membranes are relatively nonselective for cations. In particular, proton transport and back diffusion of hydroxide can consume more power for processing. Finally, the Teflon backbone structure used in most of these membranes is highly susceptible to irradiation damage at relatively low gamma doses.

Uses: Organic electrodialysis membranes have been used more than a century. In 1911, Donnan explained the manner in which negatively charged anions are repelled by the fixed charges in the polymeric membrane matrix (*termed Donnan Exclusion*) while permitting the cations to be transported.[1] These membranes have evolved into the basis for the "plate-and-frame" electrodialysis process for the chlor-alkali industry's production of caustic soda and chlorine. Buehler et al.[1] experimentally evaluated the "separation" of Na+ and Cs+ by electrodialysis with a Nafion™ 417 membrane. They found that cesium transported approximately 25% faster than sodium. Hence, in DOE's,

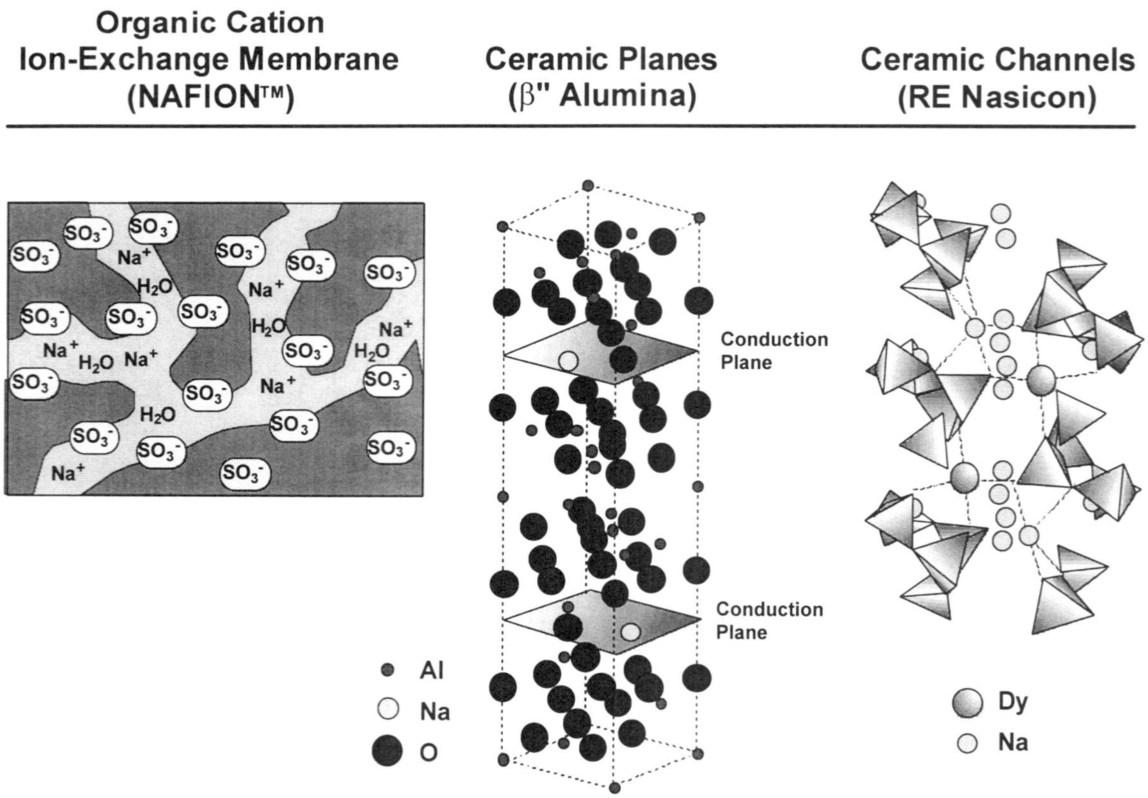

Figure 1. Schematic Comparison of the Structure of an Organic Cation Exchange Membrane (after Buehler) with the Structures of β" Alumina (after Landgrebe) and RE-NaSICON (after Maximov)

caustic tank-waste streams, where concentrations of sodium are much higher than cesium, application of electrodialysis would provide essentially no advantageous separation of the constituents.

β" Alumina

Characteristics: β" alumina is a ceramic electrolyte that permits sodium ions to reversibly be transported from pure sodium at the cathode to form sodium sulfide at the anode in sodium sulfur batteries. Operating temperatures of 300°C provides conductivities of 0.2 S/cm, which results in acceptable electrochemical efficiencies. Tubes up to 49 mm in diameter and 600 mm long have been fabricated. β" alumina has been evaluated extensively as a membrane for the sodium-sulfur battery programs throughout the world.

Uses: In 1967, early researchers at Ford Motor Co, Yu Yao and Kummer,[(4)] measured diffusional transport rates in β" alumina by tracer studies and found that sodium had a transport rate much faster than the larger, monovalent rubidium or cesium ions as shown in Figure 2. The ion selectivity of the ceramic was thought to be associated with (1) the size of the ion (2) the charge on the ion within the crystallite. For example, the Cs ion is almost twice the size of the Na ion in six-fold coordination in a crystal.[(5)]

Rassmussen[(6)] had conducted experimental electrorefining on sodium metal that had up to 4000 ppm of potassium present. The larger potassium ion concentration in the catholyte sodium metal was reduced by a factor of 5 after being electrochemically "pumped" through the β" alumina tube, even though potassium is chemically very similar to sodium.

Youngblood et al.[(7)] subsequently demonstrated the removal of nonradioactive cesium from molten sodium by β" alumina as a model for radioactive waste cleanup. After electrochemically refining cesium-doped sodium (1990 μg/gm in anolyte) through the β" alumina, the cesium concentration of the sodium on the cathode side remained at its original value of approximately 4 μg/gm during runs at 258 and 350°C. Resolution limits in the measurement of catholyte's molten sodium only permitted the decontamination factor to be defined as being greater than 500. A decontamination factor is the ratio of the element's transport rate divided be the transport rate of the sodium.

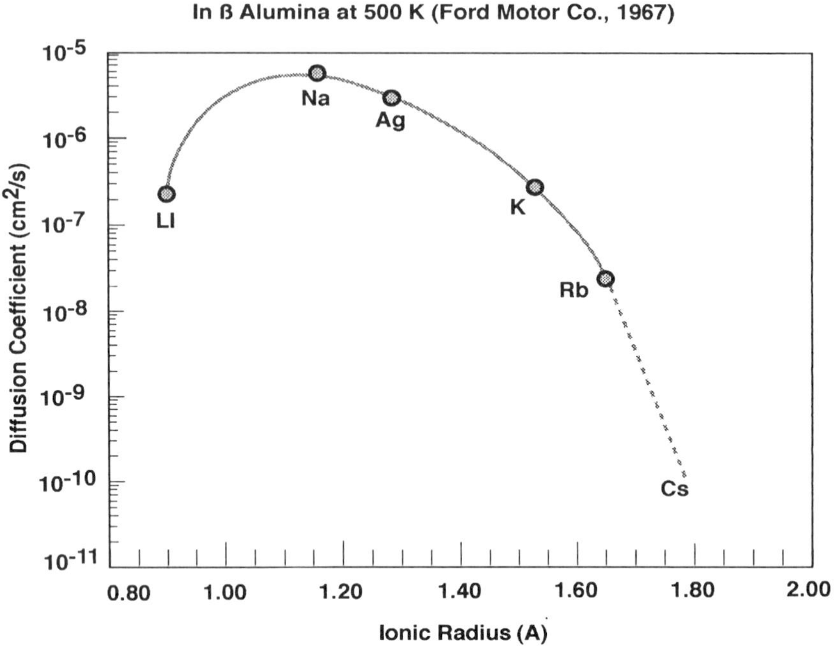

Figure 2. Effect of Ionic Radius on the Diffusion Coefficient Measured by Yao and Kuemmer as Represented by Hollenberg et al. Note that the cesium diffusion coefficient was so low that this technique could not measure it.

RECENT DEVELOPMENTS WITH NASICON SODIUM ION CONDUCTORS

Characteristics

In the 1970s, scientists searched for alternatives to β" alumina and developed ionic conductors with almost as good a conductivity as β" alumina. Goodenough et al. were perhaps the first to test $Na_3Zr_2Si_2PO_{12}$ (1976).[8] This family of sodium-zirconium-phosphate (NZP) compositions possesses a three-dimensional conduction path for sodium ions.[9] In these structures, the zirconium, silicon, and phosphate ions form a stable skeleton in which the more mobile sodium ions can be transported. Maximov et al.[3] independently reported that a variety of NaSICON compositions existed in the form of channel ion conductors with composition of $Na_5RESi_4O_{12}$ where RE represents rare earth elements such as Dysprosium, Gadolinium, Neodymium, etc. In Figure 1, the channels in these RE-Nascion membranes are shown with mobile sodium ions lined up along the c axis in sets of 4. The RE ions rest on the corners of some of the silicon tetrahedrons that surround the channels. Maximov further noted the relationship between the size of the RE ion and the mobility of the sodium ions. Larger RE ions provided larger channels that allowed a greater mobility of the sodium ions.

Uses

When it became evident that NaSICON membranes were incompatible with molten sodium metal, these materials were dropped in favor of β" alumina membranes for battery applications. However, for salt splitting in aqueous environments, the inverse in true, since β" alumina is incompatible with any type of aqueous environment.

In the past 10 years, Ceramatec, Inc. has been developing a variety of proprietary NaSICON membranes. The most advanced of these compositions are NASD, NASE, and NASG, which are synthesized by conventional processing techniques. The processed material was compacted to the desired shape and size in a die and punch assembly and then fired to achieve a dense ceramic with no surface porosity. The fired membranes were surface ground to a specific thickness consistent with the electrochemical cell design.

Performance Measurements

Table II lists a variety of efforts in recent years aimed at performance testing of the NaSICON membranes. Hollenberg,[(10)] Sutija,[(11)] Sutija,[(12)] Kurath,[(13, 14)] Stiemke,[(15)] and Hobbs[(16)] have reported tests on the use of RE NaSICON-type membranes for salt splitting of radioactive wastes. Kurath et al.[(14)] provided a simple schematic for an electrochemical cell for caustic removal from tank waste as shown in Figure 3. Schematically, a tank-waste solution is introduced as anolyte with radioactive impurities, including technetium, strontium, cesium, and nonradioactive impurities such as aluminum and iron. Sodium ions migrate to the catholyte through the NaSICON membrane as salt splitting occurs. The percentage of major constituents and radioactive impurities in these wastes varies widely with specific waste that would be processed. All of these membranes that were tested consisted of thin discs approximately 2.5 to 5 cm. in diameter that are sealed into a high-density polyethylene scaffold. This configuration was modeled after the classical plate-and-frame design used for organic membranes. Figure 4 is a photograph of the largest scaffold built to date at Ceramatec Inc., which contained 58 NASD ceramic discs with 5-cm diameters.

Table II. NaSICON Salt Splitting Testing

Author (date)	Membrane	Duration hrs	Current Density mA/cm^2	Na Kg	DF	t_{Na+}
Hollenberg (94)	NaSICON		250			0.95
Sutija (95)	RE	520	300	200	Al, Cs, Sr nondetected	Low (acid)
Sutija (96)	Coated Dry	1000	100	60		0.4 (acid)
Kurath (97)	NASD RE=Gd, Sm, Nd, DY	<1200	<500	60	Al 180 Cs 20000 K 15 Sr>4.5	1
Steimke (97)	NASD	170+140	146	4000		1
Hobbs (98)	NASD	174	--	64	See text	0.84-.99
Present	NASD	4000	25			0.9

Transport Efficiency and Conductivity: Sodium transport efficiency is important because large parasitic currents lead to poor electrical efficiencies. Although considerable work has been accomplished on RE-NaSICON membranes, recent efforts at Ceramatec have been focused on a new NaSICON rather than RE-NaSICON because of its higher conductivity and sodium-current efficiency. Kurath et al.[(13)] found that for caustic separation from simulated tank waste that sodium current efficiencies of 90% may be achieved over hundreds of hours of operation with the use of NASD membranes. In contrast, neodymium-based membranes, Nd-NaSICON, in the same study possessed current efficiencies of less than 25%. In Figure 5, a lifetime test of a NASD type of membrane is presented in which 1.5 M NaOH was used as the anolyte. The sodium current efficiency, i.e., the ratio of sodium predicted in catholyte to equivalent sodium transported as predicted from total electrical current, is 91% for the life of this test, which lasted for more than 4000 hours.

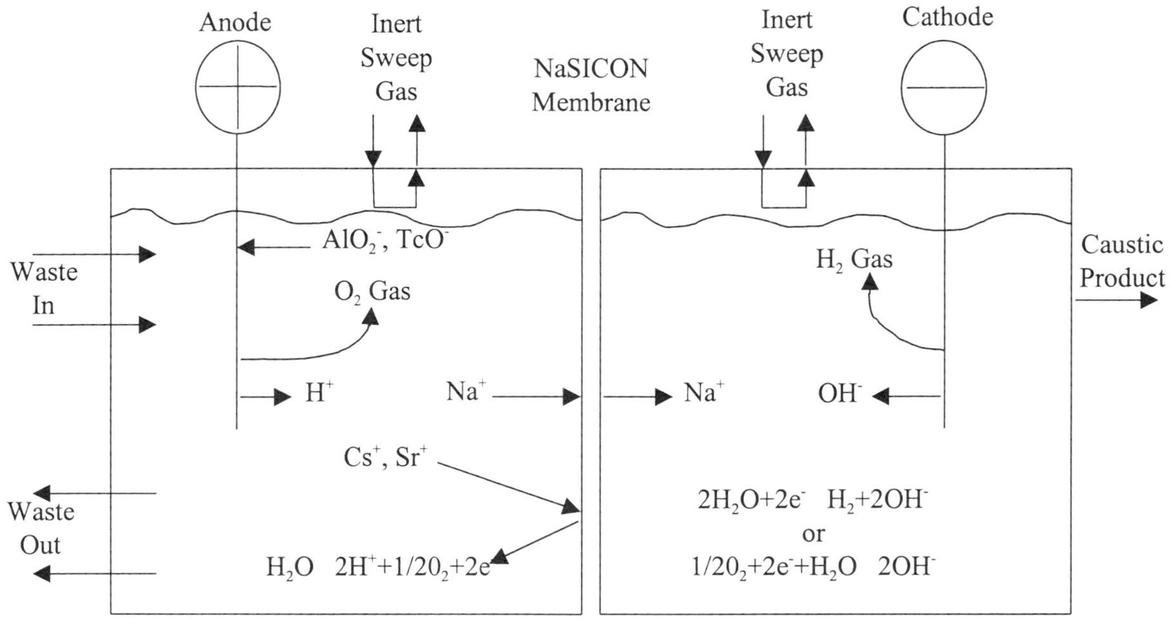

Figure 3. Schematic of an Electrochemical Process for Salt Splitting Using the NaSICON Membrane (after Kurath)

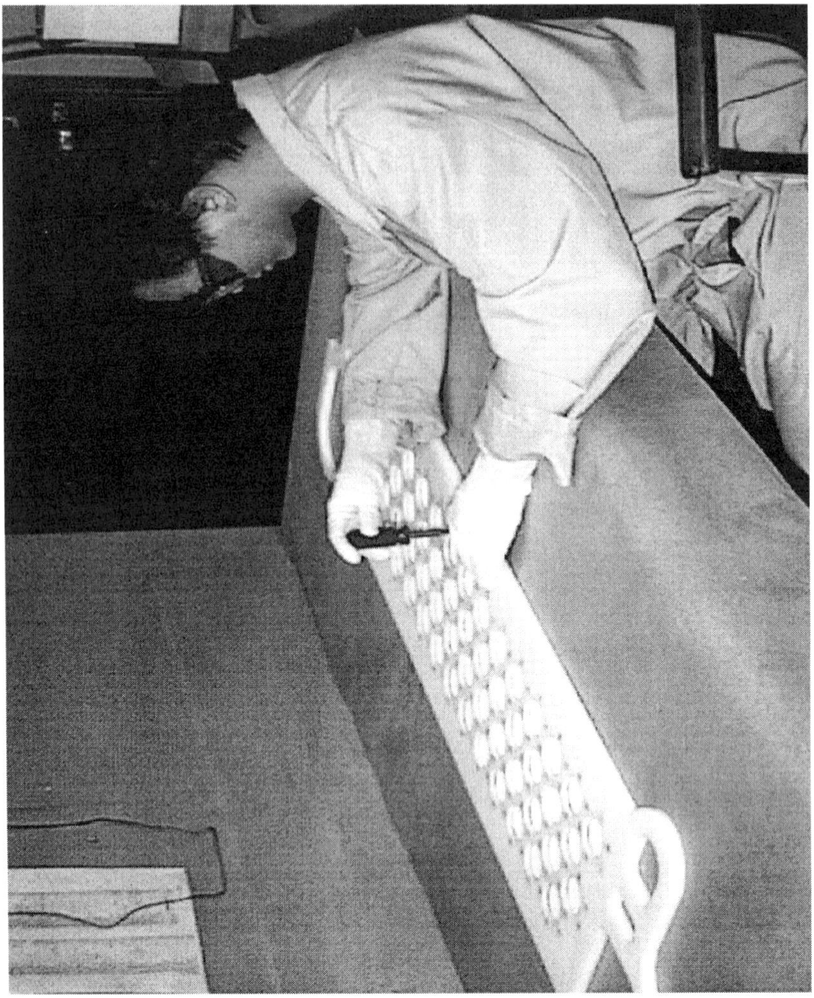

Figure 4. Ceramic Membrane Scaffold Containing 58 Disks of Approximately 5-cm Diameter Manufactured by Ceramatec Inc.

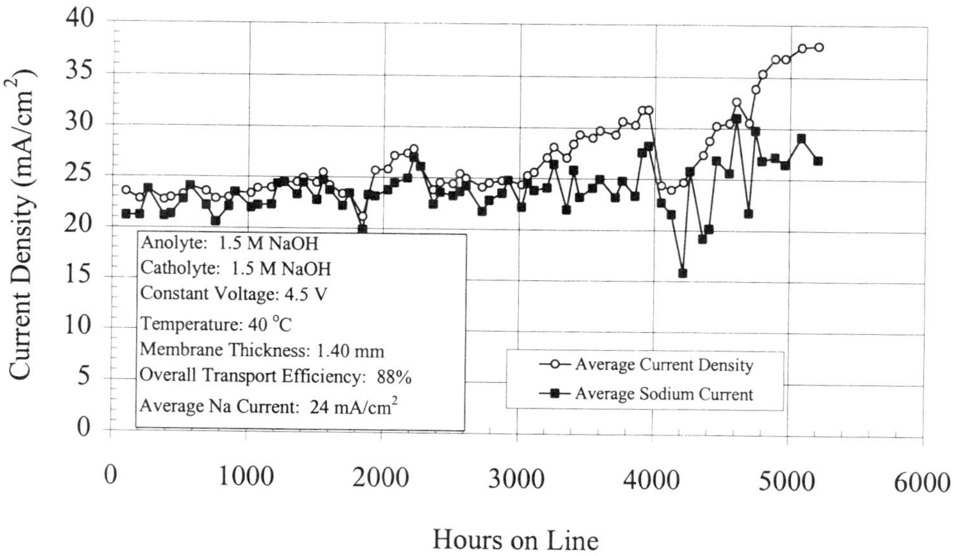

Figure 5. Lifetime Testing of an NASD-Type Membrane Demonstrating the High Sodium Current Efficiency for over 4000 Hours with a 1.5 Molar Sodium Hydroxide Anolyte and Catholyte. Overall sodium transport efficiency was 91%.

In addition, Kurath et al.[13] found that RE-NaSICON ion conductor membranes tended to microcrack extensively during operation. They explained this by an apparent 0.7% volume expansion that occurred as hydration of the crystal structure proceeds.

A new series of NaSICON compositions (NASD, NASE, and NASG) were developed at Ceramatec, which showed an order of magnitude higher sodium ionic conductivity than Dy-NaSICON at ambient temperatures.[17] The highest conductivities measured in these materials were approximately 0.02 S/cm at 64°C, which is equivalent to β" alumina at 300°C.

Fouling: Tank waste compositions present a significant potential for fouling any membrane. Kurath conducted an experiment with Nd-NaSICON using a simulated tank waste containing a caustic solution of 5 M Na^+ and 0.5 M AlO_2^- and numerous other cations and anions that are predicted to be in neutralized current acid waste in Hanford tanks.[14] In caustic aqueous environments, AlO_2^- ions are very soluble at high pH levels and relatively insoluble at neutral pH levels. In Figure 6 at approximately 475 hours, the current density dropped dramatically when sodium ions were not replenished in the anolyte compartment. Simultaneously, the pH was observed to drop below 12, which led to precipitation of large amounts of white gibbsite solids {$Al(OH)_3$}. Upon replenishing the anolyte feed stream with sodium (pH>12), the solids did not redissolve, but the cell appeared to continue to operate well with no fouling having taken place. During cell disassembly, the extensive precipitation was still apparent. Hence, under these extreme conditions, fouling did not appear to alter the performance of this ceramic membrane. This observed resistance to fouling is attributed to the high degree of ionic selectivity present in the NaSICON crystal structure.

Pilot Scale Testing: Steimke et al. used the scaffold shown in Figure 4 to conduct two pilot-scale tests of a NASD electrochemical cell on 35 gallons of solutions containing hydroxide waste simulants.[15] In the first run, an anolyte consisting of 1.3 M sodium hydroxide and 1.8 M sodium nitrate was included while in the second run a nitrate destruction process was assumed to have occurred, and 3.5 M of sodium hydroxide was present without any significant level of sodium nitrite. Each simulant also contained 0.3 M sodium aluminate and lower levels of sodium nitride, sodium sulfate, and sodium carbonate. During the first week-long run, approximately 116 $\pm$ 14 moles (2.7 Kg) of sodium were transported into the catholyte. This amount represents approximately 20% of the total sodium in the anolyte simulant. In the two runs, sodium current efficiencies as measured by catholyte-sodium-gained were approximately 100%, although the measurement technique employed only possessed an accuracy of ±13%. Although a milky precipitate of gibbsite was formed, the current and voltage were stable during the run.

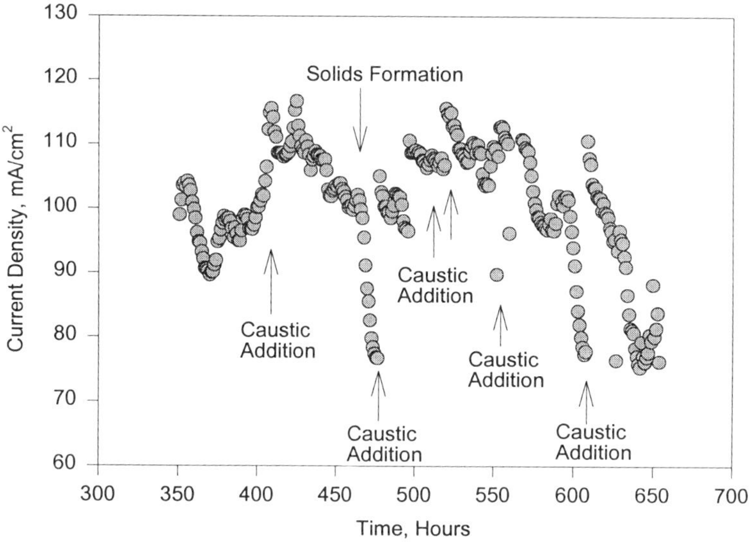

Figure 6. Current Density Versus Time for a Salt Splitting Run. Note that at 475 hours, sodium depletion in the anolyte caused the current to decrease, but after being replenished, the current regained its original level.

Radioactive Testing: Hobbs[16] has recently completed electrochemical separation of caustic from actual radioactive waste from the Salt Stone Treatment Plant with both Nafion™ and NASD membranes. Again, high electrical efficiencies were observed with the NASD membrane and very little water transport in comparison to the Nafion™ membrane. As would be expected from Buehler's work, the Nafion™ membrane transported much of the ^{137}Cs into the catholyte whereas the NASD membrane transported little of the ^{137}Cs into the catholyte as seen in Table III. The percent transported ^{137}Cs was calculated by dividing the final ^{137}Cs (total) in the catholyte by the total amount of ^{137}Cs originally in the anolyte. Hence, some error is introduced in the total quantity measured. However, it is easily observed that very small quantities of ^{137}Cs were transported with the ceramic membrane.

Table III. Caustic Recovery From Actual Low-Level Alkaline Waste

Membrane	^{137}Cs Retained in Anolyte	^{137}Cs Transported to Catholyte	Na Transported (moles)
Nafion 350	32 %	71 %	2.98
Nafion 350	13 %	128 %	4.97
NAS D	88 %	1 %	1.51
NAS D	94 %	2 %	1.29

Compatibility and Lifetime: In basic or neutral environments, like the 4000-hour test shown in Figure 4, the NASD ceramics have been observed to be stable and relatively inert. In the past few years, lifetime testing has been extended to longer and longer tests to evaluate this parameter. It is anticipated that lifetimes approaching 6000 to 7000 hours will be required to make these membranes economically attractive. However, in acidic environments, the ceramic membranes can be severely attacked. For radioactive tank waste, much of the benefit of sodium recycle can be achieved without complete removal of sodium from the waste stream. Consequently, tank waste processing need not encounter acidic conditions. However, for the paper and pulp mill applications, it is essential that sulfuric acid be one of the byproducts. Hence researchers have explored the use of coatings to obtain the selectivity of the NaSICON membranes without acidic corrosion. Sutija attempted to achieve acid compatibility by applying acid-resistant, ceramic surface coating directly to a NaSICON membrane.(12) Alternatively, Izquierdo et al. have applied a NaSICON coating onto an organic membrane by a pulsed laser deposition technique.(18)

Ceramic membranes will always be more vulnerable to brittle fracture. Within the past testing programs, cracked discs have been observed. Cracking is thought to be associated with material physical instability, as in the case for RE-NaSICON since stresses on the membrane are more controllable. But in many other ceramic applications, this inherent brittleness has been addressed in two ways:

(1) Component quality improvements by reduction of flaw size in the as-manufactured component. Even though pilot scale testing has been

conducted, it is still possible to improve the fundamental quality of each NASCION disc produced.

(2) Design modifications to reduce tensile and shear stresses in the material. Organic membranes can tolerate low stresses but relatively high strains, while ceramics are inverse to this relationship. The present scaffold design is based upon organic membrane technology and may not be as tolerant as other designs that permit the ceramic to bear homogeneous compressive loads and be less strained.

POTENTIAL APPLICATIONS OF SODIUM ION CONDUCTORS

Sodium ion conductors have a number of potentially valuable applications. Three examples of applications considered for electrochemical separation using ceramic membranes are discussed here.

Metallic Sodium Cleanup

Rasmussen[6] in the 1980s examined the use of β" alumina in the electrochemical winning or electrorefining of high purity sodium from either low grade sodium or $NaCl$-$AlCl_3$ salt mixtures. In the 21st century, sodium from sodium-cooled nuclear reactors such as the Hallam, Fast Flux Test Facility, Experimental Breeder Reactor II, and Fermi I will be treated in preparation for disposal. Although actinides, radionuclides, and hazardous (Hg) impurities may be present, ^{137}Cs is the highly radioactive impurity that is thought to be difficult to chemically separate from the host sodium. Electrorefining would provide a simple yet effective method for removing cesium and other unwanted impurities before processing waste.

Aqueous Waste Streams Dominated by Sodium Salts

In the past, sodium sulfate waste from the paper and pulp industry has either been sent to landfills or sewers for disposal. An economically viable process for sodium sulfate salt splitting and recycling does not presently exist. The NaSICON electrochemical separation would produce sodium hydroxide and sulfuric acid from the sodium sulfate waste produced by chlorine dioxide generators and other processes:

$$Na_2SO_4 + 3\ H_2O = H_2SO_4 + 2\ NaOH + 1/2\ O_2 + H_2 \qquad (1)$$

Most of these products could be recycled immediately at the mills.[12]

Radioactive Contaminated Salt Cake and Supernate

For the past half-century, nuclear fuel reprocessing facilities in the United States generated large quantities of radioactive acid waste streams. In order to store these waste streams in mild steel tanks, the streams were neutralized with nonradioactive sodium hydroxide before storage as salts or highly alkaline solutions. At the Hanford site, 330 metric tons of sodium salts are distributed within the 60,000,000 gallons of liquids and solids in 177 underground tanks. Today, the U.S. Department of Energy (DOE) is addressing the long-term isolation of this legacy by manufacturing leach-resistant (i.e., low-soda) glasses for high-level wastes. Low-level waste forms (i.e., high-soda glasses or grout with low levels of radioactivity) will be used to immobilize the soda-bearing waste. Thus, the separation of tank wastes into a low-soda, high-level waste and a high-soda, low-level-waste is a primary goal of tank remediation activities.

Many of the treatment options of these wastes from tanks to waste storage find it desirable and sometimes necessary to add sodium hydroxide. For example, sodium hydroxide is used in the caustic leaching of alumina from sludges that contain actinides. If pure sodium hydroxide can be extracted near the end of the process stream, then additional sodium need not contribute to the low-level waste processing and repository. Hence, a method for extracting sodium hydroxide from the waste streams offers an attractive alternative to adding more sodium (>15%) to the already immense low-level waste repository. The overall goal of sodium removal from radioactive wastes is the partial removal of sodium hydroxide via salt splitting. It should be noted that most of the advantage in this process can be gained without ever reducing the pH of the anolyte to less than 13.5, which is much different from the paper mill application, above, in which the anolyte becomes very acidic.

CONCLUSIONS

In the last decade, significant advances have occurred in the technology associated with the use of ceramic membranes for separating sodium from other constituents. Ceramic membranes appear to possess a high degree of ionic selectivity that is useful in reducing parasitic currents in electrochemical cells, reducing fouling, and separating sodium from radionuclides during salt splitting (i.e., producing a clean caustic product). Ceramic membranes, like any other ceramic, suffer from brittleness and lower throughput than organic membranes, which will necessitate both quality and design improvements in the future. This unique class of room-temperature ionic conductors, which is stable in certain aqueous environments, presents imaginative ceramists and chemical engineers with a new dimension on processing in a variety of sodium-based waste and chemical systems.

ACKNOWLEDGMENTS

This work was funded by the Office of Science and Technology, within the Department of Energy's Office of Environmental Management, under the Efficient Separations and Processing Crosscutting Program. Pacific Northwest National Laboratory is operated for the U.S. Department of Energy by Battelle under Contract DE-AC06-76RLO 1830.

REFERENCES

1. M.F. Buehler, W.E. Lawrence, and J.D. Norton, *Evaluation of Aqueous Na^+/Cs^+ Separation by Electrodialysis*, PNL-9022, Pacific Northwest Laboratory, Richland, WA (1993).

2. A.R. Landgrebe, R.D. Weaver, and F.K. Sen, *Sodium-Sulfur Batteries*, Volume 87-5, The Electrochemical Society, Inc., 10 S. Main St., Pennington, NJ 08534 (1987).

3. B.A. Maximov, I.V. Petrov, A. Rabenau, and H. Schulz, "X-Ray Investigations and Possible Mechanism for the Ionic Conductivity in the $NasRLSic_1O_{12}$ (RE=Y,Sc) Class of Compounds," *Sol. St. Ionics*, **6**, 195-200 (1982).

4. Y.F. Yu Yao and J.T. Kummer, "Ion Exchange Properties of and Rates of Ionic Diffusion in Beta-Alumina," *J. Inorg. Nucl. Chem.*, **29**, 2453-2475 (1967).

5. R.D. Shannon, "Revised Effective Ionic Radii and Systematic Studies of Interatomic Distances in Halides and Chalcogenides," *Acta Cryst.*, **A32**, 751-766 (1976).

6. J.R. Rasmussen, G.R. Miller, and R.S. Gordon, "Electrorefining of Sodium Using Sodium Beta-Alumina Ionic Conductors," *Electrochemical Society Meeting,* May 9-14, 1982.

7. G.E. Youngblood, D.J. Senor, D.E. Baldwin and G.W. Hollenberg, "Electrochemical Refining/Separation of Sodium Using Solid Membrane β" Alumina and NaSICON," Proceedings of *Environmental Electrochemistry Symposium*, Richland, WA 1994.

8. J.B. Goodenough, Y.P. Hong and J.A. Kafalas, "Fast Na^+ Ion Transport in Skeleton Structures," *Mater. Res. Bull.*, **11**, 203 (1976).

9. J. Alamo and R. Roy, "Crystal Chemistry of the $NaZr_2(PO_4)_3$, NZP or CTP, Structure Family." *J. Mat. Science*, **21**, 44-450 (1986).

10. G.W. Hollenberg, C. Carlson, A. Virkar, J. Gordon and A. Joshi, "Salt Splitting of Sodium-Dominated Radioactive Waste using Ceramic Membranes," *Proceedings of Spectrum '94, Nuclear and Hazardous Waste Management International Topical Meeting*, American Nuclear Society, Aug 1994.

11. D.P. Sutija, B. Webb, S. Balagopal, J.F. Smith, A.V. Virkar, and G.W. Hollenberg, "Selective Sodium Removal from Radioactive Waste Streams," Proceeding of the 1st International Symposium on Ceramic Membranes, to be published by the Electrochemical Society, Chicago, IL, Nov. 1995.

12. D. Sutija, S. Balagopal, T. Landro, and J. Gordon, "Ceramic Cleansers," *Electrochem. Soc. Inter.*, **5**, 26-30 (1996).

13. D.E. Kurath, K.P. Brooks, G.W. Hollenberg, S. Balagopal, T. Landro, and D.P. Sutija, *Caustic Recycle from Nonradioactive Tank Waste Simulants using a Ceramic Membrane Salt Splitting Process: FY1997 Experimental Results*, August 1997, Pacific Northwest National Laboratory, Richland, WA (1997).

14. D.E. Kurath, K.P. Brooks, G.W. Hollenberg, D.P. Sutija, T. Landro, and S. Balagopal, "Caustic recycle from high-salt nuclear wastes using a ceramic-membrane salt-splitting process," *Separ. Purifica. Tech.*, **11**, 185-198 (1997).

15. J.L. Steimke, D.T. Hobbs, M.D. Fowley, and T.J. Steeper, *Caustic Recovery using the TFL Electrochemical Reactor,* WSRC-TR-00355, Westinghouse Savannah River Company, Aiken, SC (1997).

16. D.T. Hobbs. *Radioactive Demonstration of Caustic Recovery from Low-Level Alkaline Waste by an Electrochemical Separation Process*, WSRC-TR-97-00363, Rev 0, Westinghouse Savannah River Company, Aiken, SC, (1998).

17. R.D. Shannon, B.E. Taylor, T.E. Gier, H.Y. Chen and T. Berkins, "Ionic Conductivity in Na_5 YSi_4O_{12} Type Silicates," Inorganic Chemistry **17** [4] 969-973 (1978).

18. R. Izquierdo, E. Quenneville, D. Triylidas, F. Girard, M. Meunier, D. Ivanov, M. Paleologou, and A. Yelon, "Pulsed Laser Deposition of NASICON Thin. Films for the Fabrication of Ion Selective Membranes, J. Electrochem. So. **144**, [12] L323-L325 (1997).

DEVELOPMENT OF MIXED-CONDUCTING CERAMIC MEMBRANES FOR HYDROGEN SEPARATION

Jie Guan, Stephen E. Dorris, Uthamalingam Balachandran
Energy Technology Division, Argonne National Laboratory, Argonne, IL 60439

Meilin Liu
School of Materials Science and Engineering, Georgia Institute of Technology, Atlanta, GA 30332

ABSTRACT

$SrCeO_3$- and $BaCeO_3$-based proton conductors have been prepared and their transport properties have been investigated by impedance spectroscopy in conjunction with open circuit voltage and water vapor evolution measurements. $BaCe_{0.8}Y_{0.2}O_{3-\delta}$ exhibits the highest conductivity in a hydrogen-containing atmosphere; however, its electronic conductivity is not adequate for hydrogen separation in a nongalvanic mode. In an effort to enhance ambipolar conductivity and improve interfacial catalytic properties, $BaCe_{0.8}Y_{0.2}O_{3-\delta}$ cermets have been fabricated into membranes. The effects of ambipolar conductivity, membrane thickness, and interfacial resistance on permeation rates have been investigated. In particular, the significance of interfacial resistance is emphasized.

Keywords
Mixed conductors, hydrogen separation, transport properties, perovskites

INTRODUCTION

Mixed ionic-electronic conductors (MIECs) have been studied for application in many solid-state electrochemical systems such as solid oxide fuel cells (SOFCs), chemical sensors, and membranes for gas separation [1-7]. One well-known class of MIECs consists of partially substituted perovskite-type oxides such as $CaZrO_3$, $SrCeO_3$, and $BaCeO_3$, in which substitution of Ce or Zr by trivalent cations introduces oxygen vacancies and other charged defects and leads to mixed conduction in atmospheres containing O_2, H_2, and H_2O vapor.

These materials exhibit significant proton conduction in a hydrogen-containing atmosphere and are used for various applications. $CaZrO_3$-based ceramics are being used as a hydrogen and humidity sensor [3]; $BaCeO_3$-based prototype SOFCs are in development and significant progress has been made [4]. Meanwhile, dehydrogenation of hydrocarbons, electrolysis of steam, and pumping of hydrogen have also been demonstrated [5].

Because oxygen ion/electron conductive ceramics are being studied intensively as membranes for oxygen separation [6], it is immediately evident that dense ceramic membranes fabricated from proton/electron-conducting ceramics could also provide a simple and efficient means of separating hydrogen from gas steams and offer an alternative method of hydrogen recovery. To be suitable for hydrogen separation in a nongalvanic mode, the ceramic membranes must have sufficient protonic and electronic conductivities to achieve high permeability and selectivity to hydrogen. In addition, the materials must exhibit high catalytic activity for oxidation and evolution of hydrogen at the solid/gas interfaces.

In this study, doped $SrCeO_3$ and $BaCeO_3$ were prepared and their transport properties were investigated. Dense $BaCe_{0.8}Y_{0.2}O_{3-\delta}$ cermet membranes exhibited high hydrogen permeation rates. The dependence of permeation rate on membrane thickness and on interfacial polarization was characterized.

MEMBRANE SEPARATION MECHANISMS

In a hydrogen pump, hydrogen can be pumped by a voltage applied to the ceramic conductor from the anode to the cathode side. The hydrogen evolution rate is governed by Faraday's law, and the current efficiency depends on the transport properties of the conductor, such as the proton transference number under the operating conditions. Alternatively, hydrogen can be pumped by shorting the external circuit, where the driving force is the electrochemical potential difference across the proton conductor.

A simple and efficient approach is to use a mixed conducting membrane without electrodes. Under the conditions of ambipolar diffusion, the current density due to the motion of protons can be expressed as [8]

$$J_{H^+} = \frac{\sigma_{amb}}{L}\left(E_N - \eta\right) \tag{1}$$

where L is the thickness of the membrane, η is the interfacial overpotential, E_N is the Nernst potential imposed across the membrane, and σ_{amb} is the ambipolar conductivity of the membrane defined as

$$\sigma_{amb} = \frac{\sigma_e \sigma_{H^+}}{\sigma_e + \sigma_{H^+}}, \tag{2}$$

if the interaction between electrons and protons is negligible. The permeation flux of hydrogen, N_{H_2}, is related to the proton current density by

$$N_{H_2} = \frac{J_{H^+}}{2F}. \tag{3}$$

It is clear that the permeation flux of hydrogen through the MIEC membrane can be improved by enhancing ambipolar conductivity and reducing the membrane thickness and interfacial overpotential.

The interfacial overpotential η and ambipolar conductivity σ_{amb} can be determined by measuring permeation rates through membranes with different thicknesses. If Eqs. 1 and 3 are rewritten as [9]

$$\frac{E_N}{2FN_{H_2}} = \frac{\eta}{2FN_{H_2}} + \frac{L}{\sigma_{amb}}, \tag{4}$$

then the interfacial polarization resistance, R_p, can be estimated from

$$R_p = \left(\frac{E_N}{2FN_{H_2}}\right)_{L \to 0}, \tag{5}$$

while ambipolar conductivity can be determined as

$$\sigma_{amb} = \left[\frac{\partial\left(E_N / 2FN_{H_2}\right)}{\partial L}\right]^{-1}. \tag{6}$$

EXPERIMENTAL

Sample Preparation

$SrCeO_3$-and $BaCeO_3$-based ceramic powders and pellets were prepared as described in Refs. 10 and 11. Pt paste was applied to polished surfaces of the pellets and fired in air at 1200°C for 12 min to form porous electrodes before electrochemical characterization. Cermet membranes were prepared by mixing metallic powder with $BaCe_{0.8}Y_{0.2}O_{3-\delta}$. After being ball-milled for 24 h, the resulting powders were pressed at 200 MPa and then sintered at 1400°C for 5 h in 4% H_2 balance argon.

Impedance Spectroscopy

The ceramic pellets with two Pt electrodes were immersed in various atmospheres at different temperatures (600-800°C). Total conductivity was measured with an impedance analyzer (HP 4192A) in the frequency range of 13

MHz to 5 Hz under open-circuit conditions. Water vapor was introduced by bubbling desired gases through deionized water at room temperatures.

Ionic Transference Numbers

Transference numbers were determined by measuring either open-circuit voltage (OCV) or water vapor evolution rates. OCVs of fuel cells and concentration cells, measured by a high-impedance digital multimeter (HP 3446A), were used to estimate the ionic and electronic transference numbers. The ionic transference numbers so obtained may be underestimated in some cases because of interfacial polarization [12]. Evolution of water vapor by discharging an O_2/H_2 fuel cell was used to determine the ratio of proton and oxide ion transference numbers to the total ionic transference numbers. The gas inlets in the setup were positioned as close as possible to the test sample. Gas flow rates were adjusted to 50-100 cm^3/min. Water vapor concentration in both cathode and anode sides was measured with a digital hygrometer (Fisher Scientific) under open- and short-circuit conditions.

Hydrogen Pumping and Water Vapor Electrolysis

A 2.09-mm-thick pellet of $BaCe_{0.8}Y_{0.2}O_{3-\delta}$ with three electrodes was affixed to a test setup (Fig. 1) with a glass sealant. Hydrogen was pumped from the anodic to the cathodic side by applying a constant DC current. The anodic side was fed with 4% H_2 balance N_2 at 86 cm^3/min and the cathodic side was swept by Ar at 43 cm^3/min. The anodic overpotentials were directly determined by measuring the potentials between the anode and the reference electrode. In water vapor electrolysis, Ar saturated with water vapor at room temperature was fed to the anode side at 86 cm^3/min, and the cathode side was also swept with Ar at 43 cm^3/min. Gas concentrations were analyzed with a gas chromatograph (HP 5830A) and a hygrometer.

Hydrogen Permeation through Cermet Membranes

Polished cermet membranes with three different thicknesses (0.095, 0.144, and 0.205 cm) were also affixed with a glass sealant to one end of Al_2O_3 tubes, as illustrated in Fig. 2. The effective membrane area for permeation was 1.4 cm^2. A sample gas (4% H_2 balance N_2) was fed to one side at a rate of 86 cm^3/min, while a carrier gas (Ar) was swept to the opposite side of the membrane at a rate of 43 cm^3/min. The gas concentrations were analyzed with a gas chromatograph that was periodically calibrated with a standard gas. Gas leakage through the setup was tested with helium as the feeding gas; typical leakage was <10% of the total permeation flux.

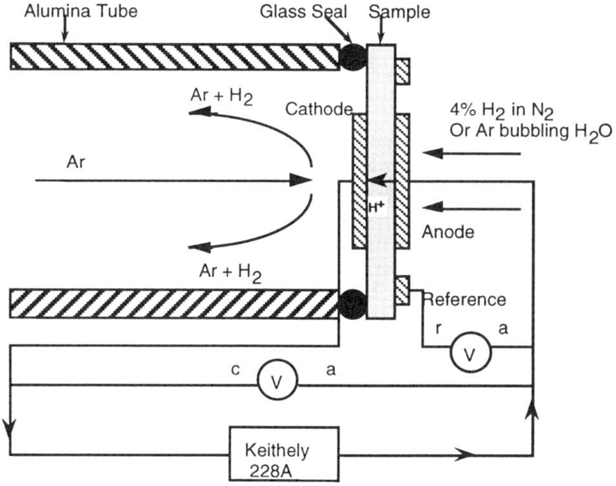

Fig. 1 Schematic illustration of hydrogen pumping and water vapor electrolysis.

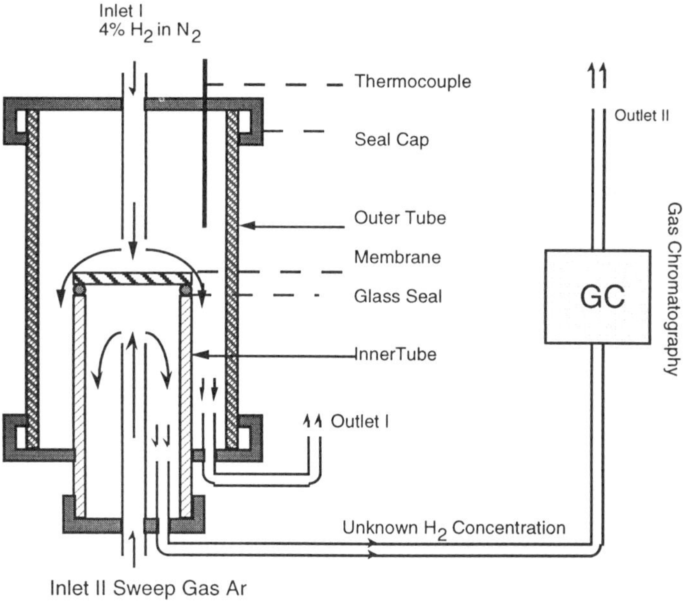

Fig. 2 Schematic illustration of setup for hydrogen separation.

RESULTS AND DISCUSSION
Conductivity

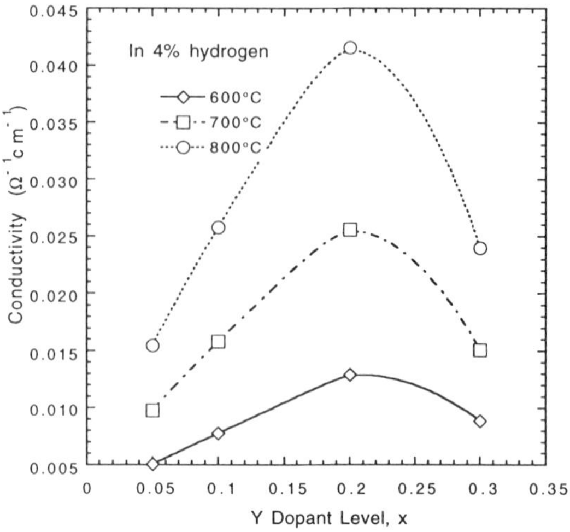

Fig. 3 Conductivity of $BaCe_{1-x}Y_xO_{3-\delta}$ in 4% hydrogen at various temperatures.

$BaCeO_3$-based materials exhibit higher total conductivities than $SrCeO_3$-based materials in both hydrogen and oxygen [11]. Conductivity of $BaCe_{1-x}Y_xO_{3-\delta}$ varied with the doping concentration x (Fig. 3). At a given temperature, conductivity measured in 4% H_2, which is dominated by ionic conduction, increased monotonically with dopant concentration from x = 0.05 to 0.2, then decreased for an Y-concentration of 0.3, suggesting that the conductivity in hydrogen reached its maximum at $0.2 \le x < 0.3$. The increase in conductivity with dopant concentration could be related to an increase in charge carrier concentration through the reaction

$$Y_2O_3 \xrightarrow{CeO_2} 2Y'_{Ce} + 3O_O^\times + V_O^{\bullet\bullet} \quad (7)$$

in the solubility limit. The decrease in conductivity for x > 0.2 was probably contributed to strong defect interaction which may lower the defect mobility.

Ionic Transference Numbers

The proton transference number of Y-doped $SrCeO_3$ and $BaCeO_3$ materials was estimated to be about 0.8-0.95 from a hydrogen concentration cell in the temperature range of 600-800°C. The ratios of proton and oxide ion transference numbers to total ionic transference numbers were further studied with a fuel cell of

$$100\%\ O_2,\ Pt \mid SrCe_{0.95}Y_{0.05}O_{3-\delta} \mid Pt,\ 4\%\ H_2 + Ar. \qquad (8)$$

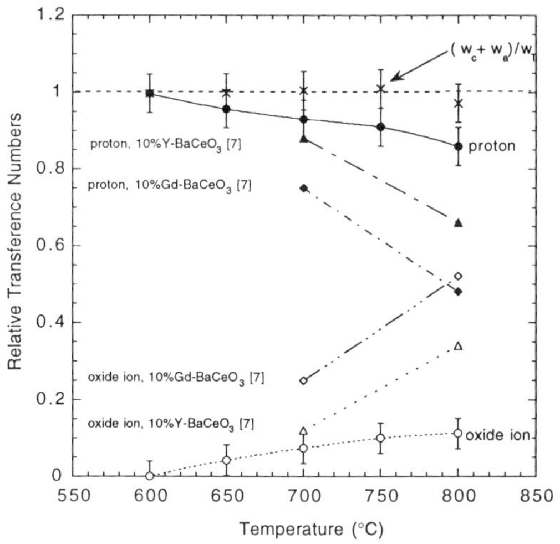

Fig. 4 Ratios of proton and oxygen ion transference number to total ionic transference number in $SrCe_{0.95}Y_{0.05}O_{3-\delta}$ and in $BaCeO_3$-based materials [7].

When the two electrodes are set at equal potentials, the current that passes through the electrolyte is approximately ionic, while electronic current is negligible. Accordingly, the ratios of (t_{H^+}/t_{ion} and $t_{O^{2-}}/t_{ion}$) can be determined as follows:

$$\frac{t_{H^+}}{t_{ion}} = \frac{w_c}{w_t}, \qquad (9a)$$

$$\frac{t_{O^{2-}}}{t_{ion}} = \frac{w_a}{w_t}, \qquad (9b)$$

and

$$t_{ion} = t_{H^+} + t_{O^{2-}}, \qquad (9c)$$

where w_c and w_a are the rates of water vapor evolution from the cathode and anode sides, respectively, and w_t is the total rate of water vapor evolution from both sides or the total rate of water evolution calculated from the short-circuit currents (when Faraday's law is applied). The rates of water vapor evolution due to the electrochemical process were calculated from the flow rates of carrier gas and humidity measured in both sides of the cell under open- and short-circuit conditions. The rate of water vapor evolution from the anode side (hydrogen side) was much slower than that from the cathode side (oxygen side), indicating that proton transport dominated the ionic transport. The total rate of water vapor evolution observed on both sides was in very good agreement with the rate calculated from short-circuit currents. As shown in Fig. 4, the ionic transference numbers are dominated by protons, even at 800°C. Unlike $BaCeO_3$-based materials [13], $SrCe_{0.95}Y_{0.05}O_{3-\delta}$ did not exhibit a conduction transition from proton to oxide ion as temperature increased from 600-800°C. In the temperature range studied, proton transference numbers were much higher than oxide ion transference numbers.

Hydrogen Pumping and Water Vapor Electrolysis

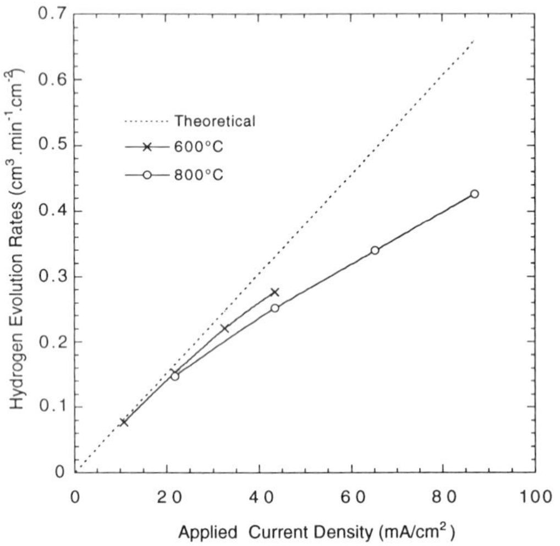

Fig. 5 Hydrogen pumping rates as a function of current density and temperatures.

Shown in Fig. 5 are the hydrogen evolution rates as a function of applied currents and temperatures. The theoretical values calculated from Faraday's law are also shown for reference. When the applied current was low (<22 mA/cm^2), the current efficiency was close to unity, which was in good agreement with the high proton transference numbers of the materials (0.8-0.95). Current efficiency dropped significantly at higher current densities, and this was more evident at high temperature (800°C) than at low temperature (600°C).

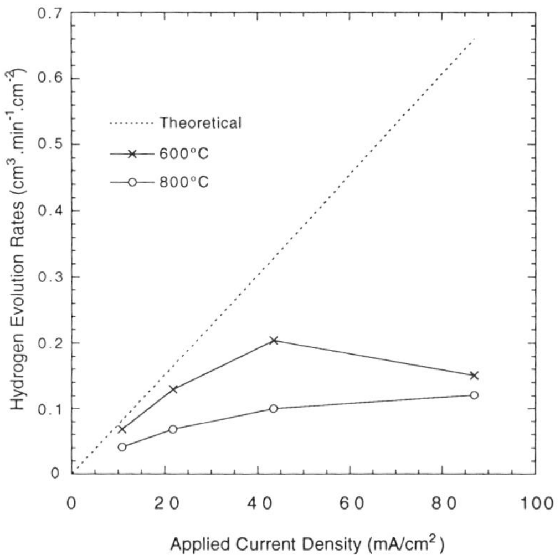

Fig. 6 Hydrogen evolution rates during water vapor electrolysis.

Very dry hydrogen (dew point < -40°C) was obtained by electrolyzing water vapor. As shown in Fig. 6, the current efficiencies were lower than those observed in hydrogen pumping. The low current efficiencies suggested that oxygen ion and electron conduction was significant in water vapor electrolysis. Electron holes could have been formed in a reaction of the oxygen vacancies with the oxygen, which may remains locally at the anodic interface with a relatively higher pressure as a by-product of the water vapor electrolysis:

$$\frac{1}{2}O_2 + V_O^{\bullet\bullet} \rightarrow O_O^{\times} + 2h^{\bullet}. \qquad (10)$$

Hydrogen Permeation through Cermet Membranes

Figure 7 shows net permeation flux through three cermet membranes. Hydrogen permeation rates increased with temperature and were approximately proportional to the inverse of membranes thickness (0.095 to 0.205 cm) at high temperature (800°C), with a slope close to $(E_N\sigma_{amb})/2F$. This suggests that the rate-limiting step is the bulk properties for samples with thicknesses between 0.095 and 0.205 cm. However, at low temperatures (<650°C), the flux data showed significant deviation from this relationship, suggesting that interfacial resistance, rather than bulk properties, is significant at low temperatures.

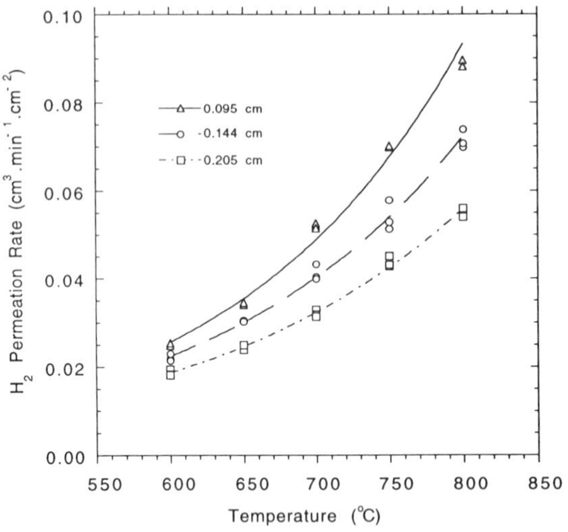

Fig. 7 Temperature and thickness dependence of hydrogen permeation rates through $BaCe_{0.8}Y_{0.2}O_{3-\delta}$ cermet membranes.

Significance of Interfacial Polarization

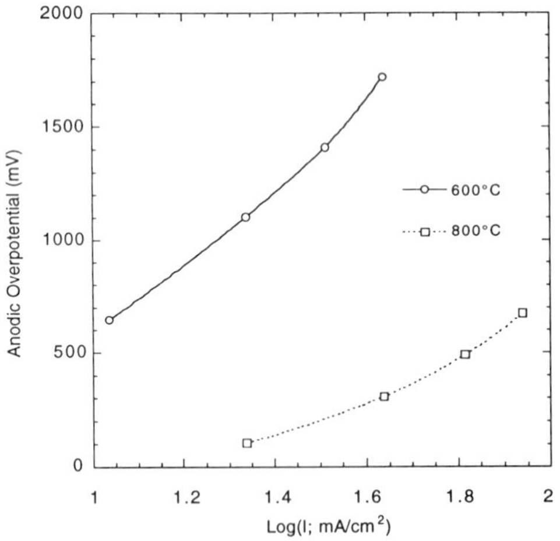

Fig. 8 Anodic overpotentials as a function of applied current density during hydrogen pumping.

As shown in Fig. 8, significant anodic overpotentials were observed in hydrogen pumping, which consumed about 50-70% of the applied potentials under the pumping conditions studied. Interfacial polarization resistances of cermet membranes were investigated through dependence of the gas permeation rate on membrane thickness, as implied by Eq. 5. The interfacial polarization resistance declined dramatically with an activation energy of about 1.03 eV when temperature was increased from 600 to 800°C (Fig. 9). At high temperature (800°C), bulk resistance is much greater than interfacial resistance when sample thickness is >0.095 cm; at lower temperatures, however, interfacial resistance is likely predominant. Thus, reducing the membrane thickness will enhance the permeation flux significantly only if the bulk transport properties are more important than interfacial polarization.

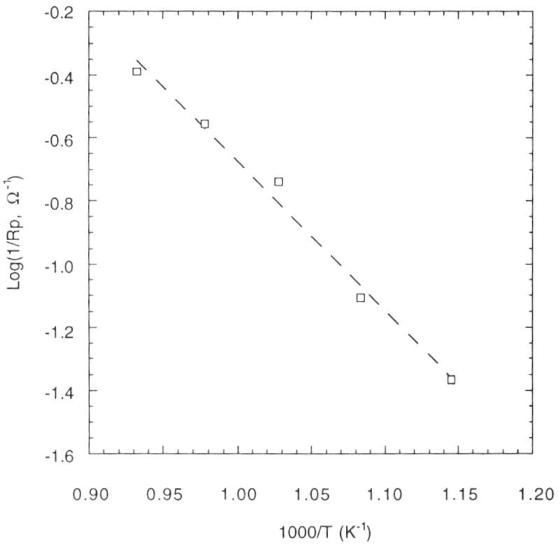

Fig. 9 Interfacial resistance of $BaCe_{0.8}Y_{0.2}O_{3-\delta}$ cermet membrane (1.4 cm^2 in area) as a function of temperature.

CONCLUSIONS

$SrCeO_3$- and $BaCeO_3$-based materials are good proton conductors in hydrogen-containing atmospheres. Hydrogen was pumped though $BaCe_{0.8}Y_{0.2}O_{3-\delta}$ by a DC current; the current efficiency decreased with an increase in temperature and applied current density. Significant interfacial polarization was observed and further reduced the energy efficiency in hydrogen pumping and steam electrolysis.

$BaCe_{0.8}Y_{0.2}O_{3-\delta}$ cermets can be used as MIEC membranes for hydrogen separation in a nongalvanic mode. Interfacial resistances were significant at low temperatures. Further reduction of membrane thickness may not improve the permeation rate effectively if interfacial polarization is more significant than bulk resistance.

ACKNOWLEDGMENTS

This work has been supported by the U.S. Department of Energy, Federal Energy Technology Center, under Contract W-31-109-Eng-38

REFERENCES

[1]. N.Q. Minh and T. Takahashi, "Science and Technology of Ceramic Fuel Cells," Elsevier Science, Amsterdam, 1995.

[2]. H.J.M. Bouwmeester and A.J. Burggraaf, in: Fundamentals of Inorganic Membrane Science and Technology, A.J. Burggraaf and L. Cot, eds., Elsevier Science, Amsterdam, 435 (1996).

[3]. N. Fukatsu, N. Kurita, T. Yajima, K. Koide, and T. Ohashi, J. Alloys and Comp., **231**, 706-12 (1995).

[4]. N. Taniguchi, E. Yasumoto, T. Gamo, J. Electrochem. Soc., **143**, 1886 (1996).

[5]. J. Langguth, R. Dittmeyer, H. Hofmann, and G. Tomandl, Chemie Ingenieur Technik, **69**, 3197 (1997).

[6]. U. Balachandran, J.T. Dusek, S.M. Sweeney, R.B. Poeppel, R.L. Mieville, P.S. Maiya, M.S. Kleefisch, S. Pei, T.P. Kobylinski, C.A. Udovich, and A.C. Bose, Am. Ceram. Soc. Bull., **74**, 71 (1995).

[7]. H. Iwahara, T. Yajima, and H. Uchida, Solid State Ionics, **70/71**, 267 (1994).

[8]. M. Liu, Proc. 1st Int. Symp. on Ionic and Mixed Conducting Ceramics, T. A. Ramanarayana and H. L. Tuller, eds., The Electrochemical Society, Pennington, NJ, Vol. **91-12**, 215 (1991).

[9]. M. Liu and A. Joshi; 1st Int. Symp. on Ionic and Mixed Conducting Ceramics, T. A. Ramanarayana and H. L. Tuller, eds., The Electrochemical Society, Pennington, NJ, Vol. **91-12**, 231 (1991).

[10]. J. Guan, S. E. Dorris, U. Balachandran, and M. Liu, Solid State Ionics, **100**, 45 (1997).

[11]. J. Guan, S. E. Dorris, U. Balachandran, and M. Liu, J. Electrochem. Soc., **145**, 1780 (1998).

[12]. M. Liu and H. Hu, J. Electrochem. Soc. **143,** L109 (1996).

[13]. N. Bonanos, K. S. Knight, and B. Ellis, Solid State Ionics, **79**, 161 (1995).

REFRACTORY CORROSION IN PROCESSING OF MIXED-WASTE GLASS MELT USNG A DC PLASMA ARC MELTER

S. K. Sundaram[a]
Pacific Northwest National Laboratory[b]
P. O. Box 999
Richland WA 99352

ABSTRACT

Materials corrosion in processing of mixed-wastes in a bench-scale DC plasma arc melter was studied in normal and reverse modes. Normal mode is defined as using the top-entering electrode as a cathode; reverse mode is using a top-entering electrode as an anode. Six different refractory candidate materials: C104, Nuline R20, Ruby, ZS 1500, Monofrax K3, and aluminum nitride (AlN) were used. The radius loss values after the test at the glass line, as well as half-down location, were reported as corrosion. The results indicated the normal mode of operation resulted in efficient and steady processing with lower plenum temperature. Electrode corrosion in reverse mode of operation was higher due to the enhanced oxidation of graphite at higher plenum temperatures. Corrosion of the refractories that survived the tests was generally less in normal mode of operation, except the case of AlN. AlN was the material least corroded under the processing conditions. Higher metal retrieved in the reverse mode of operation indicated cathodic reduction of iron in the melt.

---

[a] Member, The American Ceramic Society

[b] Pacific Northwest National Laboratory is operated for the U. S. Department of Energy by Battelle under Contract DE-AC06-76RLO 1830.

Introduction

The DC-plasma arc furnace [1] is a robust, high-temperature thermal treatment system being tested to treat mixed wastes as part of a project supported by the U.S. Department of Energy (DOE) Mixed Waste Focus Area (MWFA). In this system, applying a DC potential between a graphite electrode and the graphite hearth of the furnace creates a stable DC arc. The thermal energy produced by this arc creates and maintains a molten bath of material (glass/slag or metal) in the furnace hearth. Waste materials fed into the system are melted in the bath. Organic compounds are pyrolyzed at the high operating temperature and may be destroyed in the plenum or in a suitable afterburner. Oxide materials, including many of the hazardous and radioactive species, are immobilized in the durable glass/slag phase. As a second phase, metals separate out of the melt and are removed. A major advantage of this technology is its potential to vitrify waste streams containing a wide range of materials such as debris, trash, metals, and soil.

Refractory corrosion is one of the barriers to implementing plasma technology for the treatment of radioactive/mixed waste. High melt temperatures (in excess of 1500 °C), the variability in melt chemistry, and the operating conditions create a hostile environment for the materials in contact with or containing the melt. A preliminary evaluation[c] of refractory corrosion in plasma/torch furnaces indicated significant gaps in the materials knowledge between the data necessary to advance and develop plasma technology and the data actually available. An integrated corrosion evaluation of selected refractories was undertaken by Pacific Northwest National Laboratory (PNNL) under the MWFA program[d]. Both laboratory crucible tests, as well as bench-scale tests, were used in determining the corrosion of selected refractory candidate materials. In the bench-scale tests, the effects of process variables on the corrosion were determined. The data from these tests were compared to the data from the laboratory crucible tests. Based on these data, the performance of the refractories was ranked and suitable refractories were recommended for the application.

The arc current is a function of the quantity of electrons released by thermionic emission from the cathode surface. These electrons are accelerated in the cathode

---

[c] S. K. Sundaram, "Corrosion of Slag and Metal Contact Materials in Plasma Arc/Torch Furnaces," MWFA, TTP No. RL3-6-MW51, March 1997.
[d] S. K. Sundaram, "Refractory Corrosion Evaluation: Final Report," MWFA, TTP No. RL3-6-MW51, August 1997.

voltage drop region by the intense electric field around the cathode. The voltage drop is shown in Figure 1. The electrons acquire high kinetic energy, enabling them to ionize neutral atoms by collision and, beforehand, if the plasma generating gas is polyatomic, disassociation of the molecules. The positive ions formed are accelerated in the opposite direction and strike the cathode, releasing their energy and enabling maintenance of the thermionic emission. The polarity of operation in a DC-plasma arc furnace is crucial in optimizing and maximizing its performance. No data are available in the literature on the mode of operation of the DC plasma arc furnace. This paper summarizes the results of bench-scale tests performed to address this aspect of DC-plasma arc technology.

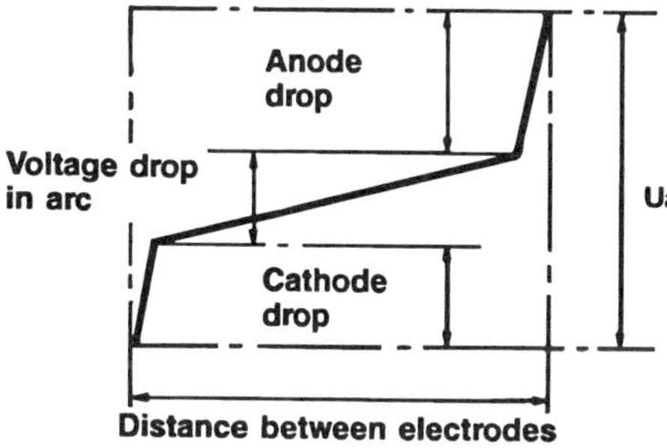

Figure 1. Voltage Drop in an Arc [1]

BACKGROUND

Refractory scoping tests were performed at the Diagnostic Instrumentation and Analysis Laboratory (DIAL) at Mississippi State University in March 1995 [2]. A 250 kW plasma torch attached to a gimbal mount to allow for material consisted of briquettes with a composition reflecting a typical low-level mixed waste heterogeneous debris waste of composition. From groups of oxides, nitrides, carbides, borides, fused cast refractories, and composites, 33 materials were tested. Two tests, a multiple cycle test and a continuous operation test, were conducted to assess the performance of refractory materials. A graphite crucible was used to contain the test materials. Because many of the samples were odd-shaped, the walls and bottom of the crucible were notched to accommodate the samples and to minimize sample movement during the tests. Because they were scoping tests, no effort was made to track the effects of cross-contamination. After the tests, only the very high chromium refractory and aluminum nitride with 0.25 wt.% yttria survived. Oxide samples, silicon carbide samples, and fused cast refractories showed evidence of thermal shock. Severe corrosion was observed at the glass line of all the refractories tested. Corrosion in samples near the bottom of the crucible was much less than near the surface of the melt. The results were qualitative and inconclusive.

Attempts have been made to evaluate refractories during or after processing wastes as a part of furnace operation. Scientific Applications International Corporation (SAIC) performed preliminary refractory testing [3] by constructing hearths of the sample materials. Tests were conducted in two different systems: the 1.2-MW torch proof-of-principle (POP) system and a 200-kW torch system located at the SAIC Science and Technology Applications Research (STAR) Center. The systems were operated on batch mode with melt temperatures typically in the range of 1600-1700°C. The materials tested were Ruby®, Korundal®, Harklase®, Nocacon®, Harbide® brick and castable, and Nuline R20® (all from Harbison-Walker), and Monofrax K3® from Monofrax Inc. Under the acidic slag (basicity of 0.38) conditions, no significant differences were noted between the performance of any of the oxide refractories. In the case of basic slag (basicity of 2.5), the Monofrax K3 performed the best, followed (in decreasing order) by Ruby, Harklase, Korundal, and Zircon. The Monofrax K3 was susceptible to thermal shock. The Ruby brick performed well and was less susceptible to thermal shock. Non-oxide materials such as silicon carbide and the Nuline refractory (containing 20% carbon) were significantly inferior in performance to the oxide materials under all slag conditions. A definite crucible size effect on refractory performance, a dramatic decrease in refractory wear with an increase in crucible volume, was observed. The general rule of thumb for glass-contact refractories is that corrosion rate doubles for each 50-100 °C increase in the interface temperature [4]. Lower corrosion rate means material conservation, lowered idle time, higher productivity, lower repair cost, lower operating cost, and less maintenance. Additional cost savings in time and labor are realized by avoiding the processing problems due to seriously corroded refractories that pollute the melt and cause problems of high viscosity, segregation, and lower product quality.

EXPERIMENTAL

Waste and glass compositions

A large number of waste streams identified for treatment reside at the Idaho National Engineering Environmental Laboratory (INEEL) and the Savannah River Site (SRS). A significant portion (60%) of the mixed transuranic waste (MTRU) at SRS is specifically designated for vitrification. The heterogeneous nature of such mixed waste streams will have a significant impact on many aspects of arc furnace operation and performance. The waste form and/or composition may affect partitioning behavior of key elements, product durability, system component durability, and operability of product removal systems, as well as overall system operability. Table I presents the composition of the SRS debris surrogate and SRS (L area) soil used in this testing program. The surrogate wastes studied in this effort support the mixed waste treatment activities at the INEEL and the SRS. The debris composition is identical to the feed material used in the tests performed by the Diagnostic Instrumentation and Analysis Laboratory (DIAL) of Mississippi State University [5]. In addition to the base composition presented in Table I, hazardous metals, radionuclides, and radionuclide surrogates are added at 1000 ppm. The debris wastes required additives for producing a low-viscosity glass at 1300 to 1500°C.

Table I. Baseline Feed Composition

Surrogate Debris (40 wt.%)		+	Soil (60 wt.%)	
Material	Wt.%		Component	Wt.%
Carbon steel	14.05		Al_2O_3	3.5
Perlite	14.05		CaO	<0.1
PVC	14.05		Fe_2O_3	1.0
Soda-lime silica glass	14.05		K_2O	<0.2
Wood	14.05		MgO	0.1
Alumina	10.18		Na_2O	0.9
Activated carbon	6.31		SiO_2	86.0
Silica	5.60		TiO_2	0.4
Portland™ cement	5.60		Others	8.1
Aluminum	2.04			
Total	100.00			100.00

Soil was chosen as a glass-forming additive because it is inexpensive and easily available. This criterion ensures that the glass can be poured from a continuously drained furnace. Lime was added to a given waste with each soil type to reduce the melting temperature of the final glass. Compared to lime, the oxides of sodium and potassium typically reduce glass-melting temperatures at lower weight percentages; however, these oxides are extremely volatile at high temperatures and would vaporize out of the glass. Based on the data from melting studies and the bench-scale non-radioactive processing tests completed using SRS soils, the baseline feed of 40% SRS debris and 60% SRS soil was selected for further investigation. All the bench-scale tests at PNNL used this baseline feed composition.

Refractory candidate materials

On the basis of the limited data available in the literature and the PNNL expertise regarding design and operation of DC-plasma arc and other furnaces [6-8], the potential candidate refractory materials were chosen. The candidate materials and manufacturers are Monofrax K3 from Monofrax, Inc., mag-carbon (Nuline R20) and Ruby from Harbison-Walker, chrome-magnesia (C104) and dense zircon (ZS1500) from Corhart Refractories, and aluminum nitride (AlN) from Advanced Refractory Technologies. Chrome and chrome-containing refractories (Monofrax K3, C-104, and Ruby) were chosen for their corrosion resistance to common silicate glasses and slags. Nuline R20 was selected for its extensive use in steel industries. Monofrax K3, Ruby, ZS1500, and AlN performed well in tests conducted at DIAL. Therefore, ZS1500 and AlN (with yttria added as a sintering aid) were also tested.

Bench-scale arc furnace testing

A schematic of a bench-scale system is shown in Figure 2. The furnace utilizes a single top-entering graphite electrode (cathode) and a fixed graphite hearth (anode). This is the normal mode of operation. In the case of reverse mode, the top entering graphite electrode is the anode and the graphite hearth is the cathode. The electrode is 1.5 cm in diameter and can be vertically adjusted using a motor-driven screw. The graphite hearth is crucible shaped; both its internal diameter and its height are 11 cm. During testing, an open bottom ceramic crucible is placed inside the graphite hearth during testing to protect the sides from oxidation. The small size of the hearth does not make it amenable to glass and metal drains. Consequently, the furnace is operated in a semi-batch mode, where it is continuously fed until full of molten product, then cooled and disassembled to remove the glass and metal phases produced. The outside of the furnace consists of a cylindrical steel vessel lined with approximately 10 cm of refractory and insulation.

During operation, a 15-kW, constant-current DC power supply provides up to 100 volts of arcing potential between the furnace electrode and hearth. A thermocouple in the furnace plenum area monitors temperature inside the furnace. This temperature, along with the off-gas temperatures, power supply voltage, and current are electronically logged by a PC-based data acquisition system. As depicted in Figure 2, dry material is fed into the furnace via a screw auger. A variable speed motor drives the auger that allows the operator to adjust the feed rate during testing. On-line feed rates cannot be measured from the system. Therefore, feed rates are calculated after each test by dividing the total mass of the feed batch by the amount of time required to feed that batch. Off-gases exit the furnace and enter a large void chamber where gas velocities are decreased to allow large particulate to settle. The gases next pass through a water-cooled heat exchanger, then through one of two 0.5 μm (absolute) polyester bag filters. In front of the bag filter, the temperature of the off-gas is increased to approximately 80°C, using heat tape wrapped around the tubing to ensure that little or no water condenses in the bag filter. Behind the bag filter, the off-gas flows through a high efficiency particulate air (HEPA) filter. All of these components can be completely dismantled so that the entire mass of off-gas solids produced can be collected, weighed, and analyzed.

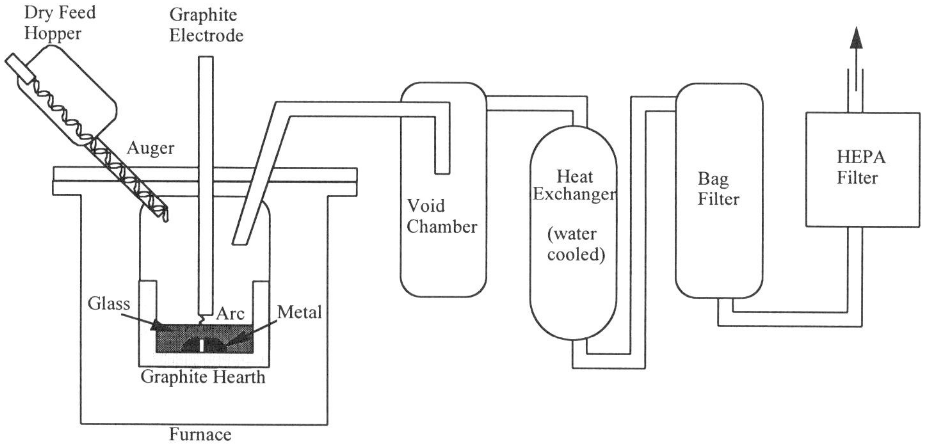

Figure 2. Schematic of Bench-Scale Arc Furnace System

The test samples were placed along the sides of the crucible opposite to the point of feeding and secured in place using carbon steel wire. Two tests were performed, one using the normal mode and one using the reverse mode. In the case of normal mode of operation, 85 -g of stainless steel was added at the bottom of the hearth before starting the test for stable arc. After completing the test, the samples were retrieved and the metal and slag were separated. The samples were sliced at the glass line and half-down positions and the dimension change was measured. The loss of material was measured using a stereoscope and an optical microscope and reported as was corrosion observed. The corrosion rate was computed by dividing the corrosion by the total processing (feeding) time. In the case of severely or fully corroded samples, no corrosion measurements could be made. But, a corrosion rate was estimated from the initial sample dimension, assuming complete corrosion of the sample, and the corrosion rate was reported to be greater than that rate. In the case of ZS1500, a reaction layer was found at the refractory-glass interface in a few samples. In these cases, the thickness of the layer was measured using a scanning electron microscope (SEM) and the corrosion rate was corrected for the thickness.

RESULTS AND DISCUSSION

Bench-scale processing

Bench-scaling processing data for normal and reverse modes of operation are presented in Figures 3 and 4, respectively. In the normal mode of operation, the average power was around 2.5 kW and the resistance was 0.8 ohm. The plenum temperature increased sharply to about 400°C within 10 minutes of operation and then gradually increased to about 700°C at the end of the processing. These data indicate the process stabilized within the first 10 minutes of operation.

The power for reverse mode of operation averaged at approximately 3 kW. The resistance gradually increased to about 0.8 ohm after 30 minutes of operation and then continued to increase to 1.0 at a slower rate through the end of the process. The plenum temperature increased gradually to about 950°C at the end of the processing. After 120 minutes of operation, the temperature was about 900°C. These data indicate the process did not stabilize until 120 minutes of processing.

Corrosion data

The corrosion of the refractory candidate materials is summarized in Table II. The half-down corrosion data indicated that C104, Nuline R20, and Ruby corroded severely in both the normal and reverse modes of operations. ZS 1500 and Monofrax K3 corroded approximately two times more in reverse mode than in normal mode. AlN corroded the least under these processing conditions. The glass line corrosion data are not indicative of the actual corrosion inside the melt and are shown for completeness.

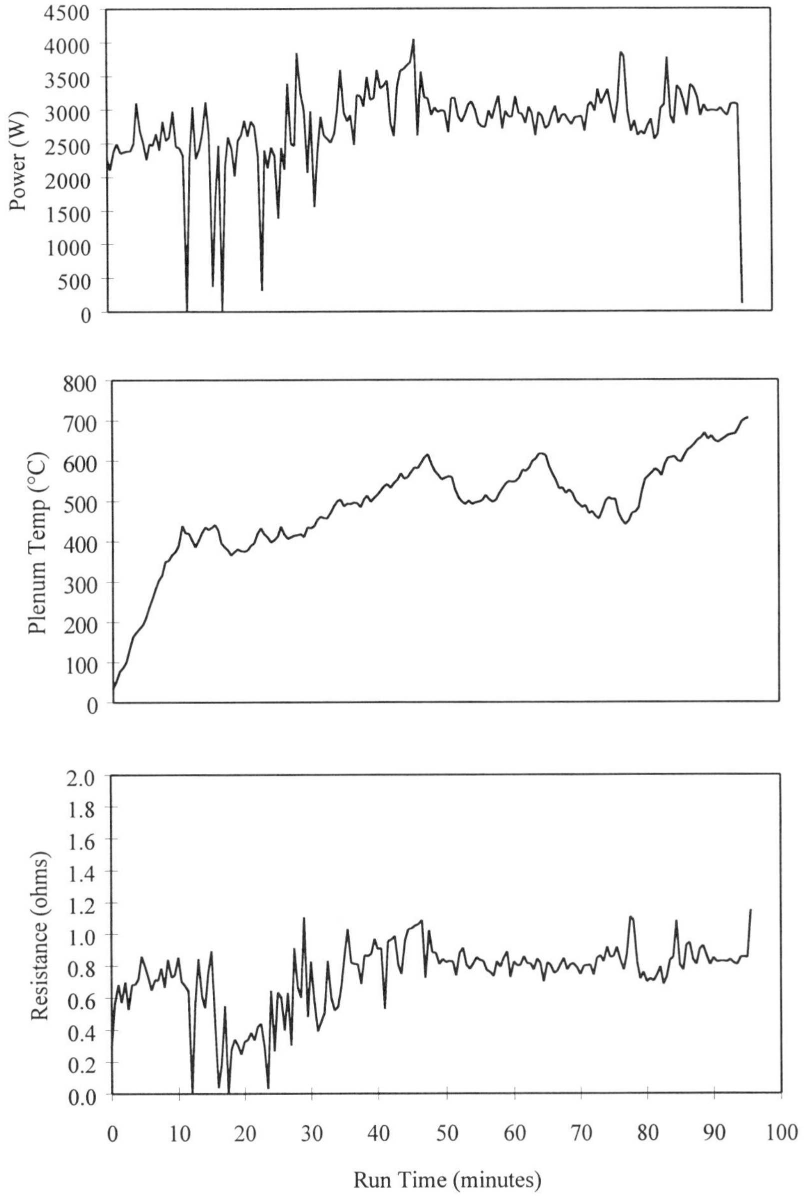

Figure 3. Processing Data for Normal Mode of Operation

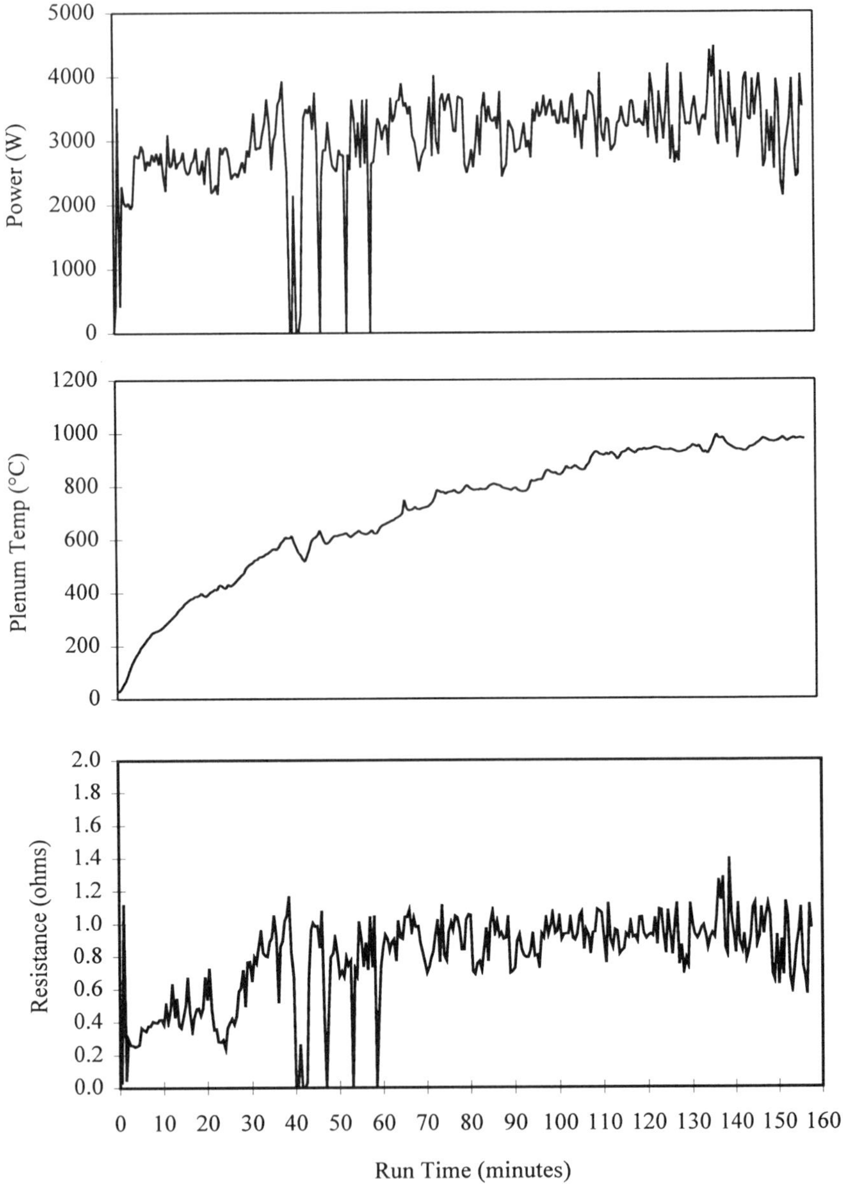

Figure 4. Processing Data for Reverse Mode of Operation

Table II. Corrosion Data

Refractories	Corrosion Rate (in./h)			
	Normal Mode		Reverse Mode	
	Glass Line	Half-Down	Glass Line	Half-Down
C104	0.0411	> 0.1525	0.0173	> 0.0948
Nuline R20	0.0269	> 0.1635	> 0.0974	> 0.0974
Ruby	0.0246	> 0.1627	> 0.0957	> 0.0957
ZS1500	0.0120	0.0438	0.0097	0.0990
Monofrax K3	0.0009	0.0135	> 0.0945	> 0.0945
AlN	0.0035	0.0113	0.0005	0.0014

Table III and Figure 5(a) and (b) compare the results of the bench-scale melter runs. The data indicated that normal mode of operation 5(a) showed efficient and steady processing and lower plenum temperature. Reverse mode of operation 5(b) resulted in erratic operation and high plenum temperature. The high plenum temperature will result in more volatilization losses and off-gas corrosion.

The electrode corrosion in normal mode was about a third of the corrosion in reverse mode (Table III). Higher plenum temperature in reverse mode of operation increased the oxidation of the electrode, leading to higher electrode corrosion, as shown in Figure 5(b). Accumulation and interaction of metal with graphite hearth was also observed in reverse mode of operation. Higher retrieved metal is attributed to cathodic reduction of Fe at the graphite hearth plate.

Table III. Summary of Normal vs. Reverse Modes

Parameters	Normal (Top electrode as cathode)	Reverse (Top electrode as anode)
Average power (watts)	2768	3001
Feed rate (g/min)	5.53	10.26
Plenum temperature	700°C	1000°C
Processing	Steady	Erratic
Metals recovered	117-g (85-g added)	58-g
Electrode corrosion (g(electrode)/g(feed))	0.020	0.058
Refractory corrosion	AlN > Monofrax K3 > ZS1500	AlN > ZS1500

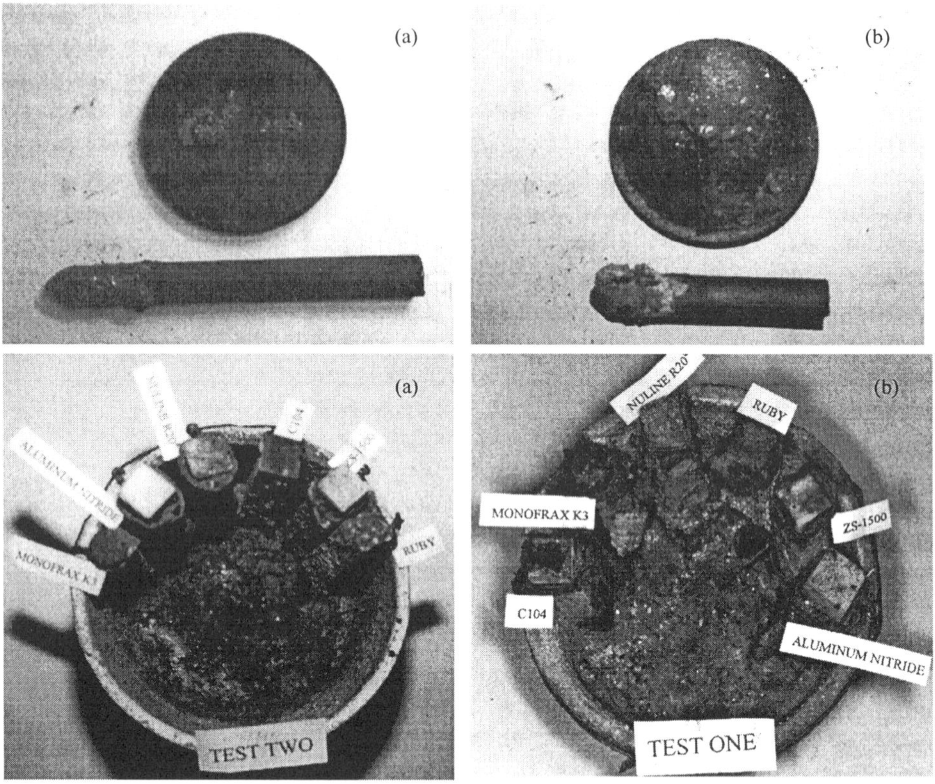

Figure 5. Graphite Hearth and Electrode (top) and Crucible (bottom) after (a) Normal and (b) Reverse Modes of Operation

CONCLUSIONS

The normal mode of operation resulted in efficient and steady processing with lower plenum temperature. Electrode corrosion in reverse mode of operation was higher due to enhanced oxidation of graphite at higher plenum temperature. Corrosion of refractories that survived the tests was generally less in normal mode of operation, except the case of AlN. AlN was the least corroded material under the processing conditions. The higher amount of metal retrieved in reverse mode of operation indicated cathodic reduction of iron in the melt.

ACKNOWLEDGMENTS

The author acknowledges the support of David A. Lamar and William F. Bonner of PNNL. The Mixed Waste Focus Area of the DOE Office of Science and Technology supported the study and preparation of this paper.

REFERENCES

1. M. Orfeuil in *Electric Process Heating Technologies/Equipment/Applications*, pp.627-667, Battelle Press, Columbus, Ohio, 1989.

2. J. C. Marra, J. W. Congdon, A. L. Kielpinski, R. F. Schumacher, A. A. Ramsey, J. Etheridge, and R. Kirkland, "Corrosion Assessment of Refractory Materials for Waste Vitrification", Proceedings of the Corrosion of Materials by Molten Glass Symposium, Ceramic Transactions, **78**, pp. 289-293 (1996).

3. G. L. Leatherman, and J. A. Batdorf, "Crucible Materials for the Plasma Treatment of Radioactive Waste", Proceedings of the Corrosion of Materials by Molten Glass Symposium, Ceramic Transactions, **78**, pp. 289-293 (1996).

4. W. Trier in Glass Furnaces (Design, Construction, and Operation), Translation by K. L. Loewenstein, Society of Glass Technology, Sheffield, UK, pp.60-75 (1987).

5. R. A. Green, B. Nail, P. Norton, P-R. Jang, J. P. Singh, W. Okhuysen, F-Y. Yueh, H. Zhang, J. Andol, B. Kirkland, J. C. Marra, R. F. Schumacher, and J. Congdon, "Refractory Lifetime Testing for Westinghouse Savannah River Company," Diagnostic Instrumentation and Analysis Laboratory at Mississippi State University, March (1995).

6. G. L. Smith, S. K. Sundaram, and P. A. Smith, "Glass-Refractory Material Interactions", PVTD-C95-02.03M, August (1995).

7. C. J. Freeman, D. A. Lamar, and S. K. Sundaram, "Evaluation of High-Temperature Electrode and Refractory Materials for Application in Hanford High-Level Waste Vitrification", PVTD-T3C-95-139, September (1995).

8. Proceedings of Materials for Vitrification Systems Workshop, Chairman: K. J. Imrich, Savannah River Technology Center, October 1-2 (1996).

SYNTHESIS OF SPINEL $Li_4Mn_5O_{12}$ CATHODES BY OXIDATION REACTIONS IN SOLUTIONS

A. Manthiram and J. Kim
Texas Materials Institute, ETC 9.104
The University of Texas at Austin
Austin, TX 78712

ABSTRACT

The commercial lithium cells use cobalt oxide cathodes and the high cost and toxicity of Co have created enormous interest to develop alternate cathodes. Manganese oxides are attractive in this regard as Mn is inexpensive and environmentally benign. But the spinel $LiMn_2O_4$ that has been investigated over the years tends to exhibit capacity fading on cycling due to structural distortions. Although the structural distortion can be suppressed by a cationic substitution of Li for Mn to give $Li_4Mn_5O_{12}$, it is difficult to achieve all Mn^{4+} without oxygen vacancies in $Li_4Mn_5O_{12}$ by conventional ceramic procedures. In this paper, a low temperature synthesis process involving the oxidation of Mn^{2+} by lithium peroxide is presented to obtain $Li_4Mn_5O_{12}$ with all Mn^{4+}. $Li_4Mn_5O_{12}$ is unstable at higher temperatures and it disproportionates to $Li_{1+y}Mn_{2-y}O_4$ (y <0.33) and Li_2MnO_3 at T > 500 °C. $Li_4Mn_5O_{12}$ obtained by this approach shows a capacity of about 160 mAh/g, which is close to the theoretical value, with excellent cyclability.

INTRODUCTION

The rapid advances in technology and the miniaturization in electronics have created an ever increasing demand for lightweight, compact, high-energy density batteries [1]. For example, popular portables such as laptop computers, camcorders, and cellular phones require small, but efficient and reliable batteries. In addition, there is enormous interest to develop advanced high-energy density batteries for electric vehicle propulsion [2]. The development of high-energy-density batteries requires the use of electrode materials that can provide both high capacity and cell voltage. Lithium batteries have become the choice in this regard as they offer higher voltage and energy density compared to other rechargeable systems. Fig. 1 compares the volumetric and gravimetric energy densities of the various rechargeable systems and lithium batteries are lighter and smaller than other systems [3]. In addition, the use of nonaqueous electrolytes in lithium batteries permits the operation over a wide range of temperatures -40 < T < 70 °C.

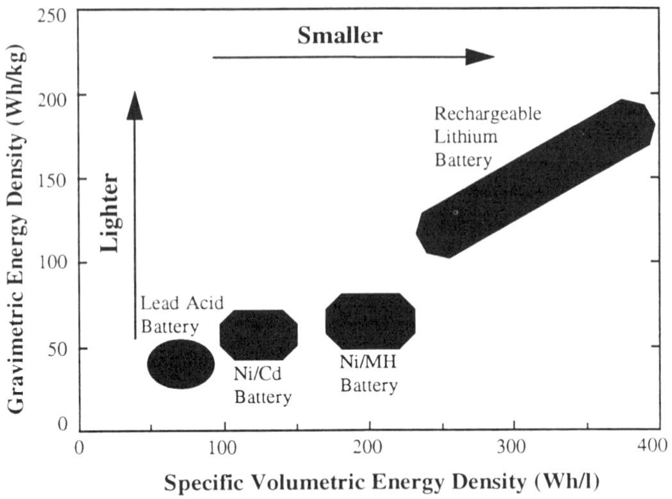

Fig. 1 Comparison of the volumetric and gravimetric energy densities of lithium batteries with other rechargeable systems (adapted from ref. 3).

A recognition of the advantages of lithium batteries has led to the development of several lithium insertion compounds as electrodes over the years [4-7]. Fig. 2 compares the electrochemical potential ranges of various insertion compounds versus metallic lithium. Although rechargeable lithium batteries were initially pursued with metallic lithium anode, the chemical reactivity of lithium metal with most of the nonaqueous electrolytes and the safety problems associated with lithium metal have forced to use lithium insertion compounds as anodes. Lithium cells employing insertion compounds as both anodes and cathodes are called "rocking-chair" or "lithium-ion" batteries. This strategy, however, requires a careful selection of electrode pairs so as to maintain an acceptable cell voltage of at least 2 V over the discharge cycle and to minimize the added weight associated with the host framework of an insertion anode. Carbon, with a discharge voltage of < 1 V versus metallic lithium and low atomic weight, has become an attractive anode for lithium-ion cells. On the other hand, the layered oxides $LiCoO_2$ [8] and $LiNiO_2$ [9] and the three dimensional spinel oxide $Li[Mn_2]O_4$ [10] having a mid-discharge voltage of about 4 V versus metallic lithium have become attractive cathodes.

However, it is difficult to synthesize $LiNiO_2$ as an ordered material due to the reduction of Ni^{3+} to Ni^{2+} and a volatilization of lithium at higher temperatures during the synthesis process [11]. The commercial lithium-ion cells are consequently made with the layered $LiCoO_2$ cathodes. But cobalt is expensive and relatively toxic, which necessitates the development of alternate cathodes. In this regard, manganese oxides are intensively pursued as manganese is inexpensive and environmentally benign.

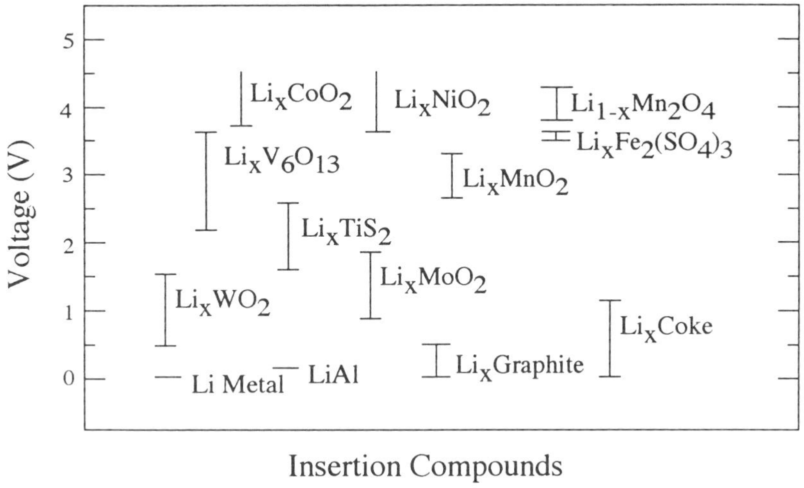

Fig. 2 Electrochemical potential ranges of some lithium insertion compounds versus metallic lithium.

MANGANESE OXIDES

Among the manganese oxides, $LiMn_2O_4$ having the spinel structure (Fig. 3) has been investigated widely. $LiMn_2O_4$ exhibits two plateaus in the voltage versus lithium content, one at 4 V corresponding to the extraction of lithium from the 8a tetrahedral sites and the other at 3 V corresponding to the insertion of an additional lithium into the 16c octahedral sites to give $Li_2Mn_2O_4$ [10]. The lithium extraction/insertion proceeds by a single phase mechanism maintaining the cubic symmetry of the spinel structure in the 4 V region. On the other hand, a transition from cubic to tetragonal symmetry occurs as the average oxidation state of manganese falls below 3.5+ and the lithium insertion/extraction proceeds by a two phase mechanism involving cubic $LiMn_2O_4$ and tetragonal $Li_2Mn_2O_4$. The tetragonal transition is due to the Jahn-Teller distortion associated with the single electron in the e orbitals of a Mn^{3+}:$3d^4$ ion. This transition is accompanied by a 16 % increase in the *c/a* ratio of the unit cell parameters, which is too severe for the electrode to maintain structural integrity during the discharge-charge cycling. As a result, $LiMn_2O_4$ exhibits poor electrochemical cyclability in the 3 V region. Moreover, the 3-dimensional spinel framework provides a lower lithium-ion diffusion compared to the layered $LiCoO_2$ and an early formation of the tetragonal $Li_2Mn_2O_4$ on the surface of the particles, particularly under higher current densities, will degrade the cyclability in the 4 V region also. These problems have plagued the commercialization of the $LiMn_2O_4$ spinel cathodes.

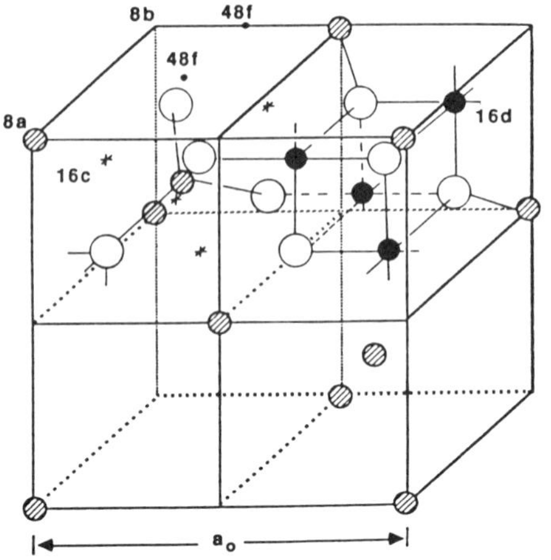

Fig. 3 Crystal structure of $LiMn_2O_4$ spinel.

One way to suppress the Jahn-Teller distortion is to increase the average oxidation state of manganese through cationic substitutions in the spinel lattice. For example, the average oxidation state of manganese can be increased progressively from 3.5+ to 4+ by a substitution of Li for Mn in $Li_{1+y}Mn_{2-y}O_4$ and the end member $Li_4Mn_5O_{12}$ with all Mn^{4+} is formed at the limiting value of y = 0.33 [12-14]. Such a cationic substitution has been found to give a better cyclability in the 3 V region, but at the expense of a monotonic decrease in capacity in the 4 V region. For instance, the $Li_4Mn_5O_{12}$ spinel exhibits better cyclability in the 3 V region compared to the $LiMn_2O_4$ spinel, but shows no capacity in the 4 V region. The better cyclability in the 3 V region is due to the preservation of the cubic symmetry to higher degrees of lithium insertion into the 16c sites of the cation substituted spinels. In fact, the cubic symmetry has been found to be preserved at least to x = 2.5 in $Li_{4+x}Mn_5O_{12}$. These considerations have made $Li_4Mn_5O_{12}$ an attractive cathode for 3 V lithium cells.

$Li_4Mn_5O_{12}$ is, however, unstable at higher temperatures and disproportionates to the more stable $LiMn_2O_4$ spinel and Li_2MnO_3 [12, 15]. Takada *et al* [16] have reported that $Li_4Mn_5O_{12}$ is stable up to about 600 °C. More recently, Thackeray *et al* [17] have reported from a careful thermogravimetric analysis (TGA) study that $Li_4Mn_5O_{12}$ begins to lose oxygen above 420 °C and the oxygen is not fully regained upon cooling even though the sample is heated only to 620 °C in TGA. $Li_4Mn_5O_{12}$ is generally synthesized by firing a mixture of manganese oxides or salts with lithium salts at higher temperatures. The diffusional limitations in such solid state reactions require higher firing temperatures T > 600 °C and often lead to incomplete reaction at lower firing temperatures. More importantly, it is difficult to achieve all Mn^{4+} in $Li_4Mn_5O_{12}$ by these conventional ceramic procedures and the samples are oxygen deficient. The maximum oxidation state achieved by Takeda *et al* [16] even after applying a high oxygen pressure of 10 atm is about 3.8+. The lower oxidation state for Mn can lead to a decrease in capacity from the theoretical value of 163 mAh/g, which corresponds to an insertion of three Li per $Li_4Mn_5O_{12}$ molecule. In addition, the oxidation state of manganese in the raw materials has been found to be critical. While raw materials with Mn^{4+} help to stabilize $Li_4Mn_5O_{12}$, those with Mn^{3+} tend to favor the formation of $LiMn_2O_4$.

We present in this paper the development of a low temperature synthesis procedure to obtain $Li_4Mn_5O_{12}$ with an objective to maximize the oxidation state of manganese and cell capacity. We use an oxidation reaction in solution involving the oxidation of Mn^{2+} with lithium or hydrogen peroxide in the presence of excess lithium hydroxide to obtain a precursor containing Mn^{4+}, which upon firing at moderate temperatures loses water and gives $Li_4Mn_5O_{12}$. Crystal chemical and compositional characterization, and electrochemical properties are presented.

EXPERIMENTAL

Required quantities of aqueous solutions of lithium hydroxide and lithium or hydrogen peroxide, as given in Table I, were first mixed together. The mixture was then added immediately to 20 ml of 0.25 M manganese acetate solution that was kept under constant stirring on a magnetic stirrer. The oxidation of Mn^{2+} by the lithium or hydrogen peroxide in presence of excess lithium hydroxide led to the formation of a precipitate, $Li_xMn_yO_z.nH_2O$. The solid formed was filtered and allowed to dry at ambient temperature in air.

The solid so formed was then fired in air at various temperatures ranging from 300 to 1000 °C for 3 days and characterized by X-ray powder diffraction. A heating and cooling rates of, respectively, 5 and 2 °C/min were used for the heat treatment procedure. Lattice parameters were calculated using NaCl as an internal standard and a least square program. Representative samples were also characterized by TGA in order to evaluate the temperature at which water or oxygen is lost. The lithium and manganese contents were determined by Atomic Absorption Spectroscopy (AAS) after dissolving the samples in dilute hydrochloric acid. The oxidation state of manganese was determined by dissolving the solid in a known excess of vanadyl sulfate in 2 N H_2SO_4 and titrating the unreacted vanadyl sulfate with standardized permanganate [18]. Electrochemical properties were evaluated with coin-type cells using an Arbin battery tester. Circular electrodes of about 2 cm^2 area were made by mixing the oxide with 25 weight percent fine carbon and 5 weight percent polytetrafluoroethylene (PTFE). The coin-type cells were assembled with the cathodes so prepared, metallic Li anodes, and $LiClO_4$ in propylene carbonate (PC)/1,2-dimethoxyethane (DME) electrolyte.

Table I. Synthesis conditions for the oxidation of Mn^{2+}

Sample number	Manganese Acetate		Lithium Peroxide		Lithium Hydroxide	
	Conc. (M)	Vol. (ml)	Conc. (M)	Vol. (ml)	Conc. (M)	Vol. (ml)
1	0.25	20	0.25	80	0.25	10
2	0.25	20	0.25	80	0.25	20
3	0.25	20	0.25	80	0.25	40
4	0.25	20	0.25	80	0.25	80
5	0.25	20	0.25	80	0.25	160
6[a,b]	0.25	20	30 vol. %	80	0.125	320
7[b]	0.25	20	0.25	80	0.125	320
8[a]	0.25	20	30 vol. %	80	0.25	160

[a] Hydrogen peroxide was used instead of lithium peroxide.
[b] Lithium Carbonate was used instead of lithium hydroxide.

RESULTS AND DISCUSSION

The oxidation experiments carried out with various concentrations and volumes of the reactants are summarized in Table I. The as-prepared samples were found to be amorphous as indicated by the absence of any discernible reflections in the X-ray powder diffraction patterns. A TGA plot of the as-prepared sample 5 in Table I is, for example, shown in Fig. 4. A loss of about 17 wt% occurring below 400 °C corresponds to both adsorbed and lattice water. The small weight loss occurring above 400 °C is due to the gradual reduction of manganese at higher temperatures. Since the as-prepared samples contain water and are amorphous, they were first heated at 500 °C in order to characterize the nature of the products formed with various amounts of the reactants. The X-ray diffraction patterns of the samples after firing at 500 °C for 3 days are shown in Fig. 5. As the volume of lithium hydroxide increases, the amount of the spinel phase $Li_{1+y}Mn_{2-y}O_4$ increases and that of Mn_2O_3 decreases. Sample 5 shows single phase $Li_{1+y}Mn_{2-y}O_4$ spinel. Samples 1-5 in Table I were prepared with a constant concentration and volume of lithium peroxide, but with an increasing volume of lithium hydroxide from sample 1 to 5. It is clear that lithium hydroxide acts as a lithium source and an excess Li^+ concentration in the reaction medium is essential to suppress Mn_2O_3 and obtain single phase $Li_{1+y}Mn_{2-y}O_4$.

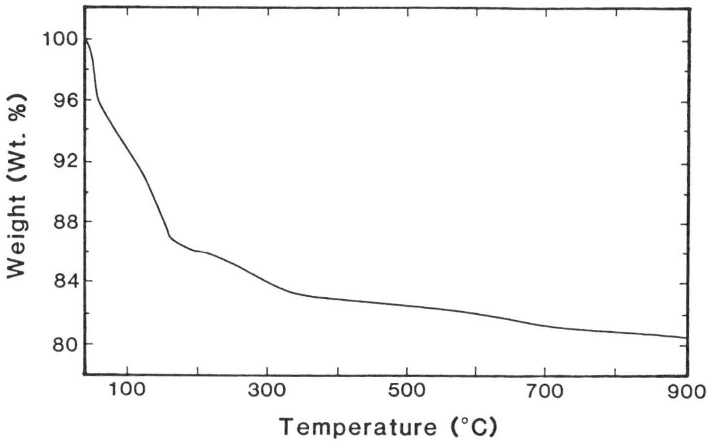

Fig. 4 TGA plot of the as-prepared sample 5 in Table I in a flowing mixture of 70% N_2 and 30% O_2 with a heating rate of 5 °C/min.

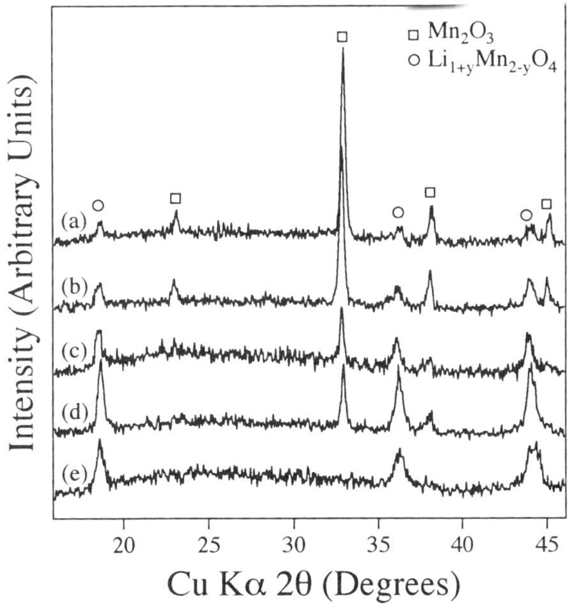

Fig. 5 X-ray diffraction patterns of the samples prepared with lithium peroxide after firing at 500 °C: (a) sample 1, (b) sample 2, (c) sample 3, (d) sample 4, and (e) sample 5 in Table I.

Fig. 6 compares the X-ray diffraction patterns of samples 6-8 with that of sample 5. Samples 6-8 used lithium carbonate and hydrogen peroxide instead of lithium hydroxide and lithium peroxide. The presence of Mn_2O_3 impurity in samples 6 and 8 reveals that lithium peroxide also acts as a lithium source in addition to lithium hydroxide. The data in Fig. 6 also reveals that the use of lithium carbonate in the place of lithium hydroxide as a lithium source does not make any significant difference as samples 5 and 7 show similar X-ray patterns.

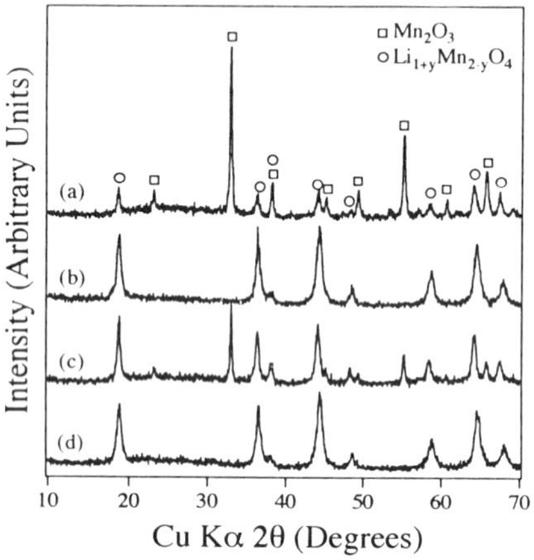

Fig. 6 Comparison of the X-ray diffraction patterns of the samples prepared with hydrogen peroxide with that prepared with lithium peroxide: (a) sample 6, (b) sample 7, (c) sample 8, and (d) sample 5 in Table I. The patterns were recorded after firing at 500 °C.

Since only samples 5 and 7 consist of single phase spinel, further characterization were carried out with sample 5 alone as sample 7 is similar to sample 5. AAS and oxidation state analysis data of sample 5 after firing at 500 °C reveal, respectively, a lithium content of 3.98 in $Li_xMn_5O_{12}$ (or $y = 0.33$ in $Li_{1+y}Mn_{2-y}O_4$) and an oxidation state of 3.97+ for Mn. Thus the chemical analysis data suggest that the spinel phase formed is $Li_4Mn_5O_{12}$.

In order to study further the influence of firing temperatures, sample 5 was fired at various temperatures in air for 3 days. The evolution of X-ray diffraction patterns as a function of firing temperature is shown in Fig. 7. The as-prepared sample does not show any discernible reflections. On the other hand, the sample fired at 300 °C begins to develop broad reflections corresponding to the spinel structure and the reflections become well defined as the firing temperature increases to 400 °C. Fig. 8 shows a magnification of the X-ray patterns over a smaller two theta range and the data reveals that the reflections move progressively to lower angles as the firing temperature increases above 500 °C. In addition, a small reflection corresponding to Li_2MnO_3 appears for firing temperatures T ≥ 700 °C. These results suggest that $Li_4Mn_5O_{12}$ begins to disproportionate at T > 500 °C, at least according to the X-ray data, to give $Li_{1+y}Mn_{2-y}O_4$ (y < 0.33) and Li_2MnO_3.

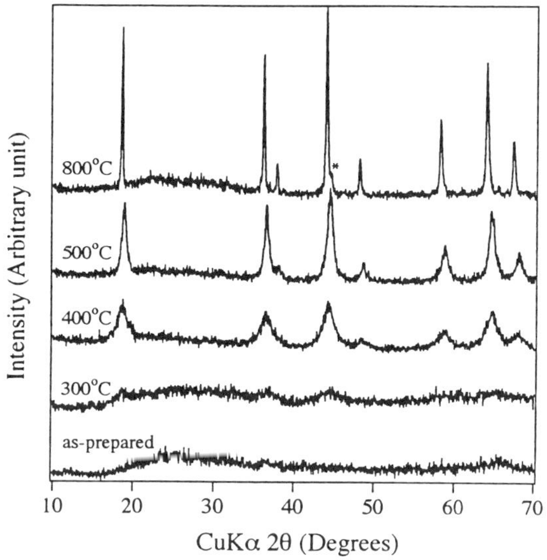

Fig. 7 X-ray diffraction patterns of sample 5 after firing at various temperatures.

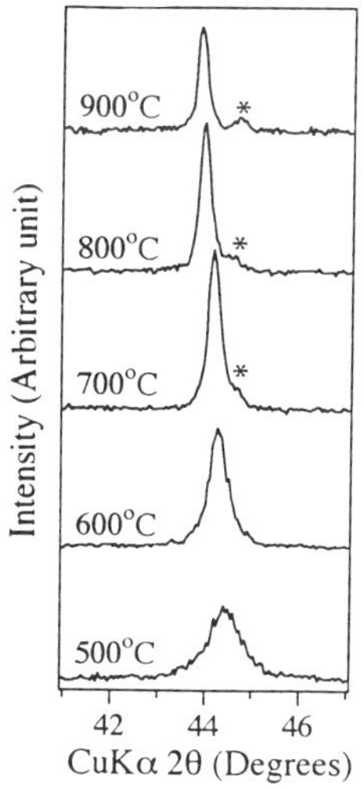

Fig. 8 X-ray diffraction patterns of sample 5 over a small two theta range illustrating the movement of peaks to lower angle with increasing firing temperature. * refers to Li_2MnO_3 reflections.

AAS and oxidation state analysis data of sample 5 after firing at 400 °C reveal, respectively, a lithium content of 3.98 in $Li_xMn_5O_{12}$ and an oxidation state of 3.98+ for Mn. The data reveal that the oxidation state of Mn does not decrease significantly on going from 400 to 500 °C. The 400 and 500 °C samples were found to have lattice parameters of, respectively, 8.121(46) and 8.131(23) Å, which are slightly smaller than the literature value of 8.137Å [12]. Although the broad diffraction peaks lead to a larger standard deviation, the smaller lattice parameter is consistent with the higher oxidation state of manganese achieved in our samples. The chemical analysis and lattice parameter values clearly indicate that the solution-based low temperature approach can access all Mn^{4+} in $Li_4Mn_5O_{12}$ without oxygen vacancies.

Since firing at T > 500 °C leads to a disproportionation of $Li_4Mn_5O_{12}$, electrochemical data were collected only with the samples fired at 400 and 500 °C. The discharge curves of the 400 and 500 °C samples are shown, respectively, in Figs. 9 and 10 for two current densities. The 400 °C sample in Fig. 9 exhibits an initial capacity of 160 mAh/g in the range 3.3 - 2.3 V, which is close to the theoretical capacity of 163 mAh/g. The 500 °C sample in Fig. 10 shows a slightly lower initial capacity of 153 mAh/g. However, the 500 °C sample exhibits better cyclability than the 400 °C. It shows excellent capacity retention with < 2% decline in capacity for 40 cycles. We believe that a better crystallinity of the 500 °C sample, as indicated by sharper diffraction peaks in Fig. 7, leads to a better capacity retention. The observed capacity and cyclability are superior to that found with samples prepared by conventional ceramic procedures [12, 13]. Furthermore, there is no significant decrease in capacity on increasing the current density from 0.1 to 0.5 mA/cm^2, indicating that the low temperature solution-based approach provides an optimum microstructure with small particle size to provide faster lithium-ion diffusion.

CONCLUSIONS

The 3 V cathode, $Li_4Mn_5O_{12}$, has been synthesized by a simple, solution-based oxidation reaction followed by firing at T ≤ 500 °C. The low temperature approach helps to achieve all Mn^{4+} without oxygen vacancies. $Li_4Mn_5O_{12}$ is unstable at higher temperatures and it begins to disproportionate above 500 °C. $Li_4Mn_5O_{12}$ samples obtained by the low temperature approach exhibits a capacity close to the theoretical value with excellent cyclability. The work demonstrates that innovative solution-based synthesis can give electrode materials with superior performance.

ACKNOWLEDGMENTS

Financial support by the Welch Foundation Grant F-1254 and the National Science Foundation Grant DMR-9401999 is gratefully acknowledged.

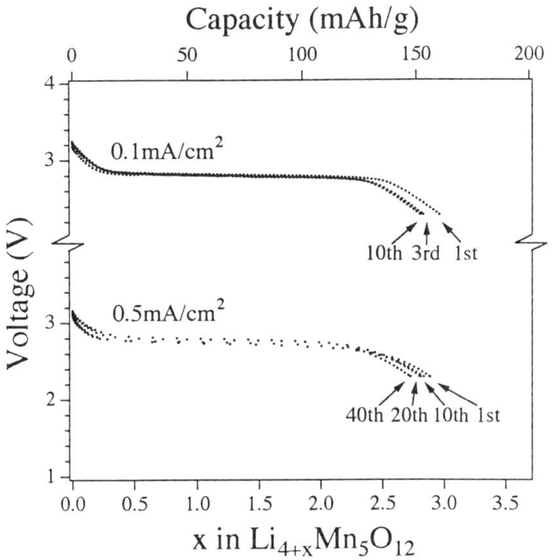

Fig. 9 Discharge curves of sample 5 ($Li_4Mn_5O_{12}$) after firing at 400 °C.

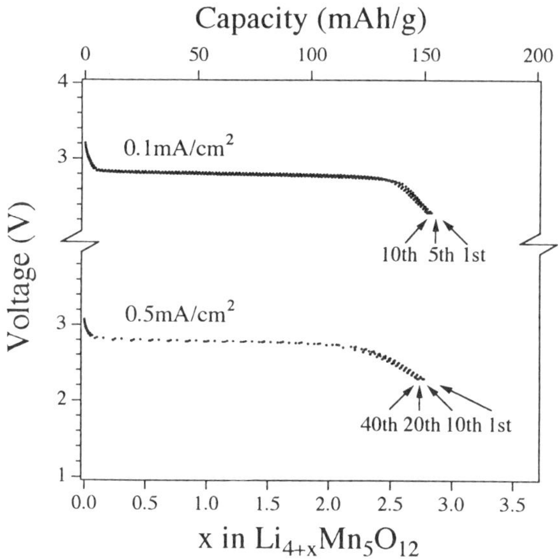

Fig. 10 Discharge curves of sample 5 ($Li_4Mn_5O_{12}$) after firing at 500 °C.

REFERENCES

1. B. Scrosati, "Challenge of Portable Power," Nature **373**, 557-558 (1995).
2. L. B. Lave, C. T. Henrickson and F. C. McMichael, "Environmental Implications of Electric Cars," Science **268**, 993-995 (1995).
3. Lithium Battery Energy Storage (LIBES) Publication, Technological Research Association, Tokyo, 1994.
4. M. S. Whittingham and A. J. Jacobson, "Intercalation Chemistry," Academic Press, New York, 1982.
5. H. V. Venkatasetty, "Lithium Battery Technology," John Wiley & Sons, New York, 1984.
6. J. P. Gabano, "Lithium Batteries," Academic Press, London, 1993.
7. D. Linden, "Lithium Cells," Handbook of Batteries, 2nd Edition, Edited by D. Linden, McGraw-Hill, New York, 1995, pp. 14.1-14.2.
8. K. Mizushima, P. C. Jones, P. J. Wiseman and J. B. Goodenough, "Li_xCoO_2 ($0 < x < 1$): A New Cathode Material for Batteries of High Energy Density," Mat. Res. Bull. **15**, 783-799 (1980).
9. J. B. Goodenough, K. Mizushima and T. Takeda, "Solid Solution Oxides for Storage Battery Electrodes," Japan. J. Appl. Phys. **19**, 305-313 (1980).
10. M. M. Thackeray, W. I. F. David, P. G. Bruce and J. B. Goodenough, "Lithium Insertion into Manganese Spinels," Mater. Res. Bull. **18**, 461-472 (1983).
11. G. Dutta, A. Manthiram and J. B. Goodenough, "Chemical Synthesis and Properties of $Li_{1-\delta-x}Ni_{1+\delta}O_2$ and $Li[Ni_2]O_4$," J. Solid State Chem. **96**, 123-131 (1992).
12. M. M. Thackeray, A. de Kock, M. H. Rossouw and D. Liles, "Spinel Electrodes from the Li-Mn-O System for Rechargeable Lithium Battery Applications," J. Electrochem. Soc. **139**, 363-366 (1992).
13. R. J. Gummow, A. de Kock and M. M. Thackeray, "Improved Capacity Retention in Rechargeable 4V Lithium/Lithium Manganese Oxide (Spinel) Cells," Solid State Ionics **69**, 59-67 (1994).
14. A. Yamada, "Lattice Instability in $Li(Li_xMn_{2-x})O_4$," J. Solid State Chem. **122**, 160-165 (1996).
15. M. M. Thackeray, M. F. Mansuetto, D. W. Dees and D. R. Vissers, "The Thermal Stability of Lithium Manganese Oxide Spinel Phases," Mater. Res. Bull. **31**, 133-140 (1996).
16. T. Takada, H. Hayakawa and E. Akiba, "Preparation and Crystal Structure Refinement of $Li_4Mn_5O_{12}$ by Rietveld Method," J. Solid State Chem. **115**, 420-426 (1995).
17. M. M. Thackeray, M. F. Mansuetto and C. S. Johnson, "The Stability of $Li_4Mn_5O_{12}$ Electrodes for Lithium Batteries," J. Solid State Chem. **125**, 274-277 (1996).
18. C. Tsang and A. Manthiram, "A New Route for the Synthesis of $LiMn_2O_4$ Cathode: Variation of Compopsition, Microstructure and Electrochemical Behavior with Synthesis Tempertature," Solid State Ionics **89**, 305-312 (1996).

ELECTRODEPOSITION AND CHARACTERIZATION OF NICKEL/SILICON CARBIDE FILMS

Maria Hepel and Christine Mahoney
Department of Chemistry
State University of New York College at Potsdam
Potsdam, NY 13676; hepelmr@potsdam.edu

ABSTRACT

Nickel/Silicon carbide (Ni/SiC) composite films were formed by potentiostatic electrodeposition from modified Watt's bath solutions on steel and Pt substrates. During the Ni/SiC film formation, SiC particles were maintained in a suspension by stirring of the solution. The effects of deposition potential, concentration of SiC in suspension and variable size of SiC on the nickel nucleation and growth transients were studied. Morphology and composition of Ni/SiC films under various conditions were investigated using Scanning Electron Microscopy (SEM) and Energy Dispersive X-Ray Analysis. Significant effect of concentration of SiC in suspension and size of SiC particles on morphological changes occurring during annealing of Ni/SiC films at 600°C were observed. Inhibition of nickel oxidation was observed for annealed Ni/SiC films at 600°C containing nanoparticles of SiC. For nickel films containing microparticles of SiC significantly less inhibition was observed.

INTRODUCTION

Metal films with incorporated inert ceramic particles or wide-bandgap semiconductor particles are composite materials of industrial interest. Composite coatings containing carbides, nitrides or oxides are promising materials with good corrosion, oxidation , and wear resistance [1-4]. Silicon carbide, as well as silicon nitride, have attracted a great attention because of their potential applications as high-temperature structural ceramics. Oxidation resistance is one of the most important properties in these applications [3].

Formation and characterization of composite films of Cu/BN, Ni/BN, Ni/TiC, and Ni/SiC by a potentiostatic electrodeposition method was reported [4-6]. Electrochemical impedance spectroscopy study of SiC particle occlusion during nickel electrodeposition was presented by Watson [7]. Effect of additives such as Coumarin and EDTA on the morphology and the amount of incorporated SiC particles was also reported [6]. Berkh et al. [8] studied the effect of a saccharin additive in the electrolyte on the composition of nickel films. They observed that the presence of saccharin considerably reduced the SiC and P content in Ni/P/SiC films [8]. The effect of stirring on the volume fraction of SiC powders incorporated into the deposit was also investigated [9,10].

EXPERIMENTAL

Chemicals. All chemicals were of reagent grade purity (Aldrich) and were used without further purification. SiC microparticles with size 0.5 μm were obtained from Johnson-Matthey Company. SiC nanoparticles with size 20 nm - 50 nm were obtained from MarkeTech International.

Procedures. The Ni/SiC films were obtained by potentiostatic electrodeposition from suspensions of silicon carbide particles in modified Watt's bath solutions containing 0.85 M $NiSO_4$ and 0.15 M $NiCl_2$. The average size of SiC microparticles was 0.5 μm and nanoparticles: 20-50 nm. Films were deposited on C-1018 carbon steel disk substrates ($A = 1.2\ cm^2$) and on a Pt substrate. They were degreased with propanol, polished with 0.05 μm alumina powder, rinsed with distilled water and ethanol, and assembled in the demountable working electrode placed in the bottom of the electrolytic cell. The depositions were carried out from deoxygenated, stirred (600 r.p.m.) suspensions, at 22°C.

High temperature annealing and oxidation of Ni/SiC films was performed in an oven Lindberg Hevi-duty with a temperature control.

Instrumentation. A Model PS-605 Precision Potentiostat/Galvanostat (ELCHEMA, Potsdam, NY) was used in measurements. The program waveform was generated by VOLTSCAN Real-Time Data Acquisition and Control System (ELCHEMA). Scanning Electron Micrographs were taken using a Model JSM-6300 SEM (JEOL) equipped with an Energy Dispersive X-ray unit for elemental analysis and a Noran Voyager X-Ray Microanalyzer.

RESULTS AND DISCUSSION

Potentiostatic Electrodeposition of Composite Ni/SiC Films

The electrodeposition of Ni under potential control conditions from Watt's bath solution in the presence and absence of SiC particles in the solution was investigated under various conditions. The current-time transients recorded for the electrodeposition of composite Ni/SiC and bare nickel films under the same macro-hydrodynamic conditions and variable deposition potential are presented in Figure 1a and Figure 1b, respectively.

In the experiments of Figure 1a, composite Ni/SiC films were formed from Watt's bath solution containing 5 g/L SiC in a suspension. When the electrodeposition potential in each experiment was changed from E = -1000 mV vs SCE to E = -1300 mV vs. SCE, a systematic increase in nucleation and growth currents was observed. Similar potential dependence was observed for electrodeposition of Ni Films without SiC particles (Figure 1b). However, current-time transients recorded in the presence of SiC particles (Figure 1a) have lower values of steady-state currents than those recorded in the absence of SiC particles (Figure 1b). The decrease in the Ni deposition currents is attributed to the blocking effect of the SiC particles, which when in contact with the electrode surface, provide less active surface for Ni electrodeposition. This blocking effect is more pronounced when concentration of the SiC particles in the suspension increases.

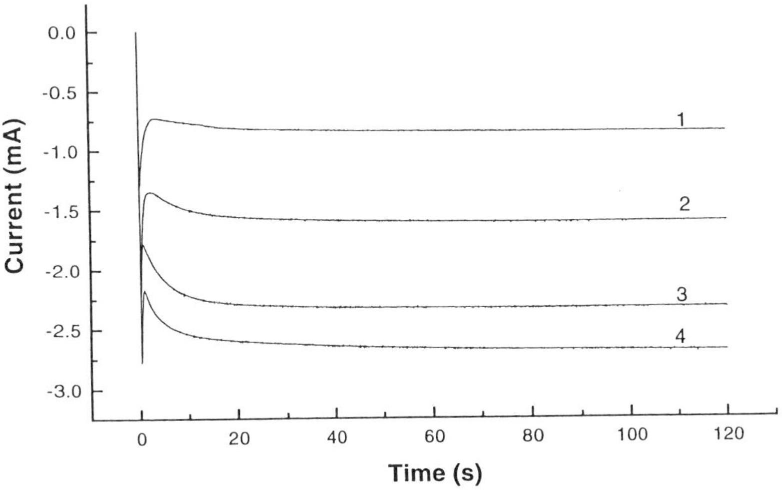

Figure 1a
Chronoamperometric current-time transients for cathodic deposition of Ni/SiC films from modified Watt's bath suspensions with suspension density 5 g/L SiC on a C - 1018 carbon steel substrate at E_{dep} [mV] vs. SCE: (1) -1000; (2) - 1100 mV; (3) -1200 mV; (4) - 1300 mV.

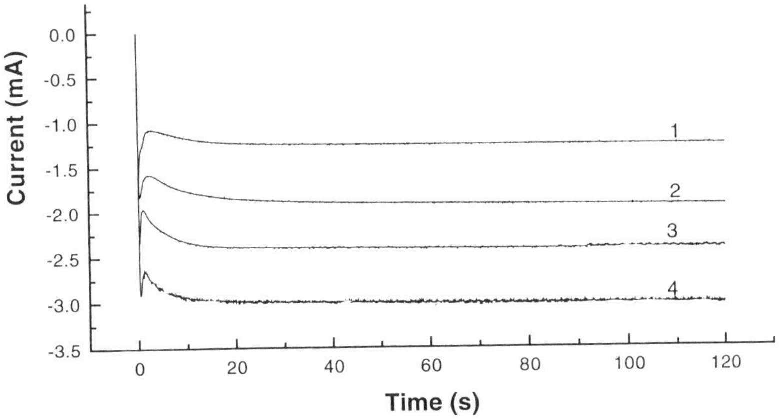

Figure 1b
Chronoamperometric current-time transients for cathodic deposition of Ni films fromWatt's bath solution on a C-1018 carbon steel substrate at E_{dep}[mV] vs SCE: (1) -1000 mV; (2) -1100 mV; (3) -1200 mV; (4) -1300 mV.

Effects of pH and Size of SiC Particles

Effect of variable pH on electrodeposition of Ni/SiC films is presented in Figure 2. Increases in the nucleation and growth currents were observed when pH of the suspension was decreased from pH = 4 to pH = 1.4 for the suspension containing microparticles or nanoparticles. However, larger increase in currents was observed for nanoparticles than for microparticles of SiC. Nucleation and growth transients recorded for the formation of Ni/SiC films from a suspension containing nanoparticles have shown higher steady-state currents (curves 1 & 2) than transients obtained from a suspension containing microsize particles (curves 3 and 4) in Figure 2. Size distribution of nanoparticles was from 20 to 50 nm.

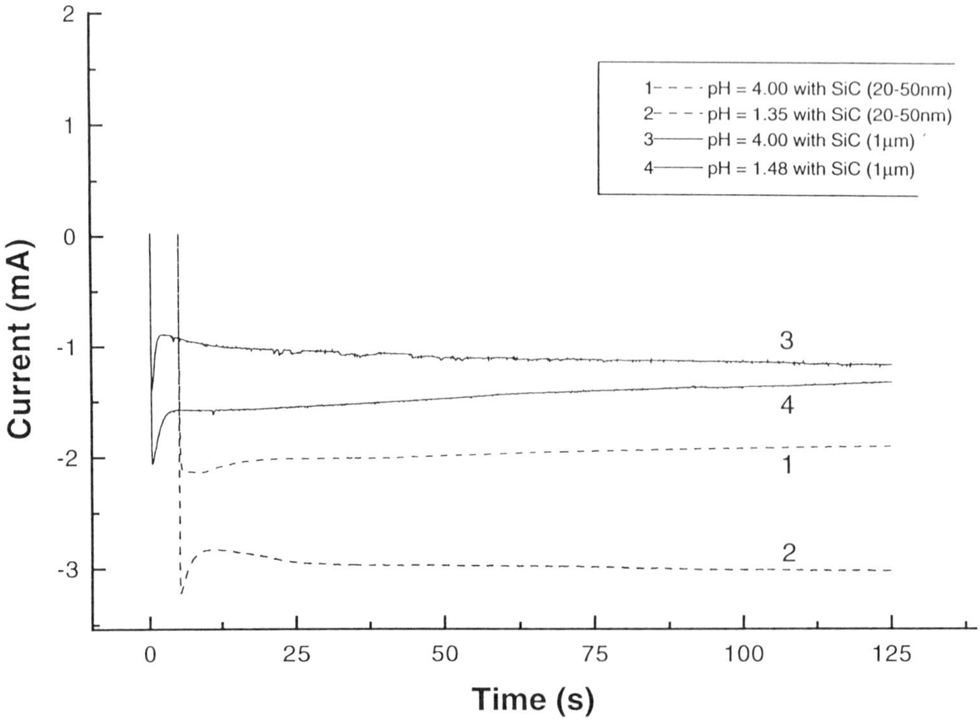

Figure 2
Chronoamperometric current-time transients for cathodic deposition at E_{dep} = -1200 mV vs. SCE of Ni/SiC films from modified Watt's bath suspensions containing nanoparticles (20 nm - 50 nm) of SiC (curve 1 & 2) and microparticles (0.5 μm) of SiC (curve 3 & 4) obtained at pH = 4.00 (curve 1 and 3) and at pH = 1.4 (curve 2 and 4).

Effect of Potential on the Morphology of Ni/SiC Films

The effect of the deposition potential on the morphology of Ni/SiC films is illustrated in Figure 3a,b. In Figure 3a, SEM image obtained for Ni/SiC film deposited at low overvoltage (E_{dep} = -1000 mV vs. SCE) from Watt's bath suspension containing 30 g/L SiC with an average 1 μm size is shown. The SEM image of Ni/SiC film deposited at a higher overvoltage (E_{dep} = -1200 mV vs. SCE) is presented in Figure 3b. The rest of the electrodeposition conditions for Figure 3b are the same as for Figure 3a. The amount of embedded SiC particles into the composite film increases when a more negative potential is applied during the film formation.

(a) (b)

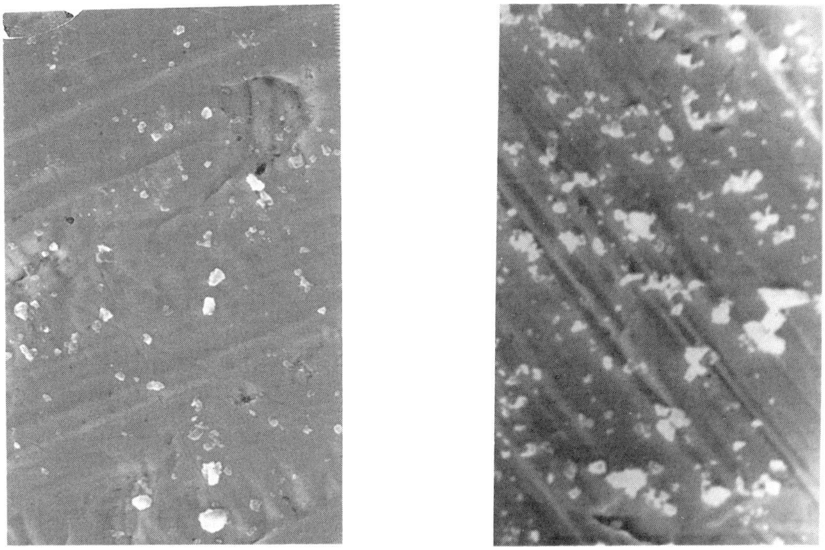

Figure 3
SEM images of the surface of composite Ni/SiC films electrodeposited on a C-1018 carbon steel substrate from Watt's bath solution containing suspension of 30 g/L SiC (0.5 μm size) at constant potential, E_{dep} [mV] vs. SCE: (a) -1000 mV; (b) -1200 mV. Magnification: 2.2kx; Scale: 1cm = 5.0μm.

Effect of Concentration of SiC in a Suspension on the Morphology of Ni/SiC Films

Films of Ni/SiC were formed for a concentration of SiC in a suspension changing from 5 g/L to 60 g/L. The effect of the variable concentration of SiC in suspension on the amount of incorporated SiC into nickel film is presented in Figure 4a, b. In Figure 4a, SEM image obtained for Ni/SiC film deposited from Watt's bath suspension containing 5 g/L SiC microparticles is shown. The SEM image of Ni/SiC film deposited from a suspension containing 30 g/L SiC microparticles is presented in Figure 4b. The electrodeposition was carried out at the same potential E = -1200 mV vs. SCE and size of SiC was 1 μm for both Figures 4a and 4b.

An increase in the amount of embedded SiC particles in the composite film was observed when the concentration of SiC particles in the suspension was increased. Also, the microhardness of Ni/SiC films was increased with the increased amount of embedded SiC particles [5].

(a) **(b)**

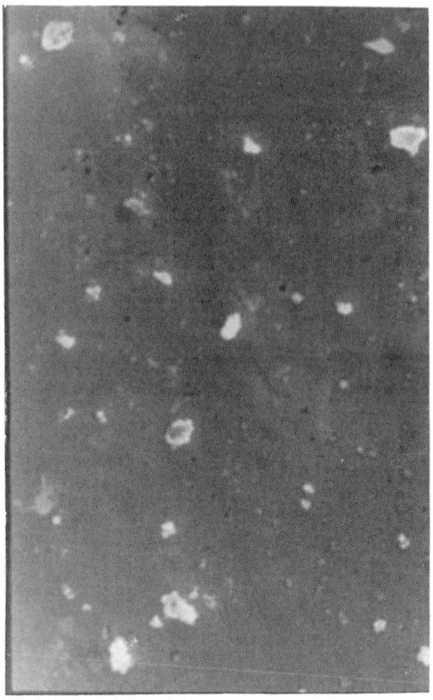

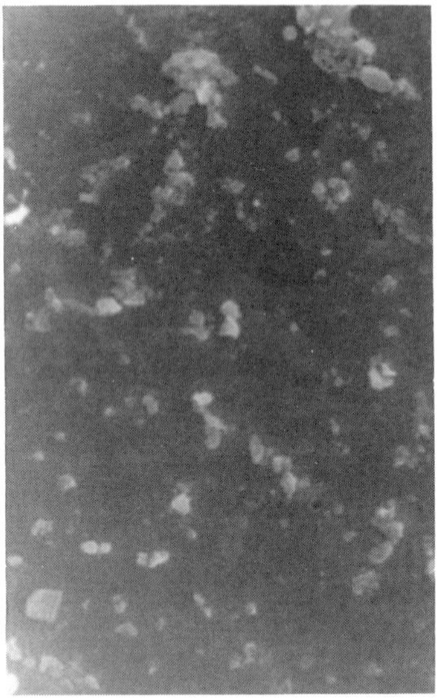

Figure 4
SEM image of the surface of composite Ni/SiC films electrodeposited at E_{dep} = -1200 mV vs. SCE on a C-1018 carbon steel substrate from Watt's bath solution containing suspension of SiC (0.5 μm size): (a) 5 g/L SiC; (b) 30 g/L SiC. Magnification: 2.0kx; Scale: 1cm = 5.0μm.

Annealing and Oxidation of Ni/SiC Films

Ni/SiC films for high temperature corrosion studies were prepared on Pt substrates from Watt's bath solutions containing 1 μm size SiC particles in suspensions with variable concentration of SiC changing from 5 g/L to 60 g/L. Films were deposited at a constant potential E = -1200 mV vs. SCE. Effect of concentration of SiC particles, 5 g/L and 60 g/L, on the film morphology after annealing at 600°C for 6 hours, is presented in Figure 5a and 5b, respectively. Less oxidation of Ni metal is observed in Figure 4a for films formed from 5 g/L SiC suspension than in Figure 5b for films formed from 60 g/L SiC suspension. The morphology of Ni film alone formed in the absence of any SiC particles and under the same set of conditions as in Figure 5a, is presented in Figure 6. This film was annealed at the same temperature T = 600°C for 6 hrs. in an air atmosphere. Film shows rounded spherical clusters of nickel oxide and is less smooth than composite Ni/SiC films in Figure 4a exposed to the same annealing conditions. For Ni/SiC films formed from 5 g/L SiC suspension and annealed at 600°C, formation of nickel oxide is slowed down. Oxidation of SiC at this temperature is rather unlikely. At temperatures higher than 700°C, a protective layer of SiO_2 is formed during the oxidation of SiC.

Significant differences in the morphology of Ni/SiC films and their resistance to high temperature oxidation were observed when nanosize SiC particles were embedded into a nickel film instead of microsize SiC particles. This effect is illustrated in Figures 7a and b.

In Figure 7a, SEM image obtained for Ni/SiC film deposited on a Pt substrate at E = -1200 mV from Watt's bath suspension containing 5 g/L SiC nanosize particles (20 nm - 50 nm). This film was annealed for 6 hrs. at 600°C.

The SEM image of Ni/SiC film obtained from suspension containing 60 g/L SiC nanosize particles (20 nm - 50 nm) and the rest conditions the same as in Figure 7a, is presented in Figure 7b.

Significant inhibition of nickel oxidation was observed when the size of embedded SiC particles was in the nanosize range (Figure 7b) even when higher concentrations of SiC particles in suspension were used.

Note that the growth of the NiO film at 600°C proceeds via paths of easy-ion movement because of the relatively low activation energy of the process (1.6 eV). The ion diffusion mechanism is consistent with the movement of doubly charged cation vacancies V''_{Ni} and adding new O^{2-} lattice sites at the oxide/air interface. The NiO scale grows outwards, which is consistent with SEM micrographs. The SiC particles do not oxidize at 600°C in dry air. At high partial pressure of water vapor, oxidation of SiC does take place with the formation of SiO_2 and CH_4 [3].

(a) (b)

Figure 5

SEM images of the surface of composite Ni/SiC films annealed at 600°C in air atmosphere for 6 hrs. Ni/SiC films were formed by electrodeposition at a constant potential E_{dep} = -1200 mV vs. SCE on a Pt substrate from Watt's bath solution containing suspension of SiC (0.5 μm size) (a) 5 g/L; (b) 60 g/L. Magnification: 6.0kx; Scale: 1cm = 1.64μm.

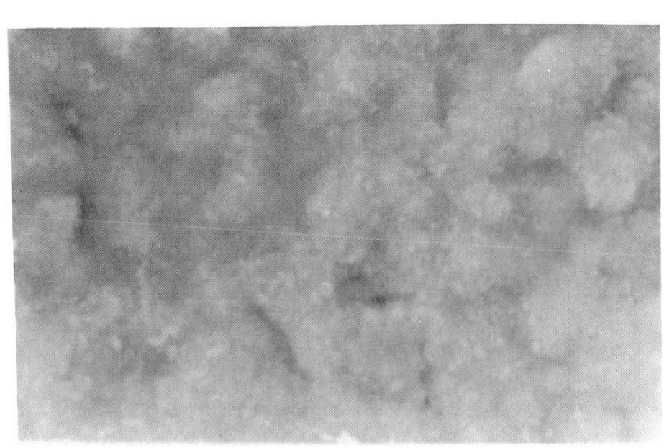

Figure 6

SEM image of the surface of Ni film annealed at 600°C in air atmosphere for 6 hrs. Ni film was formed by electrodeposition at constant potential E_{dep} = -1200 mV vs. SCE on a Pt substrate from Watt's bath solution. Magnification: 6.0kx; Scale: 1cm = 1.67μm.

(a) (b)

Figure 7

SEM images of the surface of composite Ni/SiC films annealed at 600°C in an air atmosphere for 6 hrs. Ni/SiC films were formed by potentiostatic electrodeposition at E_{dep} = -1200 mV vs SCE on a Pt substrate from Watt's bath solution containing suspension of SiC nanoparticles (20 nm - 50 nm): (a) 5 g/L; (b) 60 g/L. Magnification: 6.0kx; Scale: 1cm = 1.67μm.

Energy Dispersive X-Ray Analysis

The presence of SiC particles in Ni/SiC films deposited from SiC suspension was confirmed by performing an analysis for films obtained under different deposition conditions. The X-ray intensities of K_a lines for silicon, carbon, oxygen, and nickel were monitored along with the L lines for nickel.

Ni/SiC films annealed at 600°C for 6 hrs show more nickel oxidation on a grey image and a stronger K_α for oxygen when SiC microparticles (0.5 μm) are used instead of nanoparticles (Figure 8). Less neckel oxidation and weaker signal of K_α line for oxygen is observed in Figure 9 for a Ni/SiC film formed in suspension of 30 g/L SiC nanoparticles. Both Figures 8 and 9 confirm presence of SiC in nickel matrix.

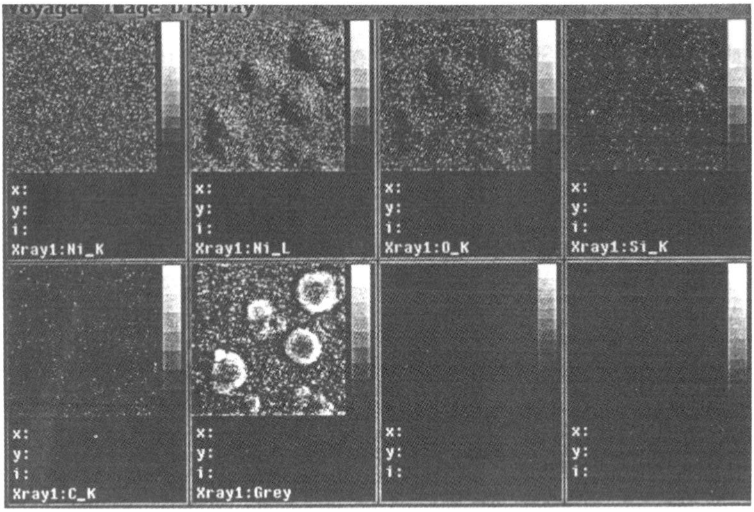

Figure 8
Energy dispersive X-ray and grey image of a composite Ni/SiC film annealed at $600^{o}C$ for 6 hrs in a dry air atmosphere. Ni/SiC film was electrodeposited potentiostatically at E = -1200 mV on a Pt substrate from Watt's bath containing a suspension of 30 g/L SiC microparticles (0.5 μm) size.

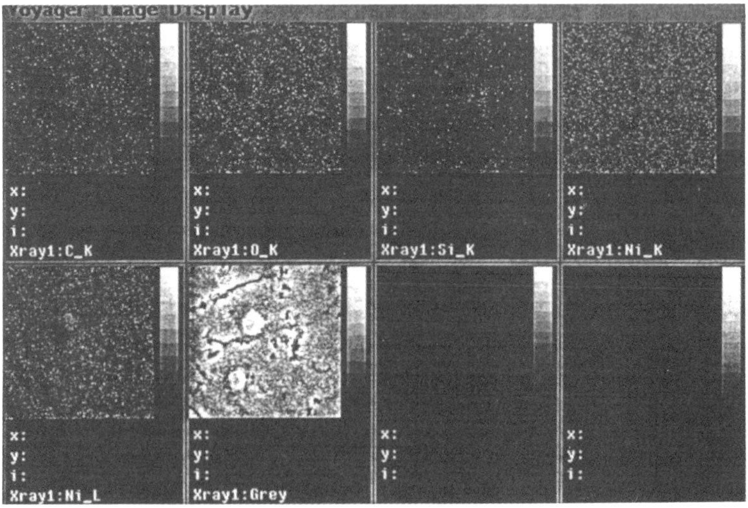

Figure 9
Energy Dispersive X-ray and grey image of a composite Ni/SiC film annealed at 600oC for 6 hrs in an air dry atmosphere. Ni/SiC film was electrodeposited potentiostatically at E = -1200 mV on a Pt substrate from Watt's bath containing a suspension of 30 g/L SiC nanoparticles (20nm - 50 nm size).

Corrosion Resistance Tests for Ni/SiC Films

Corrosion resistance measurements were performed in standard corrosive environments simulating ocean water. In Figure 10, typical anodic polarization curves obtained for Ni (curve 1) and composite Ni/SiC (curve 2) at a scan rate v = 10 mV/s, are presented. In the case of Ni/SiC films, the onset of the active anodic dissolution is shifted toward more anodic potentials in comparison to the pure Ni film. The largest anodic shift was observed for films prepared from 5 g/L SiC suspension. For films obtained from higher than 5 g/L concentration of SiC suspension, less corrosion resistance was observed.

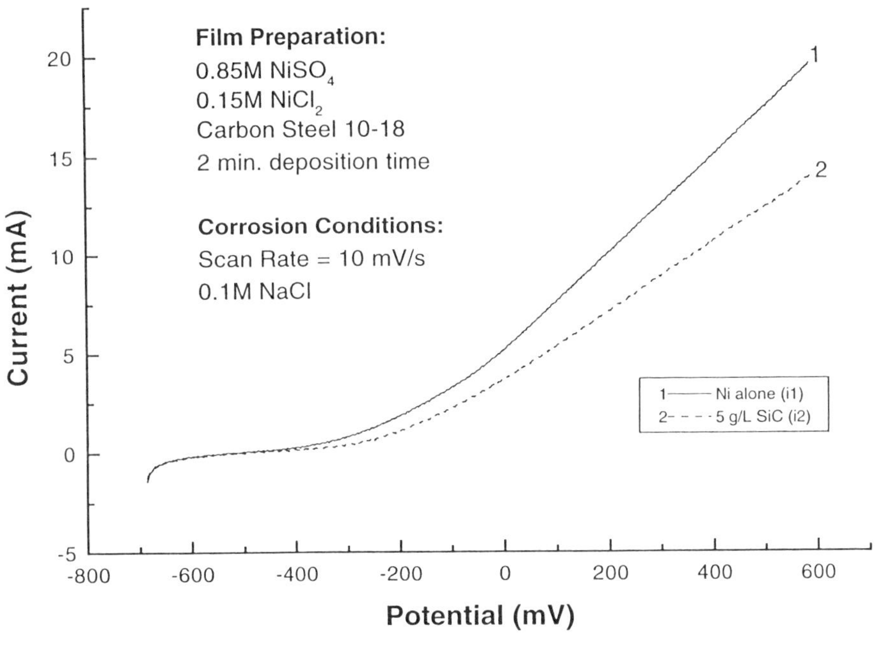

Figure 10
Potentiodynamic characteristics, current vs. potential, recorded at a scan rate V = 10 mV/s, for a C-1018 carbon steel electrode coated with bare Ni film (1) and Ni/SiC film (2) electrodeposited at E = -1200 mV vs. SCE for τ_{dep} = 160 seconds from a modified Watt's bath solution containing a suspension of 5 g/L SiC.

CONCLUSIONS

Composite Ni/SiC films can be formed by the electrodeposition method. Size of SiC particles embedded into nickel matrix significantly affects properties of composite films. Less oxidation of nickel from Ni/SiC film annealed at 600°C was observed in the presence of nanoparticles than microparticles of SiC. Ni/SiC films show improved corrosion resistance in comparison to bare nickel films.

ACKNOWLEDGEMENTS

This work was partially supported by Chema Technology Grant and Potsdam College, SUNY Grant. Help of undergraduate student, Earl Tewksbury, is appreciated.

REFERENCES

[1]E.S. Tolypin and V.N. Starovoitov, *Zh. Prikl. Khim.* **64**[2] p. 2699 (1991).

[2]F.N. Hubel, *Trans. Inst. Met. Fin.*, **56** p. 65 (1978).

[3]M. Yoshimura, J. Kase, S. Somiya, *J. Mat. Res.*, **1**[1] p. 100 (1986).

[4]M. Hepel, T. Tannahll and C. Baxter, "*Electrochemical Synthesis and Modification of Materials*" (Eds), P.C. Andricacos, S. G. Corcoran, Materials Research Society, **451** pp. 481-488 (1997).

[5]M. Hepel, C. Mahoney, and T. Tannahill, "*Fundamental Aspects of Electrochemical Deposition and Dissolution Including Modeling*" (Ed.) M. Paunovic, The Electrochemical Society, Pennington, NJ. (1998 in press.)

[6]M. Hepel, T. Tannahill, C. Baxter, "*Electrochemically Deposited Thin Films,*" (Eds.), M. Paunovic, D. A. Scherson, The Electrochemical Society, Pennington, NJ, Vol. **96-19**, 248-268 (1997).

[7]S.W. Watson, *J. Electrochem. Soc.* **140**[8] p. 2235 (1993).

[8]O. Berkh, A. Bodnevas, and J. Zahavi, *Plating & Surface Finishing*, p. 62, Nov. (1995).

[9]S.H. Yeh and C. C. Wan, *Plating & Surface Finishing*, p. 54, March (1997).

[10]G. Maurin and A. Lavanant, *J. Appl. Electrochem.* **25** p. 1113 (1995).

[11]Yu-G. Gogotsi and V. A. Lavrenko, "*Corrosion of High-Performance Ceramics*," Springer-Verlag, New York, 1986.

KEYWORD AND AUTHOR INDEX